Spatial Analysis and Modeling in Geographical Transformation Process

The GeoJournal Library

Volume 100

For further volumes:
http://www.springer.com/series/6007

Yuji Murayama · Rajesh B. Thapa
Editors

Spatial Analysis and Modeling in Geographical Transformation Process

GIS-based Applications

Springer

Editors
Prof. Yuji Murayama
University of Tsukuba
Geoenvironmental Sciences
Graduate School of Life
and Environmental Sciences
Tennodai 1-1-1
305-8572 Tsukuba, Ibaraki
Japan
mura1@sakura.cc.tsukuba.ac.jp

Dr. Rajesh B. Thapa
University of Tsukuba
Geoenvironmental Sciences
Graduate School of Life
and Environmental Sciences
Tennodai 1-1-1
305-8572 Tsukuba, Ibaraki
Japan
thaparb@gmail.com

ISSN 0924-5499
ISBN 978-94-007-0670-5 e-ISBN 978-94-007-0671-2
DOI 10.1007/978-94-007-0671-2
Springer Dordrecht Heidelberg London New York

Library of Congress Control Number: 2011921697

Printed on acid-free paper

Springer is part of Springer Science+Business Media (www.springer.com)

Preface

Interest in spatial analysis with GIS has tremendously grown in recent years in many different ways. Researchers from a variety of academic disciplines are employing geographical thinking and GIS tools to develop spatially-explicit models to understand the real world processes. Currently, spatial analysis is becoming more important than ever because enormous volumes of spatial data are available from different sources, such as GPS, Remote Sensing, and among others. Even more, non-spatial data collected are being either geo-coded or geo-referenced for converting into spatial. For example, the data related to census, traffic, patients, facilities, schools, etc. are being available at different spatial scale even in the developing countries. Therefore, there is a need to bring spatial analysis with GIS concepts and results with more empirical studies to diverse audiences.

This book is concerned with spatial analysis within a GIS framework. This book aims to provide a comprehensive discussion of spatial analysis, methods, and approaches related to human settlements and associated environment, which are based on results from empirical studies and field experiences. In this book, we planned to address a number of questions that adhere in spatial analysis such as, how can we better explain and describe geographic problems in geographic systems? How have spatial analysis and modeling been applied to real-world problems? What are the useful approaches for simulating geographic processes? What are the major issues that need to be addressed in spatial analysis? The research outputs on techniques, methods, and models with empirical experiences by the researchers are complied and make available to geographic analysts, modelers, and GIS professionals. Framing the collective efforts and practical experiences in a book will be a great contribution, especially to planners and decision makers in the developing countries. All together 16 key contributions with empirical case studies from Iran, Philippines, Vietnam, Thailand, Nepal, and Japan that applied spatial analysis including autocorrelation, fuzzy, voronoi, cellular automata, analytic hierarchy process, artificial neural network, spatial metrics, spatial statistics, regression, and remote sensing mapping techniques are collected in this edited book. A wide variety of useful results with state of the art discussion including empirical case studies is core value of this book. It will provide a milestone reference to students, researchers, planners, and other practitioners dealing the spatial problems on urban and regional issues.

University of Tsukuba, Ibaraki, Japan

Yuji Murayama
Rajesh B. Thapa

Contents

Contributors

Fatemeh Ahmadi Nejad Masouleh Department of Geography, Sinclair Community College, Dayton, OH, USA, f_ahmadinejad2000@yahoo.com

Frederic Borne CIRAD, UMR AMAP, Montpellier, France, borne@cirad.fr

Ronald C. Estoque Division of Spatial Information Science, Graduate School of Life and Environmental Sciences, University of Tsukuba, Tsukuba City, Ibaraki, Japan; Don Mariano Marcos Memorial State University, La Union, Philippines, purplebee80@yahoo.co.uk

Duong Dang Khoi Division of Spatial Information Science, Graduate School of Life and Environmental Sciences, University of Tsukuba, Tsukuba, Ibaraki, Japan, khoi_tn@yahoo.com

Akira Kitazumi Remote Sensing and GIS, School of Engineering and Technology, Asian Institute of Technology, Bangkok, Thailand, kitazumi@maia.eonet.ne.jp

Michiro Kusanagi Remote Sensing and GIS, School of Engineering and Technology, Asian Institute of Technology, Bangkok, Thailand, kusanagi@ait.ac.th; kusanagid4@gmail.com

Ko Ko Lwin Division of Spatial Information Science, Graduate School of Life and Environmental Sciences, University of Tsukuba, Tsukuba, Ibaraki, Japan, kokolwin2002@yahoo.com

Yuji Murayama Division of Spatial Information Science, Graduate School of Life and Environmental Sciences, University of Tsukuba, Tsukuba, Ibaraki, Japan, mura1@sakura.cc.tsukuba.ac.jp; mura@geoenv.tsukuba.ac.jp

Todd Wendell Rho'Dess Department of Political Science, The Ohio State University, Columbus, OH, USA, toddrhodess2003@yahoo.com

Rajesh B. Thapa Division of Spatial Information Science, Graduate School of Life and Environmental Sciences, University of Tsukuba, Tsukuba, Ibaraki, Japan, thaparb@yahoo.com; thaparb@gmail.com

Brandon Manalo Vista Department of Geography, University of Otago, Dunedin, New Zealand, brandonvista@yahoo.com

Changping Zhang Faculty of Regional Development Studies, Toyo University, Tokyo, Japan, cp-zhang@toyonet.toyo.ac.jp

Yumin Zhang School of Surveying and Land Information Engineering, Henan Polytechnic University, Jiaozuo City, Henan Province, PR China, zhangymm@hotmail.com

Yaolong Zhao School of Geography, South China Normal University, Guangzhou, Guangdong, PR China, yaolongzhao@gmail.com

Chapter 1
Spatial Analysis: Evolution, Methods, and Applications

Yuji Murayama and Rajesh B. Thapa

1.1 Development of Spatial Analysis with GIS

In a narrow sense, spatial analysis has been described as a method for analyzing spatial data, while in a broad sense it includes revealing and clarifying processes, structures, etc., of spatial phenomena that occur on the Earth's surface. Ultimately, it is designed to support spatial decision-making, and to serve as a tool for assisting with regional planning and the formulation of government policies, among other things. The world of GIS includes such terms as spatial data manipulation, spatial data analysis, spatial statistical analysis, and spatial modeling. While there are admittedly slight differences in the definitions of these terms (O'Sullivan & Unwin, 2003), they are subsumed in this chapter, which will examine spatial analysis in a broad sense.

If we think of GIS in simple terms as a computer system for entering → managing → operating → outputting spatial data, then spatial analysis would apply to the "operating" part. Viewed historically, GIS has been developed to help national and local governments to store and manage enormous amounts of geographical information. For that reason, there has been much research on the functions for entering, managing and outputting spatial data, so that today, it is reaching the mature stage in a practical sense. However, operating functions including attribute search, geometric operation and mapping are still in a state of development, in that they are still at a level that cannot satisfactorily meet users' needs. Against this backdrop, society's needs for spatial analysis have increased greatly in the 21st century, and with advances in geographical information science, there is growing enthusiasm for enhancing the operational functions of GIS.

Y. Murayama (✉)
Division of Spatial Information Science, Graduate School of Life and Environmental Sciences, University of Tsukuba, Tsukuba, Ibaraki, Japan
e-mail: mura1@sakura.cc.tsukuba.ac.jp; mura@geoenv.tsukuba.ac.jp

Y. Murayama, R.B. Thapa (eds.), *Spatial Analysis and Modeling in Geographical Transformation Process*, The GeoJournal Library 100, DOI 10.1007/978-94-007-0671-2_1,

1.1.1 Genealogy of Spatial Analysis

From a theoretical perspective, the evolution of spatial analysis is generally traced to four original current displines of geographic thought. The first current is that of quantitative geography. As it is well known, the quantitative revolution began in the United States in the 1950s as a movement away from individual descriptive thoughts (idiographic) toward empirical law-making (nomothetic) geography. Its nerve center was the geography department at Washington University. Under the leadership of Prof. William Garrison, and disciples such as Richard Morrill, Brian Berry, Michael Dacey, Duane Marble, John Nystuen, and Waldo Tobler, quantitative geographers gathered epoch-making results from a wide range of geographical areas including central place theory, factorial ecology, urban rank size distribution, regionalization (modeled), network analysis, and map conversion. Quantitative geographers called the nomothetic framework within which they searched for order, regularity, etc., in spatial patterns "spatial analysis" (Berry & Marble, 1968). This methodological paradigm shift quickly spread among North American geographers. By the late 1960s, research had shifted from an emphasis on (static) spatial patterns to (dynamic) spatial processes, with special focus on such themes as spatial diffusion, spatial behavior, mental mapping, and urban systems.

For GIS development, quantitative geography not only provided sophisticated analytical methods and refined spatial models, it also made enormous contribution to the basic concepts related to the fundamentals of GIS architecture. The geographical matrix proposed by Berry led to the attribute table concept of GIS. The abstraction of geographical space based on the point/line/plane concepts proposed by Peter Haggett led to the idea of vector-type GIS that controls geographical planes with nodes, arcs, and polygons. The methodology of Tobler's geographical conversion contributed to the development of analytical cartography, and paved the way for computer mapping.

We must also remember the contributions of the Lund school, which was pioneered by Torsten Hägerstrand in Sweden. In the 1950s and 1960s, the Lund school produced countless groundbreaking research results that laid the foundation of today's GIS, including digitizing methods, geocoding concepts, spatial reference methods, point-in-polygon operations, shortest route searches, mesh/grid analysis, and automatic image creation methods (3D maps, dot maps, theme maps). They developed NORMAP, which can be called the prototype for today's analysis-oriented GIS. The utilization of this software for corroborative research on population potential, location allotment, shortest routes, central place theory, population distribution zoning, etc., has been of great benefit to urban planners, regional policymakers, and among others (Nordbeck & Rystedt, 1972).

The second original current of spatial analysis is regional science founded by Walter Isard of Pennsylvania University (USA) (Isard, 1956). Based on regional economics, regional science looks at socioeconomic phenomena within a regional context in an effort to clarify spatial mechanisms and develop associated theories. In contrast to quantitative geography, which attempts to inductively construct models with a practical understanding of socioeconomic phenomena as a starting point, regional science makes theoretical interpretations of socioeconomic phenomena and attempts to construct spatial models in a normative and deductive manner. In short, we can say that the approach of quantitative geography is data-driven, while regional science is model-driven.

Since the late 1950s, after the establishment of Regional Science Association in 1954, the development of regional science has been accompanied by the development of countless numbers of innovative quantification methods and theoretical models. Among these are the inter-regional input-output model, industrial location analysis, population movement forecasting, commercial flow analyses, game theory, transport planning, distance decay theories, shortest distance problems (Dykstra method), theoretical trend line analysis (logistics curves, Gombertz curves, etc.), urban dynamics, gravity models, potential models, spatial interaction models, and the inter-regional linear programming model. Regional science became highly acclaimed for its applications in regional planning and regional policy, and it gradually solidified its position as a practical science.

In other words, regional science can be said to be a type of "toolbox" that contains methods and models for regional analyses. The ability to utilize this toolbox through GIS became the goal of the second generation of regional scientists. In the 1980s, spatial econometrics grew into the largest branch of regional science, especially with the series of results by Luc Anselin enabled concepts of spatial effects (spatial dependency and differences) to be manipulated with GIS, and which helped to build the foundation of a new spatial analysis theory in the early 1990s (Anselin, 1988).

The third original current is that of spatial statistics, which was established by statisticians. In spatial analysis, the analytical units of point (facilities, etc.), line (roads, etc.), and plane (administrative zones, etc.) are derived from geographical space that is generally a contiguous expanse. The interest of statisticians lies in the methods for deriving and setting analytical units. The problem is that the structural elements that are derived (set) have spatial mutual dependency, and there are points where statistical independence cannot be guaranteed. In the fields of quantitative geography and regional science, in the 1950s and 1960s, this problem tended to be ignored. During that period, regional analyses involving multivariate analytical methods (factor analysis, multiple regression analysis, cluster analysis, etc.) were widely used, but statisticians criticized that their analytical framework ignored spatial dependency, which would likely lead to contorted interpretations and erroneous conclusions.

In addition, there were warnings that the problems brought about by spatial autocorrelation were more serious than those of temporal autocorrelation, time-series analyses for example, which were being vigorously studied at the time. Against this backdrop, a far-reaching argument for spatial autocorrelation was published by Cliff and Ord (1973). This work led to increased interest in statistical estimation and testing, probability theory, error theory, etc., that could not be avoided when making spatial analyses, forcing quantitative geographers and regional scientists in a wave of more vigorous discussion and debate. Even when addressing the same problem, different disciplines take different approaches. Let us look at the example at how residuals in regression analysis are approached by statisticians and geographers. Statisticians view residuals as errors, and pour their efforts into examining significant differences and minimizing residuals by adding variables. In contrast, geographers view residuals as having meaning and look for factors that induce deviation from the predicted value (regression line). For statisticians, spatial autocorrelation is a bias and should be eliminated, while geographers attempt to use spatial autocorrelation as a basis for interpreting local structure.

In the early 1980s, Ripley (1981) published a comprehensive explanation of spatial statistics. This work, which systematically discussed spatial sampling, point distribution patterns, spatial interpolation/extrapolation, surfaces, random closed sets spatial regression, stereology, etc., was highly acclaimed by neighboring disciplines. With this work as a stimulus, interest in spatial statistics grew dramatically in the 1980s, but there was also another side emerging remote sensing, for example. Around this time, it was becoming possible to utilize high-resolution satellite image data, and it became an urgent task to develop refined methods for making statistical analyses, such as pattern recognition and image classification. In this way, spatial statistics secured a place for itself as a branch of (applied) statistics.

The latter half of the 1980s saw the development of geostatistics. This field utilizes statistical analysis methods, probability theory, etc., to investigate spatial laws and mechanisms of spatial phenomena that occur on the earth's surface. Unique techniques such as the Kriging method, variograms, drift, etc., were developed to support earth sciences (such as soil mechanics, geology, and mineralogy), and their utility has been verified through actual use, such as in exploration for mineral resources.

The fourth original current is that of computational geometry which was led by information scientists. Computational geometry uses computers to design algorithms for analyzing geometrical attributes of planar data. Making topological concepts such as neighborness, intersect, inclusion, etc., operational with GIS is the result of work by information scientists. If an effective algorithm cannot be developed, then it would not be possible to process complex planar information. The ability to process, in a short computational time, complicated problems involving point distribution patterns, Voronoi diagram, Delaunay triangulation, optimum networks, traveling salesman problems, shortest route, optimal allocation, point location decision-making, mesh analysis, minimum spanning trees, hidden line processing, convex hull, etc., is the fruit of computational geometry.

In spatial analysis, enormous amounts of real-number calculations cannot be avoided, but information scientists have developed an algorithm for avoiding computational error, and this plays an important role in processing map distortion, linkages with map images, and so on. The history of computational geometry is relatively recent, and it was not until the mid-1970s that it was given academic recognition in the West. In Japan, however, it is notable that research connecting geographic information processing with computational geometry began in the early 1970s.

This was the activity of the "Committee for Investigating and Developing Basic Algorithms for Processing Geographic Information" which was established in the *Operations Research Society of Japan* in 1971. As it incorporated methods of operations research (OR), this group sought to develop efficient methods for processing geographic information, and its results were on the cutting edge in the world at that time. According to Masao Iri, who was the chief investigator of the committee, computational geometry involved both topology/combination geometry (connective relationships, inclusive relationships, etc.) and quantitative geometry (quantitative attributes such as point coordinates, length of line segments, area of planes, etc.). It can be said that using the power of the computer, these two aspects were integrated into a GIS technology which enables the comprehensive processing (geometric computations) of planar data.

Originally, the target of computational geometry was limited to planar figures, but advances in GIS gradually expanded interest to include spatial figures as well. Applied research is being promoted which includes 3rd order Voronoi diagrams and arrangements, and which incorporates time axes, i.e., creating dynamic geometric algorithms, time-space Voronoi diagrams, and so on.

1.1.2 New Spatial Analyses

The remarkable success of "spatial analysis using GIS" has not only enabled the four original currents, it has also made it possible to incorporate results from neighboring research fields. In terms of methods for acquiring and analyzing quantitative data, it is essential to have collaborative relationships with many other academic disciplines, including regional science, computational geometry, mathematical ecology, operations research, computational science, etc., that have lent their hand to the development of effective algorithms for spatial statistics, visualization technologies, spatial manipulations, and so on.

With improvements in computer processing capacity and advances in GIS, a new academic discipline was formed in the early 1990s which is called New Spatial Analysis. This was not the result of one particular academic field, but rather the product of research groups having varied backgrounds and multiple areas of expertise. They formed close working relationships, and poured their efforts into collecting information and holding workshops and symposia. The number of opportunities for researchers from different fields and specialties to conduct joint research is being grown, as the numbers of academic papers penned by co-authors from different research fields are available.

In 1991, the National Center for Geographic Information and Analysis in the United States gathered together cartographers, statisticians, geographers, etc., to launch the Geographical Information Systems and Spatial Analysis Project. The results were published under the title "Spatial Analysis and GIS" in 1994 (Fotheringham & Rogerson, 1994). At the same year, a workshop titled "Spatial Modelling and GIS" was held at Bristol University in the United Kingdom under the auspices of the Economic and Social Research Council (ESRC), among others. The discussion there resulted in numerous suggestions for future spatial analysis research. Some of the results can be found in "Modelling in a Spatial Analysis- GIS Environment", which was published in 1996 (Longley & Batty, 1996).

In addition, the European Science Foundation (ESF) started the GISDATA program in 1993, which included the establishment of "Spatial Models and GIS," as a society of specialists. Numerous human resources were assembled there, including geographers, regional scientists, and spatial statisticians. The results of the GISDATA were published in two volumes: "Spatial Analytical Perspectives on GIS" (Fisher, Scholten, & Unwin, 1996) and "Spatial Models and GIS: New Potential and New Models" (Fotheringham & Wegener, 2000). Other organizations that were vigorously engaged in activities at that time included the Working Group on Mathematical Models of the International Geographical Union (IGU) in the field of geography, and the European Congresses of the Regional Science Association in regional science.

In the latter half of the 1990s, interest in spatial analysis further increased, and large-scale conferences were organized. In 1996, the international GeoComputation conference, centered on the UK, held its first biannual meeting, and since 2000 the international GIScience symposium has been held biannually, in the United States and elsewhere. These meetings are attended by a wide array of researchers, from human geographers and social scientists, to physical scientists and engineers. In the past few years, many technical papers with "spatial analysis" in their title have been published. It can even be said that today, spatial analysis is in a sort of boom period.

The number of national projects involved with spatial analysis, and the number of countries allocating priority funds to research, are growing. For example, in Ireland the National Institute for Regional and Spatial Analysis was established at the Maynooth campus of the National University of Ireland in 2001, and the National Centre for GeoComputation was established within the Institute in 2004. The core of these organizations is formed by geographers and computational scientists, among others, who are undertaking theoretical and corroborative research related to the processing of massive amounts of spatial data.

1.1.3 Why Is Spatial Analysis Conducted with GIS?

Why spatial analysis is conducted with GIS instead of statistical analysis packages (such as SAS and SPSS) or mathematical analysis tools (such as Mathematica)? The reason is simple. GIS manages planar data and attribute data in an integrated manner, which enables the spatial data to be manipulated in all directions. We can summarize its advantages as follows:

1.1.3.1 Calculating

GIS can eliminate the difficulty and tediousness of making massive amounts of geographical calculations, such as measurements of distance, area, mass (capacity), center of gravity, and circumference. For example, a number of indices for distance exist, such as straight lines, roads, time, cost, recognition, and psychological indices, but the use of GIS can almost instantly identify shortest route, accessibility, etc., based on these indices. It is the same way with measurements of mass (capacity); the volume of a reservoir behind a dam, for example, can be readily calculated by incorporating GIS into Digital Elevation Model (DEM). This can easily simulate changes in reservoir volume resulting from a rise or fall in the water level.

1.1.3.2 Mining

The amount of customer information, empirical data, etc., is constantly increasing. However, due to advances in GIS technology and improvements in computer processing capacity, we can represent this enormous amount of geographical information to "geocoding" ("address matching") to turn it into a spatial database which can be incorporated into GIS without difficulty.

"Mining" refers to the process of shifting out underlying patterns, causal relationships, etc., from enormous amounts of data. At one time, the use of GIS without having an outlook, hypothesis was derided as merely being "garbage in, garbage out". However, today these mountains of "garbage" (which are actually mountains of treasure) are being viewed as "diamonds in the rough", and GIS, as a powerful tool for constructing hypotheses rather than proving them, is attracting attention for its versatility. Until now, it had often been used as a mining tool for conducting multivariate analyses. While this utility has not changed, it is now being applied to a wide range of innovative methods such as fuzzy logic, neural networks, genetic algorithms, pattern recognition, and fractal analysis, among other uses.

1.1.3.3 Visualizing

One of the selling points of GIS is its outstanding ability to turn data into images. With complete freedom, the user can create an array of highly impressive visual maps, graphs, and movies, such as mesh maps, choropleth maps, dot maps, cartograms, bird's-eye view maps, kernel density distribution maps, contour maps, flow line maps, area maps (planar expression maps), time maps, time-space maps, two-variable maps, 3D maps, and animations.

Visualization helps with the construction of models, and is also useful for cultivating spatial thinking. Let us look at the example of creating a choropleth map. It is a delicate process, involving decisions on how to classify hierarchies, how to group numerical values, how to arrange scales, colors, etc. All of these are difficult tasks. However, GIS enables the map-making process to be done through trial-and-error that will ultimately result in the map that best meets the user's needs.

There are many methods for classifying hierarchies, such as equal size method, equal interval method, equal area method, and SD method, but with GIS, they can be freely changed instantly. For example, if the user wants to express a rough structure, the number of divisions can be small, but if the user wants to highlight local/community differences, then the number of divisions can be increased accordingly. Map conversion enables latent principles, spatial order, etc., in the background to be searched and discovered, which is useful for mental mapping and recognition distance as well.

1.1.3.4 Creating Data

GIS has superb database management functions. Using topology, theoretical (subordinate) relations as a "handrail", attribute data can be easily combined and changed, making GIS a very attractive tool for reconstructing data.

Let us look at the example of a geographical matrix of a certain urban area (where rows are districts and columns are arrangement of socio-economic features). With the urban core as the center, point buffers are constructed for every *x* meters of radius by concentric circle zones, or it can reconstruct a geographical matrix for each sector, divided into 8 directions. Then by combining the concentric circles with the sectors, it becomes possible to edit spatial data in greater detail. If this is overlaid with DEM data, it will be possible to create new geographical matrices that are divided by elevation.

Recently, it has become possible to utilize various types of individual data, such as POS (point-of-sale), person-trips, real estate transaction data, and observational data (temperature, water quality, air pollution, etc.). Non-tabulated (micro) data are being distributed at an accelerating pace. What should be particularly noted here is that these data normally come with location information and time information, so by utilizing the attribute search functions and spatial manipulations offered by GIS, the user can freely set time and space boundaries to create original geographical matrices. It is also possible to analyze non-tabulated spatial data without modification, and the power of GIS can also be manifested in modifiable area unit problems (MAUP).

Using positional information (latitude, longitude, address, etc.) as a "hand rail", panel data can be created by using geocoding to link different types of non-tabulated data, allowing the user to try panelizing study data from different years. Panelization can also be applied effectively to government statistics (national censuses, industrial statistics, etc.) that have accumulated over the years. This resolves problems with security and confidentiality, so the national government needs to promote the use of panelization to display non-tabulated (micro) data.

1.1.3.5 Handling Relations

Another distinguishing feature of GIS is its ability to reveal wholeness, localness, relativity, etc., based on the relationship among compositional elements (points, lines, and planes). Neighboring relationships, inclusive relationships, intersecting relationships, etc., can be freely handled. It is not an exaggeration to stress that GIS is what has stimulated ideas about positioning and relativizing local events within an entire region.

In 1970, Tobler pointed out that "Everything is related to everything else, but near things are more related than distant things" (Tobler, 1970). This simple, articulated phrase is called "the First Law of Geography". When conducting spatial analysis, this hidden quality of spatial data has an important meaning, but in the 1970s, it was essentially taken for granted and it was not able to attract academic interest in GIS fields.

However, in the 1990s, advances in geographical information science and associated qualitative principles were basking in the spotlight. This was because it had become possible to quantitatively examine spatial mutual relationships – how relationships are affected by the distance between them – using GIS. Concepts such as spatial dependency proposed by Anselin, spatial heterogeneity, etc., can trace their roots back to this, and geographically weighted regression (GWR), local spatial autocorrelations, kernel density estimation, etc., are also closely related to these principles. Today, global-versus-local research is growing into an important pillar of spatial analysis.

1.1.3.6 Understanding Processes

Spatial phenomena that occur on the Earth's surface change over time. Accordingly, if we can trace the movements of spatial phenomena in the past, we can acquire clues that will help us to unravel today's structures and formation mechanisms.

Turning their attention to the location of convenience stores, researchers made an effort to obtain results of random distribution by measuring point distribution patterns of a certain region. Unfortunately, the results were not able to provide reasons for why random distribution was indicated. However, if we can understand the dynamic spatial phenomena and trace how convenience stores have come to be located where they are, then we may be able to find some clues that can help to reveal the spatial factors that have caused changes in point distribution. In fields such as mathematical ecology and epidemiology, an increasing number of approaches are being taken from a dynamic perspective based on such concepts. GIS, which specializes in repeated computations, has also contributed to the analysis of dynamic processes. In recent years, geosimulation research, with its use of cell automata, multi-agent models, etc., has been at the forefront (Benenson & Torrens, 2004). With the increasing distribution of non-tabulated data that includes location and time, we can expect spatial process research to evolve even further. Hereafter, "spatial analysis" will undoubtedly develop into "time-space analysis" which specifically incorporates a time axis.

1.1.4 Methodological Shift

As reported above, the development of GIS theory and technology has brought new spatial analysis theories to the forefront, and perspectives, concepts, and methodologies of traditional spatial analyses are undergoing drastic changes.

1.1.4.1 From Aggregate Thinking to Non-aggregate Thinking

"Aggregate thinking" is a way of thinking that is based on total amounts and average values (amounts per unit), while "non-aggregate thinking" emphasizes individual characteristics without tabulating data.

In non-aggregate thinking, no significance is given to average values such as population densities, unemployment rates, green area rates, and average income. For example, a municipality that achieves a green cover rate of 90% is considered to have only a high proportion of greenery and is not interpreted as being "rich in greenery". Instead, the concern is how much greenery there is in built-up areas where most of the people live. It is the same way with average income – if there is one person who is earning tens of millions of dollars, the average income of that area will increase remarkably.

Most aggregate data is tabulated in units of area or district and provided to users. Therefore, the user is only able to discuss regional similarities and differences at the tabulated or aggregate level. In other words, this is thinking in the world of average values. However, in recent years there has been a growing amount of non-aggregate data, such as POS, real estate transaction data, person-trips, and empirical data. The distribution of non-aggregate data is prompting a withdrawal from "average value-ism".

In 2007, new statistical law was established in Japan, and restrictions were relaxed on the use of micro data outside of the intended purpose, which had been prohibited as a matter of principle. Various types of government statistics, such as the Population Census, Census of Agriculture and Forestry, Establishment and Enterprise Census, etc., were a storehouse of geographical information. If it is possible to use non-aggregate data that contain location information, then there will undoubtedly be major advances in spatial analyses that utilize GIS.

This is especially important for regional divisions. There are a wide variety of user needs in a unit area. At the community level, various types of local divisions can be assumed, including, blocks, school districts, postal code zones, firefighting jurisdictions, autonomous organization zones, garbage collection zones, grids, and so on. In the ideal case, it would be possible for users to undertake the tabulation work while giving sufficient consideration to privacy and confidentiality. This could be expected to make a significant contribution to the development of spatial data mining methods and research on MAUP.

1.1.4.2 From Model-Driven to Data-Driven

"Hypotheses are established in the brain, and models are constructed deductively. Models are calibrated, the data are ascertained to behave according to the theory, and the propriety and effectiveness of the model are asserted". This has been the general research style of conventional social sciences. However, recently there has been growing awareness of the existence of approaches which attempt to inductively build models starting from data acquisition. This trend has been seen even in theoretical social science and quantitative behavioral science which were far removed from empirical or observational science.

Against this backdrop, GIS is starting to be manifested as a powerful tool not for proving hypotheses, but for constructing them. This might be thought of as GIS lending its prowess to finding a diamond hidden in a mountain of garbage. Such thinking resembles the concepts of field work, which seeks individual facts at a site, finds order in them, and structuralizes them. The idea is to "explain phenomena with data". Constructing models and methods in one's head and applying corresponding data to them makes it difficult to make new discoveries, or to identify new principles and rules.

In the background of what has enabled this type of investigative approach is the remarkable development of geographical information science and computer technology. By using parallel computers, superfast computers, etc., it has become possible to process tremendous amounts of geospatial data. The approach of this angle is called "GeoComputation", and it has been extensively studied, especially in the United Kingdom.

1.1.4.3 From an Understanding of Pattern and Processes to Prediction, Control, and Management

In the late 1950s in the field of social sciences, there began a quantitative revolution which spurred a methodological shift from idiographic to nomothetic geography, and spatial analysis flourished. Spatial analysis started from an analysis of static spatial patterns during a certain period in time, but soon took on a dynamic perspective which led to the development of spatial processes research.

In the 1990s, the development of GIS made it easy to process large amounts of geographical spatial data, which led to the successive construction of detailed models for inductive spatial forecasting, which were applied to corroborative research. Among the multitude of methods developed were analytic hierarchy process, multi-criteria evaluations, genetic algorithms, cell automatons, and agent-based models.

1.2 Contemporary Spatial Analysis and Modeling

With the automated computer aided geographical techniques, spatial analysis has experienced remarkable growth in recent years in terms of theory, methods, and applications. Growth in these areas fueled by the significant achievements of spatial data acquisition technologies, for example, Global Positioning System (GPS) and Remote Sensing (RS), and manipulative and geocomputational powers of GIS which have provided new opportunities to explore human environment relations at desired spatiotemporal domain. Spatial analysis is exclusively developed for analyzing location-based digital information. It includes the use of topologies, geometry, logic, and multidimensional reasoning capabilities that are all directed towards the spatial domain. Spatial analysis is emerging to solve various location-based problems and useful to all scales from individual household to worldwide explorations and interpretations. When spatial analysis emerged with different theories and methods it produces a different horizon of applications. Several methods and applications, varied by academic disciplines, of spatial analysis are available to date. However, in this section, we discussed some of the methods and applications that are frequently used to explore urban and regional problems and provide solutions.

1.2.1 Spatial Autocorrelation

Spatial autocorrelation is a general geographical phenomenon in nature, which indicates spatial association and spatial dependence of geographic object. Objects in natural systems have some degree of spatial clusters. Spatial aggregation of objects produces a variety of distinct spatial patterns that can be characterized by the size and shape of the aggregations, and can be quantified according to the degree of similarity between the objects in their attributes values. Spatial autocorrelation is often measured by examining how the similarity of objects varied at a location and nearby. If the objects located represent homogenous spatial patterns, then the pattern refers to the positive spatial autocorrelation. When the features located in geographic space are heterogeneous then the spatial patterns of the features are analyzed as negative spatial autocorrelation. Zero autocorrelation exists when attributes values are independent of location. A mosaic of patches with different spatial autocorrelation structures is normal in natural system. Spatial structures consists of their own magnitude and size that make them distinct which are usually easy to detect and interpret. Methods for analyzing patterns of spatial attributes include global and local spatial autocorrelation tests. Global spatial autocorrelation tests can verify the relationship between the attribute values and location of spatial objects through the tests of significance of statistics such as Moran's I and Geary's c. While these statistics are effective in measuring the spatial autocorrelation of spatial attributes over an entire region, they may not be practical for finding individual local clusters within a region or for searching for heterogeneous regional patterns (Ord & Getis, 1995). To remedy this situation, Getis and Ord (1992) developed a local spatial autocorrelation statistic $Gi(d)$ introducing a distance parameter d to a weight coefficient w_{ij} to measure spatial proximity of spatial objects. A significant number of applications of spatial autocorrelations are available to date (Zhang & Murayama, Chapter 3; Zhao & Murayama, Chapter 4; Anselin, Syabri, & Kho, 2010; Fortin & Dale, 2009).

1.2.2 Geographically Weighted Regression

Geographically weighted regression (GWR) is a technique of spatial statistical modeling used to analyze spatially varying relationships between variables. It is based on the non-parametric technique of locally weighted regression developed in statistics for curve-fitting and smoothing applications, where local regression parameters are estimated using subsets of data proximate to a model estimation point in variable space (Fotheringham, Brunsdon, & Charlton, 2002; Nakaya, 2008). The procedure yields a separate model for each spatial location in a study area with all models generated from the same data set using a differential weighting scheme. The spatial coordinates of the data points are used to calculate inter-point distances, which are input into a kernel function to calculate weights that represent spatial dependence between observations. Part of the main output is a set of location-specific parameter estimates and associated t statistics which can be used to test hypotheses about individual model parameters. Therefore, unlikely the traditional regression framework,

GWR allows local rather than global parameters to be estimated. GWR has been presented as a method to conduct inference on spatially varying relationships, in an attempt to extend the original emphasis on prediction to confirmatory analysis. GWR approach is applied in several studies, for instance, regional analysis of wealth and land cover (Ogneva-Himmelberger, Pearsall, & Rakshit, 2009), evaluation of land use change factors (Thapa & Murayama, 2009), examination of cancer mortality (Páez & Wheeler, 2010), driving forces behind the deforestation (Jaimes, Sendra, Delgado, & Plata, 2010), and afforestation (Clement, Orange, Williams, Mulley, & Epprecht, 2009).

1.2.3 Spatial Metrics

Spatial metrics were developed in both information theory and fractal geometry based on a categorical, patch-based representation of the landscape (Mendelbrot, 1983). Patches are defined as homogenous regions for a specific landscape property of interest, such as "industrial land", "park" or "high-density residential area" (Zhao & Murayama, Chapter 10). Landscape metrics are used to quantify the spatial heterogeneity of individual patches, all patches belonging to a common class, and the landscape as a collection of patches. When applied to multi-scale or multi-temporal datasets, the metrics can be used to analyze and describe the changes in the degree of spatial heterogeneity (Thapa & Murayama, Chapter 15). The interest in using spatial metric concepts for the analysis of urban environments is starting to grow. Geoghegan, Wainger, and Bockstael (1997) explored spatial metrics in modeling land and housing values. Alberti and Waddell (2000) substantiate the importance of spatial metrics in urban modeling by proposing specific spatial metrics to model the effects of the complex spatial pattern of urban land-use and land-cover on social and ecological processes. Parker, Evans, and Meretsky (2001) summarize the usefulness of spatial metrics with respect to a variety of urban models and argue for the contribution of spatial metrics in helping link economic processes and patterns of land-use. The integration of remote sensing and spatial metrics in spatiotemporal analysis of urban growth, land use and land cover change (Thapa & Murayama, Chapter 15) has further strengthen the importance of spatial metrics in spatial analysis.

1.2.4 Voronoi Method

Voronoi method decomposes a set of objects in a spatial space to a set of polygonal partitions which are also called Voronoi diagram. The concept is simple. Given a finite set of distinct, isolated points in a continuous space, we can associate all locations in that space with the closest member of the point set. This process partitions the space into a set of regions, i.e. Voronoi diagram (Okabe, Boots, Sugihara, & Chiu, 2000).

The basic Voronoi concept involves tessellating an m-dimensional space with respect to a finite set of objects by assigning all locations in the space to the closest member of the object set. This concept can be operationalised in a variety of ways by considering different methods for determining "closeness", subsets of objects rather than individual objects as generators, moving objects, and different types of spaces, as well as various combinations of these. According to Okabe et al. (2000), a second order diagram can be constructed from the Voronoi diagram in m-dimensional space by joining those points whose regions share an $(m-1)$-dimensional face. This dual diagram refers to Delaunay tessellation.

Voronoi diagram can handle the k-th order variant, where each point has k closest generating points. Several different distance metrics can be applied to these diagrams, the most popular of which are Euclidean and Manhatten distances. These versatilities have resulted in Voronoi constructions being used increasingly in a wide range of disciplines for spatial data manipulation, spatial interpolation, modelling spatial structures and spatial processes, pattern analysis, and locational optimization.

An ordinary Voronoi diagram assumes that all points have the same weight. But in reality points may have different weights reflecting their variable properties; for example, the population size of a settlement, the number of functions in a shopping centre, the amount of emissions from a polluter, and so forth. If weights are taken into account, then weighted Voronoi diagrams are generated. Depending on the variation methods they can be either multiplicatively, additively, or compound weighted diagrams (Ahmadi Nejad Masouleh, 2006; Mu, 2004; Okabe et al., 2000).

Voronoi is often referred to Voronoi diagram, Voronoi foam, Thiessen polygon, Dirichlet tessellation, areas of influence polygons, plant polygons, multiplicatively weighted, and so on (Mu, 2004; Ahmadi Nejad Masouleh et al., Chapter 8). A number of applications of the Voronoi with spatial analysis are available to date. For instance, UK's postcode boundary generation (Boyle & Dunn 1991); location optimization (Okabe et al., 2000), fire area delineation (Mu, 2004), and school administrative boundary delineation (Ahmadi Nejad Masouleh, 2006; Ahmadi Nejad Masouleh et al., Chapter 8) are some of them.

1.2.5 Multi-criteria Decision Making and Analytical Hierarchy Process

Decision making itself is broadly defined to include any choice or selection of alternative course of action. Therefore, it is important in both social and natural sciences including GIS. The types of decision problems that interest geographers and spatial planners typically involve a large set of feasible alternatives and multiple conflicting and incommensurate evaluation criteria. The alternatives are usually evaluated by a number of people, i.e. managers, decision makers, stakeholders, or interest groups, who are often characterized by unique preferences with respect to the relative importance of criteria on the basis of which the alternatives are evaluated. Many real-world spatial planning and management problems give rise to spatial analysis based multi-criteria decision making. Spatial multi-criteria decision analysis provides a platform with a collection of techniques for analyzing geographic events where the results of the analysis depend on the spatial arrangements of the events. There had been several studies that were able to demonstrate the application of GIS along with multi-criteria decision techniques on a wide range of area of interests; from the technical aspect of integrating GIS with multi criteria evaluation, land suitability mapping for industry, peri-urban agriculture, and bee keeping area, land allocation for shipping canals, evaluating habitat for endangered species, and so on as discussed by Thapa et al. in Chapter 12, Estoque and Murayama in Chapter 13, and Thapa et al. in Chapter 14.

Analytical Hierarchy Process (AHP) is one of the core methods of multi-criteria decision making currently emerging in spatial analysis and getting much attentions within the geographers. It consists of a sound mathematical foundation, which has been widely used in several socioeconomic and engineering applications (Saaty, 1980; Thapa & Murayama, 2010a). It is one of the best known comparative ratings of the importance of various factors, which are extracted from stakeholders and then analyzed to produce a set of consensus weights using a fairly common form of matrix algebra. The technique is based on a pairwise comparison of elements (attributes or alternatives), and results in a comparison matrix in which the relative importance of each element or factor is determined numerically. The technique also has a scientific basis through which the consistency of the driving factor evaluation can be judged. The AHP is based on the assumption that the relevant dominance of one attribute over another can be measured by a pairwise comparison of preferences, systematically made on each level of a hierarchy of factors.

In using AHP to model a problem, one needs a hierarchical structure to represent that problem as well as pairwise comparisons to establish relations within the structure. The AHP has been used with absolute or relative types of comparisons to derive ratio scales of measurement. In absolute comparisons, alternatives are compared with a standard that has been developed through one's experience. In relative comparisons, alternatives are compared in pairs according to a common attribute. The relative measurement w_i, $i = 1,\ldots, n$, of each of n elements is a ratio scale derived by comparing the elements with each other in pairs. In paired comparisons, two elements i and j are compared with respect to common property. The i is used as the unit and the j is estimated as a multiple of that unit in the form $(w_i/w_j)/1$, where an estimate of the ratio w_i/w_j is taken from a fundamental scale of absolute values between 1 and 9 (Saaty, 1980). The numbers are used to represent how many times the larger of two elements dominates the smaller with respect to a common property or criterion (Thapa & Murayama, 2010a). AHP is used in spatial analysis for evaluating factor weights, for example, land evaluation for peri-urban agriculture (Thapa et al., Chapter 12), investigating driving forces of urban growth (Thapa & Murayama, 2010a), land suitability analysis for bee keeping (Estoque & Murayama, Chapter 13), and deforestation modeling (Khoi & Murayama, Chapter 11).

1.2.6 Fuzzy Logic

Fuzzy logic is a mathematical approach that reasons on uncertainties, vagueness, and judgments (Zadeh, 1965). Fuzzy logic has great potential to address uncertainty and imprecision by extending beyond the classical binary representations. In fuzzy logic, a statement can assume any real value between 0 and 1 which is different from classical logic (0 or 1). This logic utilizes fuzzy set theory. Because of the handling capability of partial truth in decisions making process of real world problems, fuzzy set and fuzzy logic have drawn much attention in a variety of disciplines including GIS and spatial analysis. Two kinds of spatial uncertainty inherent in spatial data, i.e., ambiguity and vagueness can be addressed by fuzzy logic and fuzzy sets. The ambiguity often occurs when one does not have unique criteria for making a decision. For example, consider the spectral determination of a pixel in a satellite image that represents a residential land or not (Thapa & Murayama, Chapter 9). As in many image processing tasks, spectral signatures are difficult to precisely delineate. Fuzzy approach helps to reduce mixed pixel problems of heterogeneous earth surface. Thus, the notion of a residential land may best be represented by a fuzzy set and probability, and fuzzy measures may be used to deal with that kind of uncertainty. Vagueness arises from the inability to make precise distinctions in the real world boundary, existence of rough boundary. This is particularly true when using linguistic descriptions of real world phenomena, for example, metropolitan area, rural-urban fringe, mountain, suitability, or steepness. Measures of fuzziness have been developed to address these kinds of uncertainty. For example, research evidence that humans' vague language can be represented by fuzzy theory is integrated with a field-based qualitative spatial reasoning while providing users more choices and degree of satisfaction as discussed by Zhao et al. (Chapter 2).

Thus, ambiguity and vagueness problems in spatial data led many to use techniques based on fuzzy set theory (Stefanakis, Vazirgiannis, & Sellis, 1999; Zhao et al., Chapter 2; Thapa & Murayama, Chapter 9). More than one set of elements representing spatial phenomena can be classified by fuzzy set theory as a member varying degrees of memberships in each class. A membership functions measures the fractional value in a statement, for example, membership of an element x in a set A, is defined by a characteristic functions that indexes the degree to which the object in questions is in the set. The index ranges from 0 (for full non-membership) to 1 (for full memberships). There is an increase in the literature looking at integrating fuzzy reasoning and fuzzy query into GIS and building dynamic simulation models (Davidson, Theocharopoulos, & Bloksma, 1994; Hagen, 2003; Liu, 2009).

1.2.7 Cellular Automata

A cellular automaton is a discrete dynamic system in which space is divided into regular spatial cells, and time progresses in discrete steps. Each cell in the system has one of a finite number of states. The state of each cell is updated according to local rules, that is, the state of a cell at a given time depends on its own state and the states of its neighbours at the previous time step (Wolfram, 1984). Cellular automata (CA) are an example of mathematical systems constructed from many identical components and capable of complex behaviors earth systems. CA with the spatial analysis develops specific models for particular system and abstract general principles applicable to a wide variety of complex system.

Models based on the principles of CA are developing rapidly. CA approach provides a dynamic modeling environment which is well suited to modeling complex environment composed of large number of individual elements. Therefore, CA based urban models usually pay more attention to simulating the dynamic process of urban development and defining the factors or rules driving the development. Different CA models have been developed to simulate urban growth, land use/cover change, and deforestation over time. The differences among various models exist in modifying the five basic elements of CA, i.e., the spatial tessellation of cells, states of cells, neighborhood, transition rules, and time. CA models have demonstrated to be effective platforms for simulating dynamic spatial interactions among biophysical and socio-economic factors associated with land use and land cover change (Clarke, Hoppen, & Gaydos, 1997; Li & Yeh, 2001; Thapa & Murayama, 2010b). The land use change and urban growth process are often compared with the behavior of a cellular automaton in many aspects, for example, the space of an urban area can be regarded as a combination of a number of cells, each cell taking a finite set of possible states representing the extent of its urban development with the state of each cell evolving in discrete time steps according to some local transition rules (Liu, 2009; Tang, Wang, & Yao, 2007). Zhao and Murayama in Chapter 5 stressed that an important component in CA based urban geosimulation models is the local spatial interaction between neighborhood land use types. The neighborhood interaction is often addressed based on the notion that urban development can be conceived as a self-organizing system in which natural constraints and institutional controls, for example, land use policies temper the way in which local decision-making processes produce macroscopic urban form. Different processes can explain the importance of neighborhood interaction.

1.2.8 Artificial Neural Network

Artificial Neural Network (ANN) is a mathematical model based on biological neural networks. It consists of an interconnected group of artificial neurons, and it processes information using a connectionist approach to computation. The neural network contains three types of data layers: input data layers, output data layers, and some hidden layers staying in between the input and output layers (Yuan, Van Der Wiele, & Khorram, 2009; Khoi & Murayama, Chapter 11). Similar to the human brain, the ANN model can be trained by sample data to learn, think, and react to stimulus. During this training process, the initial model parameter values are modified repeatedly until the model can generate acceptable outcomes that match the targeted output values. Li and Yeh (2001) formalized a model to incorporate in the transition functions of the cellular automata using ANN. A three-layer neural network with multiple output neurons was designed to calculate conversion probabilities for multiple land uses. The ANN was trained using data extracted from the local site in GIS.

Through a set of training processes, the neural network was able to generate a number of parameter values automatically, which were imported into the cellular automata model to simulate the multiple land-use change progress. Recently, Khoi and Murayama (Chapter 11) studied deforestation process integrating multilayer perceptron neural network and Markov model and confirms that the neural network framework allows the integration of the driving factors of forest change, whereas the Markov model controls the temporal dynamics of forest change. Furthermore, a multi-regression approach often performs poorly when the relationships between variables are non-linear and some variables must be transformed while the neural network models are good at dealing non-linear relationships and do not require the transformation of variables. However, a fundamental issue in using the neural network approach is that it is essentially a black-box type of model; what happens inside the black box is unknown to the modelers and the users.

1.2.9 Weight of Evidence

Weight of Evidence (WE), entirely based on the Bayesian approach of conditional probability, is traditionally used by geologists to point out areas favorable for geologic phenomena, such as seismicity and mineralization (Bonham-Carter, 1994). The WE method can combine spatial data from variety of sources to describe and analyze interactions, provide evidences for decision making and make predictive models (Soares-Filho et al., 2004; Thapa & Murayama, 2010b). In spatial context, this approach detects the favorability of a certain event, for example, an event of land use change from agricultural area to built-up surface in relation to potential evidences (proximity to urban centers, roads, water, etc.) often called driving factors of change. Weights are estimated from the measured association between the land use change occurrences and the values on the driving factors maps which are to be used as predictors. According to Bonham-Carter (1994), this method is objective that avoids subjective choice of weighting factors. It can combine multiple maps of evidence with a model that is straightforward to program with a modeling language. WE method can accommodate input maps with missing data into the model with the possibility of incorporating multiclass input maps, where each map class may associate with a weight or likelihood ratio. It can model the uncertainty due to variances of weights or missing data. However, the combination of inputs maps assumes that the maps are conditionally independent of one another with respect to the response variable (Thapa & Murayama, 2010b). The testing for conditional independence is only possible where the method is applied in a data driven mode, since it requires overlay data between pairs of evidence maps.

1.2.10 Markov Chain

The outcome of a given experiment can affect the outcome of the next experiment. This type of process is called a Markov Chain (MC). MC follows a set of states where the process starts in one of the states and moves successively from one state to another. Each move is called a step. If the chain is currently in state s_i, then it moves to state s_j at the next step with a probability denoted by p_{ij}, and this probability does not depend upon which states the chain was in before the current state. The probabilities p_{ij} are called transition probabilities. Because of the calculation power of transition probability p_{ij} MC is emerging rapidly as a tool in spatial analysis especially in land change science.

In MC models, the land use/cover change is assumed as a stochastic process, and different categories are the states of a chain. Thus, a chain is defined as a stochastic process having the property that the value of the process at time t, depends only on its value at time t–1 and not on the sequence of values t–2. Although MC model is a typical spatial transition model, early Markovian analysis is used to be a descriptive tool to predict land use change on a local or regional scale (Bell, 1974). Several applications of spatial analysis based on MC are existed in literatures, i.e., Murayama (2000) analyzed land use change pattern in Tokyo using MC; Lopez, Bocco, Mendoza, and Duhau (2001) used MC to simulate the relationships among a set of urban and social variables in predicting land use/cover change in the urban fringe of Morelia city, Mexico; Tang et al. (2007) integrated the MC with genetic algorithm to simulate urban landscape change in Daqing City; and Khoi and Murayama (Chapter 11) simulated the deforestation in Tam Dao National Park in Vietnam incorporating the MC with artificial neural network.

1.2.11 Agent Based Model

An agent based model (ABM) consists of autonomous decision-making entities (agents), an environment through which agents interact, rules that define the relationship between agents and their environment, and rules that determine sequencing of actions in the model. Autonomous agents are composed of rules that translate both internal and external information into internal states, decisions, or actions. While cellular models are focused on landscapes and transitions, ABMs focus on human actions. Agents usually consist of some degree of self-awareness, intelligence, autonomous behaviors, and knowledge of the environment and other agents as well; they can adjust their own actions in response to environmental changes. ABMs are usually implemented as multi-agent systems, a concept originated in the computer sciences that allows for a very efficient design of large and interconnected computer programs (Parker, Manson, Janssen, Hoffman, & Deadman, 2003). The ABM method follows a bottom up approach that has been specifically employed to deal with complexity, especially when coupled with GIS and spatial analysis (An, Linderman, Qi, Shortridge, & Liu, 2005). ABM predicts or explains emergent higher-level phenomena by tracking the actions of multiple low-level agents that constitute the system behaviors. ABM is distinguished from statistical modeling approaches in its focus on the ways in which macro-scale spatial patterns, e.g., urban settlement patterns, result from processes and behaviors of micro-scale actors, e.g., households and firms, and by its ability to represent nonlinear interactions. The agent based spatial analysis has been used to model various land use and land cover change related processes, for instance, reforestation in the Yucatan peninsula of Mexico (Manson, 2000), technology diffusion and resource utilization related to agricultural land use in Chile (Berger, 2001), spatial planning in Netherlands (Ligtenberg, Bregt, & van Lammeren, 2001), ex-urban development in Maryland (Irwin & Bockstael, 2002), agricultural household land-use decision making in the US Midwest (Hoffman, Kelley, & Evans, 2002; Laine & Busemeyer, 2004), agricultural land use decision making by colonist households in Brazilian Amazon (Deadman, Robinson, Moran, & Brondizio, 2004), and migration and deforestation in Philippines (Huigen, 2004).

1.3 Future Perspectives on Spatial Analysis and Modeling

As we discussed in the previous sections, with its historical development over time, the spatial analysis and modeling encompass a wide range of techniques for analyzing, computing, visualizing, simplifying, and theorizing about geographic data. Methods of spatial analysis and modeling can be as simple as taking measurements from a map or as sophisticated as complex geocomputational procedures based on numerical analysis. As we represent real world phenomena in point, line, and area patterns geometrically, how these phenomena pattern themselves and interact with one another have come to be an important element of scientific inquiry. Hypotheses include conjectures about the mapped patterns of poverty and incidence for example, the pattern of residuals from regression, the tendency for some phenomena to cluster or disperse, the differences among patterns, the spatial relationship between a given observation and other designated observations, and perhaps most important, how defined points, lines, and areas, interact with one another over time and space. In this regards, data exploration and analytical tools such as spatial autocorrelation, geographically weighted regression, and spatial metrics can help at first step.

Today, environmental sustainability is becoming main issue among the planners and researchers. Accurate forecast models are taking center stage and there is an increasing amount of research including the perspective of controlling or managing local phenomena. With attention on sustainable development, investigations are being made of measures and policies that steer this research in an appropriate direction. Scenario analysis, which utilizes GIS's high level of simulation functions, is becoming a strong tool for formulating and strengthening measures for such things as community development, regional and urban planning. Better techniques of spatial analysis and modeling of geographical transformation will have applications that span a vast range of societal concerns, for instance, land use planning, traffic and transportation planning, boundary management, hydrology and flood control, landscape characterization and measurement, ecosystem and species diversity management, spatial econometric analysis, climate change, and so on. To solve any of these concerns, we need to handle a diversified data from qualitative to quantitative information. For example, qualitative information may be in linguistic form, that acquired from field work or other sources can be quantified using fuzzy logic or multi-criteria decision making framework such as using AHP which then be considered as an input parameter in other sophisticated modeling frameworks. Voronoi approach provides an excellent platform to solve the boundary problems. Moreover, cellular automata and agent based models are getting much attention in modeling communities for understanding the existing complex environment and extrapolating the future geographical context. Probability theories such as artificial neural network, weight of evidence, fuzzy, Markov chain, genetic algorithm and mixture of these are being emerged with CA and agent based frameworks under various hypotheses and scenarios for projecting the future situation of the real world.

In this beginning of the 21st century, due to rise on global and local environmental change, industrial and economic development, and rapid urbanization process, the spatial analysis and modeling are being conceived as an important tool. At the same time, with the development of highly capable computer technologies and automated techniques, we are being benefitted by availability of unprecedented amounts of finer spatial scale data on aspects of the Earth environment from remote sensing, demographic, social, and economic sources. Although several efforts on the development of spatial analysis techniques for better understanding of the real world phenomena were made by the leading researchers in the past, the spatial analysis and modeling field is becoming more challenging and thrilling even than in before. Spatial and space-time inquires at higher spatial resolution are becoming very important to understand the daily growing complexities created by the human-Earth interactions. It is true that our ability to extract meaning and make useful decisions has not yet kept in pace. We must become better equipped to unleash the power of the new technology for testing and developing theories, identifying important processes, finding meaningful patterns, creating more effective visualizations of data, and for making important societal decisions, which are possible through improving spatial analysis and modeling techniques, and providing more empirical studies and geographical experiences.

1.4 Overview of the Book

This book presents an overview of recent research practices in spatial analysis with GIS. The contributions range from theoretical discussion to model testing and quantitative interpretation of urban and regional social-environmental consequences. The book is substantiated by empirical studies that are originated from academic and applied researches which are grouped into five coherent parts with 16 chapters including the introduction chapter. Chapter 1 provides a synopsis of spatial analysis development in the last century and discusses the contemporary theories, methods, and applications in practice.

The first part of the book consists of four chapters that highlight the major issues in spatial scale, autocorrelation, and neighborhood analysis. Chapter 2 presents a field-based fuzzy qualitative spatial reasoning framework in the case of constraint satisfaction problems and confirms through the empirical test that the humans' vague language can be represented with field-based model in GIS. Chapter 3 verified the validity of local spatial autocorrelation statistics based on k-order neighbour measurement used for pattern analysis of spatial attributes experimenting to the population distribution of elderly residents of Ichikawa City, Japan. The spatial scale effect on urban land use pattern analysis with a case study of Tokyo is presented in Chapter 4. It is found that reducing number of urban land-use types may diminish the loss of information of land-use area across the range of scale and the effect of spatial scale on urban land-use pattern analysis in certain extent. In Chapter 5, a method for modeling and calibrating neighborhood interaction in CA based urban geosimulation is proposed which provides a theoretical framework for presenting neighborhood effect of CA. An experiment of the model is presented in Tokyo Metropolitan Area.

The second part consists of three chapters highlighting zonation and population in urban area. Chapter 6 presents new dasymetric mapping technique based on GIS estimated building population which was computed from building footprints, census tract, and LIDAR derived digital volume model. An empirical case study is shown in Tsukuba City. A GIS approach using the Areametric and Volumetric methods for estimating building population based on census tracts and building footprint datasets is presented in Chapter 7. The results are evaluated using actual building population data by visual, statistical and spatial means, and validated in Tsukuba City for micro-spatial analysis. The authors provide an example by developing a standalone GIS tool called "PopShape GIS" for generating new building footprint with population attribute information based on user-defined criteria. Chapter 8 presents the school choice in the absence of defined attendance areas led to the outcome Public Female Junior High School students in Rasht city in Iran taking longer commutes than necessary. The authors applied multiplicatively weighted Voronoi diagram method to delineate school attendance areas and proposed shorter and more convenient commutes on foot for the students.

The third part of the book is devoted to land use and land cover change analysis comprising of three chapters. Chapter 9 examines four mapping approaches (unsupervised, supervised, fuzzy supervised, and GIS post-processing) for satellite image and their accuracies to extract land use/cover information. The authors present example with an experiment in Tsukuba City, Japan and claim that the spatial statistics of land use/cover derived from remotely sensed images mostly depend on the adaptation of mapping approaches and the GIS post-processing approach. The usefulness of spatial metrics to interpret urban dynamics at two different scales of land-use classification with a case of Yokohama city is discussed in Chapter 10. The authors found that some metrics show similar characteristics in representing the process of urban land-use categories while some are not sensitive in representing urban dynamics. An application of neural network-Markov model to deforestation modeling can be found in Chapter 11. The modeling is conducted in Tam Dao

National Park in Vietnam. The authors present simulated result of the forest cover for 2014 and 2021 highlighting the areas vulnerable to conversion of primary and secondary forest within the national park.

The fourth part of the book covers the multi-criteria GIS analysis with three empirical studies conducted in East Asia. Chapters 12 and 13 show the usefulness of Analytical Hierarchical Process techniques with GIS. The former one identifies the suitable land for peri-urban agriculture in Hanoi province in Vietnam while later one evaluates the potential sites for beekeeping in La Union province, Philippines. The authors in Chapter 14 attempt to assess potential sites for a shipping canal in Kra Isthmus, Thailand based on geophysical parameters, i.e., elevation, sea charts, geology, soils, and river systems. The case study reveals five suitable geographic sites including the required length within the subset of multi-criteria decisions for potential shipping canal on the Isthmus.

The last two chapters that explore the socio-environmental applications of spatial analysis are grouped into fifth part of this book. Chapter 15 shows a method to detect spatiotemporal patterns of urbanization from remotely sensed images and analyze the changing patterns of urbanization with the spatial metrics measures. The authors claim that the urban area in Kathmandu is increasing rapidly which enhances fragmented and heterogeneous landscape development in recent years. The last chapter, Chapter 16, shows an application of spatial analysis for examining poverty determinants in the Philippines. The authors argue that the spatial variation in the incidence of poverty is mainly caused by disparities on access to road infrastructure which is further exacerbated by loopholes and geographical bias in fiscal decentralization priorities and deficiency in agrarian reform implementation. Moreover, proximity to major cities where there is a high concentration of development and economic activities and differences in agro-climatic features, particularly, elevation, slope, and rainfall, also proved to be significant determinants to poverty and suggest the presence of geographically disadvantageous areas within the pilot site.

The first part of this Chapter (Sect. 1.1) is based on Murayama, Y. (2006) which is written in Japanese.

References

Ahmadi Nejad Masouleh, F. (2006). A geographical study of school attendance areas using multiplicatively weighted Voronoi diagrams: A case of Rasht City, Iran. *Geographical Review of Japan, 79*, 700–714.

Alberti, M., & Waddell, P. (2000). An integrated urban development and ecological simulation model. *Integrated Assessment, 1*, 215–227.

An, L., Linderman, M., Qi, J., Shortridge, A., & Liu, J. (2005). Exploring complexity in a human–environment system: An agent-based spatial model for multidisciplinary and multiscale integration. *Annals of the Association of American Geographers, 95*, 54–79.

Anselin, L. (1988). *Spatial econometrics: Methods and models*. Dordrecht: Kluwer.

Anselin, L., Syabri, I., & Kho, Y. (2010). GeoDa: An introduction to spatial data analysis. In M. M. Fischer & A. Getis (Eds.), *Handbook of spatial data analysis* (pp. 73–89). Berlin: Springer.

Bell, E. J. (1974). Markov analysis of land use change: An application of stochastic processes to remotely sensed data. *Socio-Economic Planning Sciences*, *8*, 311–316.

Benenson, I., & Torrens, P. M. (2004). *Geosimulation*. Chichester: Wiley.

Berger, T. (2001). Agent-based spatial models applied to agriculture: A simulation tool for technology diffusion, resource use changes and policy analysis. *Agricultural Economics*, 25, 245–260.

Berry, B. J. L., & Marble, D. F. (1968). *Spatial analysis: A reader in statistical geography*. Englewood Cliffs, NJ: Prentice-Hall.

Bonham-Carter, G. (1994). *Geographic information systems for geoscientists: Modeling with GIS*. New York: Pergamon.

Boyle, P. J., & Dunn, C. E. (1991). Redefinition of enumeration district centroids: A test of their accuracy using Thiessen polygons. *Environmental Planning A*, *23*, 1111–1119.

Clarke, K. C., Hoppen, S., & Gaydos, L. J. (1997). A self-modifying cellular automaton model of historical urbanization in the San Francisco Bay area. *Environment and Planning B*, *24*, 247–261.

Clement, F., Orange, D., Williams, M., Mulley, C., & Epprecht, M. (2009). Drivers of afforestation in Northern Vietnam: Assessing local variations using geographically weighted regression. *Applied Geography*, *29*, 561–576.

Cliff, A. D., & Ord, J. K. (1973). *Spatial autocorrelation*. London: Pion.

Davidson, D., Theocharopoulos, S., & Bloksma, R. (1994). A land evaluation project in Greece using GIS and based on Boolean and fuzzy set methodologies. *International Journal of Geographical Information Systems*, *8*, 369–384.

Deadman, P., Robinson, D., Moran, E., & Brondizio, E. (2004). Colonist household decision making and land-use change in the Amazon rainforest: An agent based simulation. *Environment and Planning B*, *31*, 693–709.

Fisher, M., Scholten, H. J., & Unwin, D. (1996). *Spatial analytical perspectives on GIS, new potential and new models*. London: Taylor & Francis.

Fortin, M.-J., & Dale, M. R. T. (2009). Spatial autocorrelation. In A. S. Fotheringham & P. A. Rogerson (Eds.), *The SAGE handbook of spatial analysis* (pp. 89–103). London: Sage.

Fotheringham, A. S., Brunsdon, C., & Charlton, M. (2002). *Geographically weighted regression: The analysis of spatially varying relationships*. New York: Wiley.

Fotheringham, A. S., & Wegener, M. (2000). *Spatial models and GIS*. London: Taylor & Francis.

Fotheringham, S., & Rogerson, P. (1994). *Spatial analysis and GIS*. London: Taylor & Francis.

Geoghegan, J., Wainger, L. A., & Bockstael, N. E. (1997). Spatial landscape indices in a hedonic framework: An ecological economics analysis using GIS. *Ecological Economics*, *23*, 251–264.

Getis, A., & Ord, J. K. (1992). The analysis of spatial association by use of distance statistics. *Geographical Analysis*, *24*, 189–206.

Hagen, A. (2003). Fuzzy set approach to assessing similarity of categorical maps. *International Journal of Geographical Information Science*, *17*, 235–249.

Hoffman, M., Kelley, H., & Evans, T. (2002). Simulating land cover change in South-central Indiana: An agent-based model of deforestation and afforestation. In M. E. Janssen (Ed.), *Complexity and ecosystem management: The theory and practice of multi-agent systems* (pp. 218–247). Cheltenham: Edward Elgar.

Huigen, M. G. A. (2004). First principles of the MameLuke multi-actor modeling framework for land-use change, illustrated with a Philippine case study. *Journal of Environmental Management*, *72*, 5–12.

Irwin, E. G., & Bockstael, N. E. (2002). Interacting agents, spatial externalities and the evolution of residential land use patterns. *Journal of Economic Geography*, *2*, 31–54.

Isard, W. (1956). *Location and space-economy*. New York: Wiley.

Jaimes, N. B. P., Sendra, J. B., Delgado, M. G., & Plata, R. F. (2010). Exploring the driving forces behind deforestation in the state of Mexico (Mexico) using geographically weighted regression. *Applied Geography*, *30*, 576–591.

Laine, T., & Busemeyer, J. (2004). Comparing agent-based learning models of land-use decision making. In C. L. M. Lovett, C. Schunn, & P. Munro (Eds.), *Proceedings of the 6th international conference on cognitive modeling* (pp. 142–147). Mahwah, NJ: Lawrence Erlbaum Associates.

Li, X., & Yeh, A. G. (2001). Calibration of cellular automata by using neural networks for the simulation of complex urban systems. *Environment and Planning A*, *33*, 1445–1462.

Ligtenberg, A., Bregt, A. K., & van Lammeren, R. (2001). Multi-actor-based land use modelling: Spatial planning using agents. *Landscape and Urban Planning*, *56*, 21–33.

Liu, Y. (2009). *Modelling urban development with geographical information system and cellular automata*. Boca Raton, FL: Taylor and Francis.

Longley, P., & Batty, M. (1996). *Spatial analysis: Modelling in a GIS environment*. Cambridge: GeoInformation International.

Lopez, E., Bocco, G., Mendoza, M., & Duhau, E. (2001). Predicting land cover and land use change in the urban fringe: A case in Morelia city Mexico. *Landscape and Urban Planning*, *55*, 271–285.

Manson, S. M. (2000). *Agent-based dynamic spatial simulation of land use/cover change in the Yucatan peninsula, Mexico*. Proceedings of the 4th international conference on integrating GIS and environmental modeling (GIS/EM4): Problems, prospects and research needs, Banff, AB.

Mendelbrot, B. B. (1983). *The fractal geometry of nature*. New York: W. H. Freeman.

Mu, L. (2004). Polygon characterization with the multiplicatively weighted Voronoi diagram. *Professional Geographer*, *56*, 223–239.

Murayama, Y. (2000). Land use change in Tokyo. In Y. Murayama (Ed.), *Japanese urban system* (pp. 227–236). Dordrecht: Kluwer.

Murayama, Y. (2006). Kukan-bunseki to GIS (Spatial analysis with GIS). In A. Okabe, & Y. Murayama (Eds.), *GIS de Kukan-bunseki (Spatial analysis using GIS)* (pp. 1–20). Tokyo: Kokon-shoin in Japanese.

Nakaya, T. (2008). Geographically weighted regression (GWR). In K. K. Kemp (Ed.), *Encyclopedia of geographic information science* (pp. 179–184). London: Sage.

Nordbeck, S., & Rystedt, B. (1972). *Computer cartography: The mapping system NORMAP: Location models*. Lund: Studentlitteratur.

Ogneva-Himmelberger, Y., Pearsall, H., & Rakshit, R. (2009). Concrete evidence & geographically weighted regression: A regional analysis of wealth and the land cover in Massachusetts. *Applied Geography*, *29*, 478–487.

Okabe, A., Boots, B., Sugihara, K., & Chiu, S. N. (2000). *Spatial tessellations: Concepts and applications of Voronoi diagrams*. Chichester: Wiley.

Ord, J. K., & Getis, A. (1995). Local spatial autocorrelation statistics: Distributional issues and application. *Geographical Analysis*, *27*, 286–306.

O'Sullivan, D., & Unwin, D. J. (2003). *Geographic information analysis*. Hoboken, NJ: Wiley.

Páez, A., & Wheeler, D. C. (2010). Geographically weighted regression. In M. M. Fischer, & A. Getis (Eds.), *Handbook of spatial data analysis* (pp. 461–486). Berlin: Springer.

Parker, D. C., Evans, T. P., & Meretsky, V. (2001). *Measuring emergent properties of agent-based landuse/landcover models using spatial metrics*. Proceedings of 7th annual conference of the international society for computational economics, Yale University.

Parker, D., Manson, S., Janssen, M., Hoffman, M., & Deadman, P. (2003). Multiagent system models for the simulation of land-use and land-cover change: A review. *Annals of the Association of American Geographers*, *93*, 316–340.

Ripley, B. (1981). *Spatial statistics*. Chichester: Wiley.

Saaty, T. L. (1980). *The analytic hierarchy process*. New York: McGraw-Hill.

Soares-Filho, B. S., Alencar, A., Nespad, D., Cerqueira, G. C., Dial, M., Del, C., et al. (2004). Simulating the response of land-cover changes to road paving and governance along a major Amazon Highway: The Santarem-Cuiaba corridor. *Global Change Biology*, *10*, 745–764.

Stefanakis, E., Vazirgiannis, M., & Sellis, T. (1999). Incorporation fuzzy set methodologies in DBMS repository for the application domain of GIS. *International Journal of Geographical Information Science*, *13*, 657–675.

Tang, J., Wang, L., & Yao, Z. (2007). Spatio-temporal urban landscape change analysis using the Markov chain model and a modified genetic algorithm. *International Journal of Remote Sensing*, *28*, 3255–3271.

Thapa, R. B., & Murayama, Y. (2009). Land use change factors in Kathmandu valley: A GWR approach. In B. G. Lees, & S. W. Laffan (Eds.), *Proceedings of the 10th international conference on geocomputation* (pp. 255–260). Sydney, NSW: The University of New South Wales.

Thapa, R. B., & Murayama, Y. (2010a). Drivers of urban growth in the Kathmandu valley, Nepal: Examining the efficacy of the analytic hierarchy process. *Applied Geography*, *30*, 70–83.

Thapa, R. B., & Murayama, Y. (2010b). Urban growth modelling of Kathmandu metropolitan region, Nepal. *Computers, Environment and Urban Systems*. DOI: 10.1016/j.compenvurbsys.2010.07.005.

Tobler, W. (1970). A computer movie simulating urban growth in the Detroit region. *Economic Geography*, *46*, 234–240.

Wolfram, S. (1984). Cellular automata as models of complexity. *Nature*, *311*, 419–424.

Yuan, H., Van Der Wiele, C. F., & Khorram, S. (2009). An automated artificial neural network system for land use/land cover classification from Landsat TM imagery. *Remote Sensing*, *1*, 243–265.

Zadeh, L. A. (1965). Fuzzy sets. *Information Control*, *8*, 338–353.

Part I
Spatial Scale, Autocorrelation and Neighborhood Analysis

Chapter 2
Field-Based Fuzzy Spatial Reasoning Model for Constraint Satisfaction Problem

Yaolong Zhao, Yumin Zhang, and Yuji Murayama

2.1 Introduction

While geographical information systems (GIS) have attracted many attentions for its strong functions of managing spatial information and providing spatial decision-making tools to users, there has been little change in the functionality of the systems (Guesgen & Albrecht, 2000). The way in which they perform spatial reasoning, i.e., the extraction of new information from stored spatial data, has been intuitively quantitative in nature. On the other hand, humans often prefer a qualitative analysis over a quantitative one, as this is more adequate in many cases from the cognitive point of view (Clementini, Felice, & Hernández, 1997). For instance, in the sentence *to find a parcel of wasteland which is not far from a reservoir*, extension space of the word *not far* is so vague that current spatial analysis functions of GIS software are difficult to solve such analogous problems. Dealing with these problems entails qualitative spatial representation and reasoning approaches.

Spatial decision-making under the support of GIS can be conceived as technique extension of traditional map analysis function. As most spatial expressions in natural language are purely qualitative, the lack of function of effective qualitative spatial reasoning like Artificial Intelligence (AI) in software inevitably limits the application of GIS in the spatial decision-making process. Stefanakis, Vazirgiannis, and Sellis (1999) discussed in detail the significance of providing efficient tools in GIS packages to decision-makers who use GIS, and mainly emphasized the uncertainty of geographical phenomena in GIS.

Two options are available for extending the functions of GIS: to extend the functions of spatial analysis of GIS software repository with qualitative spatial

Y. Zhao (✉)
School of Geography, South China Normal University, Guangzhou, Guangdong, PR China
e-mail: yaolongzhao@gmail.com

This chapter is improved from "Yaolong Zhao, Yumin Zhang, and Yuji Murayama (2005), Field-based fuzzy spatial reasoning model for geographical information systems: Case of constraint satisfaction problem, Theory and Applications of GIS, 13, 21–31", with permission from GIS Association, Japan.

Y. Murayama, R.B. Thapa (eds.), *Spatial Analysis and Modeling in Geographical Transformation Process*, The GeoJournal Library 100, DOI 10.1007/978-94-007-0671-2_2,

reasoning and to implement qualitative spatial analysis (reasoning) functions in special application systems associated with GIS through extension programming. But both of the options must face the difficulties of describing uncertainty, especially fuzzy uncertainty, of real-life geographical phenomena and ambiguity of human language. All the confronted problems influence the implementation of qualitative spatial reasoning function in GIS. Fuzzy uncertainty of real-life geographical phenomena mainly affects the representation and store of spatial knowledge. For examples, the boundary between woodland and grassland, and between urban and rural areas, may be gradual through a transition zone rather than a crisp boundary. Some researchers attempt to find alternative approaches to represent the type of phenomena (Molennar & Cheng, 2000; Cheng, Molenaar, & Lin, 2001) and topological relations of these special objects (Bjørke, 2004), whereas the approaches have yet not been effectively embedded in current commercial GIS software. Ambiguity of human language is more complex as it is difficult to model human natural language which usually contains vague instructions. Laudably, some literatures in other disciplines are looking at the problems (Bloch & Ralescu, 2003; Claramunt & Thériault, 2004; Renz, 2002). Fuzzy sets show superiority in representing qualitative phenomena. This research proposes a framework of field-based spatial reasoning using fuzzy set theory through which qualitative description usually encountered in spatial analysis function can be handled quantitatively.

The remainder is organized as follows. The next section reviews the definition of fuzzy sets and options of fuzzy membership functions. Section 2.3 discusses object-based and field-based data models applied to represent spatial entities in GIS, and put forward field-based model for qualitative spatial knowledge. This is followed by proposing field-based fuzzy spatial reasoning models in the case of constraints satisfaction problems. Then a case study compares the approach with traditional spatial reasoning process. Sections 2.6 and 2.7 give some discussions and summarize the results.

2.2 Fuzzy Sets

2.2.1 Fuzzy Set Theory

In classical set theory (Boolean logic) an individual is a member or not a member of any given set. The membership degree to which the individual z belongs to the set A is expressed by the membership function, μ_A, which can take the value 0 or 1, i.e.,

$$\mu_A(z) = \begin{cases} 1, & c_1 \leq z \leq c_2 \\ 0, & z \geq c_1 \; or \; z \leq c_2 \end{cases} \tag{2.1}$$

where, c_1 and c_2 define the boundaries of set A. For example, if the boundaries between the direction "north", "east", "south" from object O_1 to object O_2 were to

Fig. 2.1 Definition of azimuth θ from object O_1 to O_2

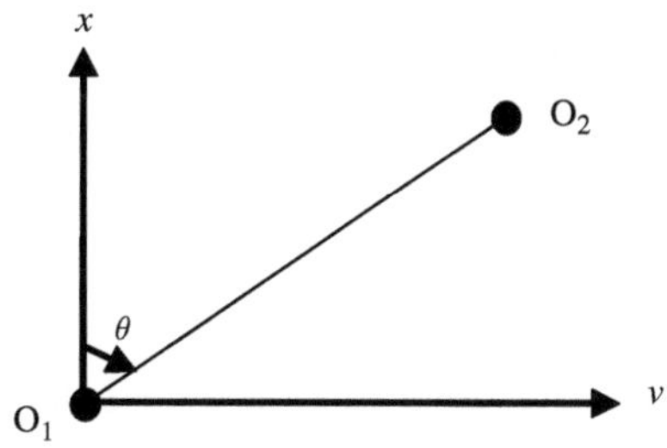

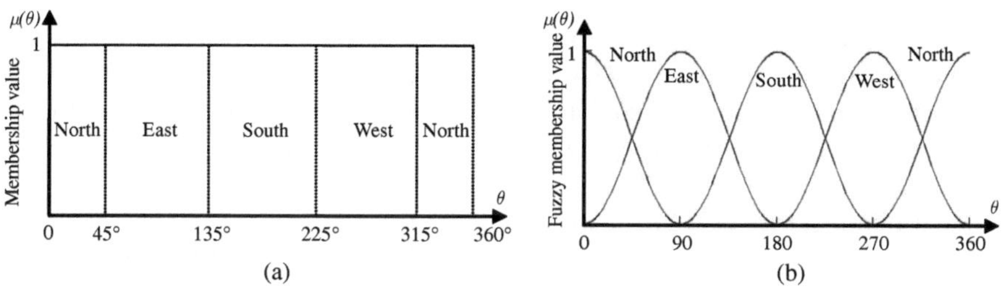

Fig. 2.2 An example of the classic (**a**) and fuzzy (**b**) classification for direction. θ stands for the azimuth from object O_1 to O_2

be set at azimuth θ_1=45° and θ_2=135° from O_1 to O_2 (Fig. 2.1), then the membership function defines the direction "east" (Fig. 2.2a). Note that classical sets allow only binary membership functions (i.e., TRUE or FALSE).

Fuzzy set theory (Zadeh, 1965) is an extension of classical set theory. A fuzzy set A in a universe Z of discourse is characterized by a membership function $\mu_A(z)$ which associated with a real number in the interval [0, 1], representing the "degree of membership of z in A". Thus, the nearer the value of $\mu_A(z)$ to unity, the higher the grade of membership of z in A. That is to say, $\mu_A(z)$ of z in A specifies the extent to which z can be regarded as belonging to set A. The fuzzy sets can be represented as a set of ordered pairs:

$$A = \{z, \mu_A(z)\}, z \in Z \tag{2.2}$$

2.2.2 Fuzzy Membership Function

The choice of fuzzy membership function, i.e., its shape and form, is crucial in fuzzy sets application. In correspondence to classical set theory, two options are available for choosing the membership functions for fuzzy sets (Burrough, 1996): (*a*) through an imposed "expert" model; and (*b*) by a data driven multivariate procedure.

The first approach uses a priori membership function, based on expert knowledge, with which individual entities can be assigned a degree of membership

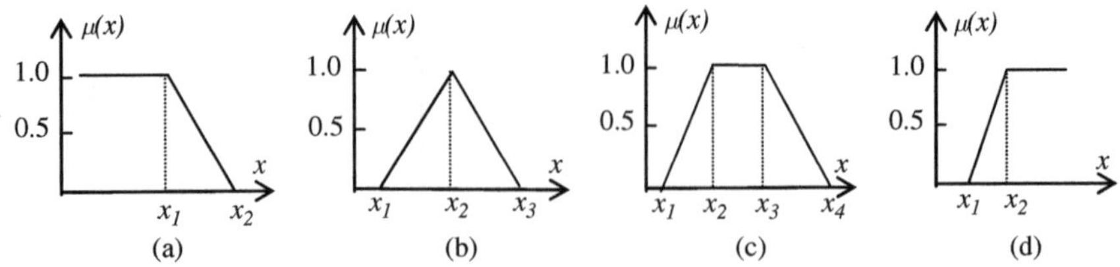

Fig. 2.3 Several conventional membership functions. (**a**) L Function; (**b**) Λ Function; (**c**) Π Function; (**d**) Γ Function

regarding a lexical value characterizing a theme. This method is known as the semantic import approach or model. Each of these functions has its own characteristics and its behavior may simulate better or worse various physical phenomena. Several conventional linear models which can be used as membership function are shown in Fig. 2.3.

Some nonlinear models also can be applied to represent fuzzy phenomena in terms of their characteristics. In direction example, we choose $\sin^2(\theta)$ and $\cos^2(\theta)$ referring to relative position relation functions proposed by Miyajima and Ralescu (1994) to describe the direction [north, east, south, west] from object O_1 to O_2 (Fig. 2.1) as follows.

$$\mu_{\text{north}}(\theta) = \begin{cases} \cos^2(\theta), & \frac{3\pi}{2} \leq \theta \leq 2\pi \; or \, 0 \leq \theta \leq \frac{\pi}{2} \\ 0, & \frac{\pi}{2} \leq \theta \leq \frac{3\pi}{2} \end{cases} \tag{2.3}$$

$$\mu_{\text{east}}(\theta) = \begin{cases} \sin^2(\theta), & 0 \leq \theta \leq \pi \\ 0, & \pi \leq \theta \leq 2\pi \end{cases} \tag{2.4}$$

$$\mu_{\text{south}}(\theta) = \begin{cases} \cos^2(\theta), & \frac{\pi}{2} \leq \theta \leq \frac{3\pi}{2} \\ 0, & \frac{3\pi}{2} \leq \theta \leq 2\pi \; or \, 0 \leq \theta \leq \frac{\pi}{2} \end{cases} \tag{2.5}$$

$$\mu_{west}(\theta) = \begin{cases} \sin^2(\theta), & \pi \leq \theta \leq 2\pi \\ 0, & 0 \leq \theta \leq \pi \end{cases} \tag{2.6}$$

where, θ stands for the azimuth from object O_1 to O_2. Figure 2.2b illustrates the fuzzy classification of direction values. Note that the classic way (Fig. 2.2a) to classify direction involves discrete classes with special ranges, while fuzzy classification captures the graduate transition between classes, providing a better way to categorize imprecise concepts, such as north and east direction. It is more "human-like" or "cognitively adequate" than the classic way.

In the second approach the choice of the membership functions is data-driven in the sense that they are locally optimized to match data set. This method is analogous to cluster analysis and numerical taxonomy (Kaufman & Rousseuw, 1990) and is known as the natural classification model.

Whichever approach is chosen, the ultimate form or shape of the function should be "human-like" and close to the reality.

2.3 Field-Based Qualitative Spatial Representation

2.3.1 Field-Based Models

To implement spatial decision-making for users, GIS have to provide the ability of representing spatial phenomena which intersperse in the space under certain data models. As pointed out by Goodchild (1992), the GIS data models are divided into two broad categories. First, entity-based or object-based data models which represent geographic data set conceived as collections of discrete objects littering an otherwise empty space, and able to overlap freely. Second, field-based models which represent variation conceived as being spatially continuous, such that for every point in the plane there is exactly one value of the field.

Spatial entity refers to a phenomenon that can not be subdivided into like units (Laurini & Thompson, 1992). A house is not divisible into houses, but can be split into rooms. The discrete object view represents the world as objects with well-defined boundaries in empty space (Longley, Goodchild, Maguire, & Rhind, 2001). Entity-based data models catch the characteristics of categories of spatial phenomena and make it easy to represent spatial phenomena in vector data structure. Therefore, the models show superiority in representing spatial entity with crisp (well-defined) boundaries. But they are not suited to mapping poorly defined phenomena (Zhang & Stuart, 2001) and qualitative description.

Field-based data models come from field view of representing geography. For example, while we might think of terrain as composed of discrete mountain peaks, valleys, ridges, slopes, etc., there are unresolvable problems of definition for all of these objects. Instead, it is much more useful to think of terrain as a continuous surface, in which elevation can be defined rigorously at every point. Such continuous surfaces form the basis of the common view of geographic phenomena, known as the field view. In this view the geographic world can be described by a number of variables, each measurable at any point on the Earth's surface, and changing in value across the surface (Longley et al., 2001).

In digital representation, field-based data models can be represented as the following continuous two-order relationship on 2-D plane N^2 (Zhao, Zhao, & Wang, 1999):

$$R = \int_{(x,y)} \frac{\mu_R(x,y)}{(x,y)} \quad x, y \in N^2 \tag{2.7}$$

where, fuzzy membership value $\mu_R(x,y)$ represents attribute density of a surface feature character on the point (x,y). That is to say, it stands for the extent to which a point belongs to one class (object). If $\mu_R(x,y)$ equals anyone of both numbers $\{0, 1\}$, it indicates that all the objects in real-life have crisp boundaries. If $\mu_R(x,y)$ is a numerical value in the interval [0, 1], R becomes a fuzzy sets (Zadeh, 1965) and the model can represent fuzzy geographical phenomena.

The relationship can be expressed with a 2-D matrix in which row and column numbers are the coordinates of the spatial surface feature. For example, urban area can be represented as following Fig. 2.4:

The value of cells in Fig. 2.4 stands for the extent to which the cell belongs to urban area. 1.0 indicates that the cell belongs to the classification of urban area entirely; $0 <$ the value < 1.0 means that the cell belongs to the urban area partly; 0, the cell does not be characterized by urban area at all. The two-order relationship reflects the field view of geographic phenomena.

2.3.2 Field-Based Qualitative Spatial Representation

In geographical space, reasoning on spatial entities is supported by natural language representations that involve direction, topological, ordinal, distance, size and shape relationships (Pullar & Egenhofer, 1988). In qualitative spatial reasoning, it is common to consider the main spatial aspects: topology, direction, distances and to develop a system of qualitative relationships between spatial entities which cover this spatial aspect to some degree and which appear to be useful from an application or from a cognitive perspective (Renz, 2002). Many traditional commercial GIS soft wares include the functions of representing topological relationships of spatial

0	0	0	0	0	0.1	0.1	0.1	0	0	0	0
0	0	0	0.1	0.1	0.2	0.2	0.2	0.1	0.1	0	0
0	0.1	0.2	0.2	0.3	0.4	0.4	0.4	0.3	0.3	0.2	0.1
0.1	0.2	0.4	0.5	0.5	0.7	0.7	0.6	0.5	0.5	0.3	0.2
0.1	0.3	0.4	0.6	0.7	0.9	0.9	0.8	0.7	0.6	0.4	0.3
0.3	0.4	0.6	0.7	0.9	1.0	1.0	1.0	0.8	0.7	0.5	0.3
0.3	0.5	0.7	0.9	1.0	1.0	1.0	1.0	1.0	0.9	0.7	0.5
0.4	0.6	0.8	1.0	1.0	1.0	1.0	1.0	1.0	1.0	0.8	0.6
0.4	0.6	0.8	1.0	1.0	1.0	1.0	1.0	1.0	1.0	0.8	0.6

(a)

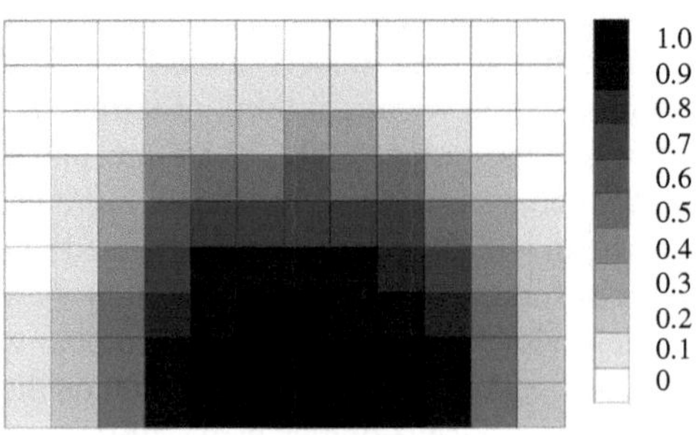

(b)

Fig. 2.4 Fuzzy representation of spatial object – urban area. (**a**) In numerical form; (**b**) in *grey* form. *Grey values* hierarchy is shown in the *right* of (**b**) entities. Therefore in this study we just consider the representation of other two important spatial aspects: direction and distance.

When we try to describe the relationships of spatial entities we always select one of spatial entities as a reference object. The relationships of other spatial entities with the reference object are associated with the locations of them. That is to say, the relationships can be considered as a field surrounding the reference object. Therefore, the spatial aspects also can be represented using the field-based models as continuous two-order relationship on 2-D plane N^2 (see Eq. 2.7). Here, the fuzzy membership value $\mu_R(x,y) \in [0, 1]$ represents the degree to which the point (x, y) belongs to the relationship representation of between spatial entities according to their location. We name the approach as field-based qualitative spatial representation structure.

2.3.2.1 Representation of Direction

Direction – also called orientation – relationships of spatial entities with respect to other spatial entities is usually given in terms of a qualitative category like "to the north of" rather than using a numerical expression like "12 degrees" (which is certainly more common in technical communication like in aviation). It is important and common-sense linguistic and qualitative property used in everyday situations and qualitative spatial reasoning (Frank, 1996). Direction of spatial entities is a ternary relationship depending on the located object, the reference object, and the frame of reference which can be specified either by a third object or by a given direction. In the literature one distinguishes between three different kinds of frames of reference, extrinsic ("external factors impose a direction on the reference object"), intrinsic ("the direction is given by some inherent property of the reference object"), and deictic ("the direction is imposed by the point of view from which the reference object is seen") (Hernández, 1994). Given the frame of reference, direction can be expressed in terms of binary relationships with respect to the frame.

Most approaches to qualitatively dealing with direction are based on points as the basic spatial entities and consider only two-dimensional space. Frank (1991) suggested different methods for describing the cardinal direction of a point with respect to a reference point in a geographic space, i.e., directions are in the form of [north, east, south, west] depending on the granularity. Without loss of generality, here we consider all objects even that with irregular shape and size as a point and follow frank's suggestion about direction (centroid-based method, where the direction between two objects is determined by the angle between their controids). According to Fig. 2.1, the azimuth θ from object O_1 to object O_2 is computed. This angle, denoted by $\theta(O_1, O_2)$, takes values in $[0, 2\pi]$, which constitutes the domain on which primitive directional relations are defined.

We choose $\sin^2(\theta)$ and $\cos^2(\theta)$ as fuzzy membership functions to describe the direction [north, east, south, west] referring to relative position relation functions proposed by Miyajima and Ralescu (1994) (Eqs. 2.3, 2.4, 2.5 and 2.6 and Fig. 2.2b). Miyajima and Ralescu (1994) have used the square trigonometric function to illustrate relative position relations [above, right, below, right] of segmented images. The square trigonometric functions are also suitable for the direction in the form of [north, east, south, west]. For instance, in Fig. 2.1, if $\theta = 50°$, then the direction relationship is [0.4132, 0.5868, 0, 0] in the form of [north, east, south, west] according to Eqs. (2.3), (2.4), (2.5), and (2.6). It means that object O_2 is located to the north of object O_1 with 0.4132 of membership degree and to the east with 0.5868 of membership degree. That is, $\mu_{north}(O_1, O_2)$= 0.4132, $\mu_{east}(O_1, O_2)$= 0.5868, $\mu_{south}(O_1, O_2)$= 0, $\mu_{west}(O_1, O_2)$= 0, and $\mu_{north}(O_1, O_2)$+ $\mu_{east}(O_1, O_2)$+ $\mu_{south}(O_1, O_2)$+ $\mu_{west}(O_1, O_2)$=1. Therefore, the fuzzy membership functions not only show the characteristics of transition of direction relationship but also ensure the integrality of definition of direction for any target object.

In GIS, the direction relations can be seen as field view in order to represent them. An example is given to illustrate the representation of direction relations using field-based qualitative spatial representation structure. In the example, we require representing the linguistic sentence "to the north of a city". According to Eq. 3, we get the field-based representation in raster structure like Fig. 2.5a as numerical form and Fig. 2.5b as grey form. Here, the city is abstracted as a point.

In Fig. 2.5, the marked point with "C" stands for the city. The number in Fig. 2.5a or the grey grade in Fig. 2.5b denotes the extent to which the location belongs to the linguistic sentence "to the north of a city".

2.3.2.2 Representation of Distance

In spatial decision-making process, distance relation between spatial entities always plays a key role. Dealing with distance information is an important cognitive ability in our everyday life (Renz, 2002). When representing distance, we usually use qualitative categories like "A is close to B" (binary constraint) or qualitative distance comparatives like "A is closer to B than to C" (ternary constraint), but also numerical

0.7	0.7	0.8	0.9	0.9	1.0	0.9	0.9	0.8	0.7	0.7
0.6	0.7	0.7	0.8	0.9	1.0	0.9	0.8	0.7	0.7	0.6
0.6	0.6	0.7	0.8	0.9	1.0	0.9	0.8	0.7	0.6	0.6
0.5	0.6	0.7	0.8	0.8	1.0	0.8	0.8	0.7	0.6	0.5
0.4	0.5	0.6	0.7	0.8	1.0	0.8	0.7	0.6	0.5	0.4
0.4	0.4	0.5	0.6	0.8	1.0	0.8	0.6	0.5	0.4	0.4
0.3	0.4	0.4	0.5	0.7	1.0	0.7	0.5	0.4	0.4	0.3
0.1	0.2	0.3	0.4	0.5	1.0	0.5	0.4	0.3	0.2	0.1
0	0	0	0	0	C	0	0	0	0	0

(a)

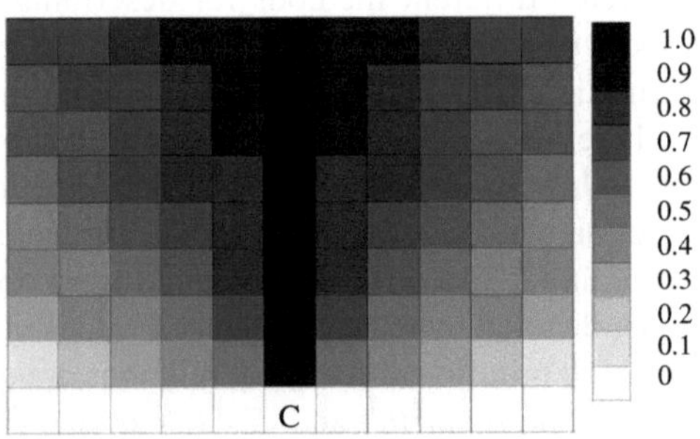

(b)

Fig. 2.5 Field-based representation of "to the north of city C". (**a**) In numerical form; (**b**) in *grey* form. *Grey values* hierarchy is shown in the *right* of (**b**) values like "A is about 20 m away from B". One can distinguish between absolute distance relations (the distance between two spatial entities) and relative distance relations (the distance between two spatial entities as compared to the distance to a third entity) (Guesgen & Albrecht, 2000; Renz, 2002). The choice which relations should be used depends on the application domain and the requirements posed by decision-makers. In this work we restrict ourselves to absolute distance relations, i.e., binary constraints. For two individual locations A, B, which are abstracted as points in general, the Euclidean distance is given as follows:

$$d(A,B) = \sqrt{(x_A - x_B)^2 + (y_A - y_B)^2} \tag{2.8}$$

where (x_A, y_A) and (x_B, y_B) denote the coordinates of two locations A, B, respectively.

Qualitative absolute distance relations are obtained, e.g., by dividing the real line of distance into several sectors such as "very close", "close", "commensurate", "far", and "very far" depending on the chosen level granularity (Hernández, Clementini, & Felice, 1995). In practice we usually use one of the sectors. For instance, Fig. 2.6 represents the "close" degree from a point on the map to a city.

Where,

$$\mu_{\text{close}}(x) = \begin{cases} 1, & x \leq 5 \\ (20 - x)/15 & 5 \geq x \leq 20 \\ 0, & x \leq 20 \end{cases} \tag{2.9}$$

x denotes the distance (in kilometer) from a location to the city. The division values such as 5 and 20 km are designed arbitrarily by decision-makers according to the understanding of definition of "close" degree.

In field-based model, distance relations are represented as a field surrounding the reference spatial entity. Every location in the field has a membership value that means the extent to which the location belongs to the qualitative absolute distance relations. Herein we use field-based representation of above example to illustrate the model (Fig. 2.7). Where, the marked point with "C" stands for the city. The number in Fig. 2.7a or the grey grade in Fig. 2.7b denotes the grade of fuzzy membership, the extent to which the location belongs to the linguistic sentence "close to a city".

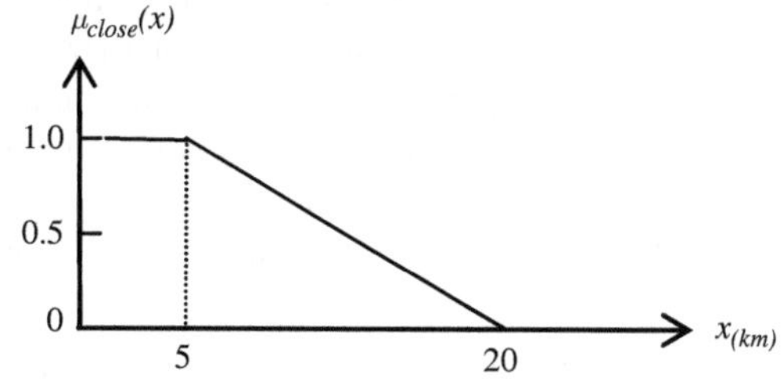

Fig. 2.6 Membership function of "close to a city"

0	0	0	0	0	0	0	0	0	0	0
0	0	0	0	0.1	0.1	0.1	0	0	0	0
0	0.1	0.1	0.2	0.2	0.3	0.2	0.2	0.1	0.1	0
0.1	0.2	0.3	0.3	0.4	0.5	0.4	0.3	0.3	0.2	0.1
0.2	0.3	0.4	0.5	0.6	0.7	0.6	0.5	0.4	0.3	0.2
0.3	0.5	0.7	0.7	0.8	0.9	0.8	0.7	0.5	0.5	0.3
0.4	0.6	0.7	0.9	1.0	1.0	1.0	0.9	0.7	0.6	0.4
0.5	0.7	0.8	1.0	1.0	1.0	1.0	1.0	0.8	0.7	0.5
0.5	0.7	0.9	1.0	1.0	C	1.0	1.0	0.9	0.7	0.5

(a)

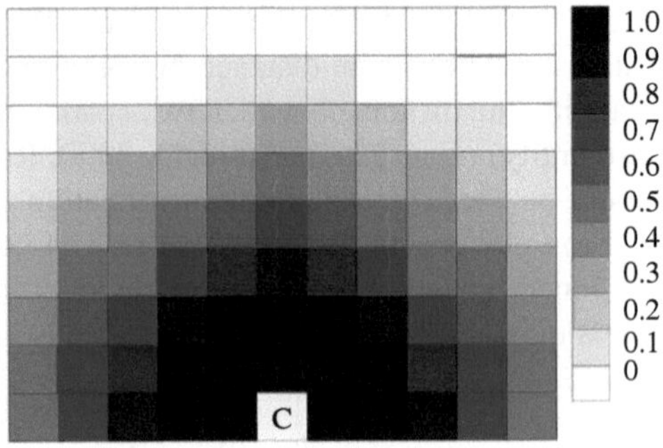

(b)

Fig. 2.7 Field-based representation of "close to a city C". (**a**) In numerical form; (**b**) in *grey* form. *Grey values* hierarchy is shown in the *right* of (**b**)

2.4 Field-Based Fuzzy Spatial Reasoning

Spatial qualitative reasoning is an approach to deal with commonsense spatial knowledge. While knowledge about spatial entities or the relationships between spatial entities are often given in the form of constraints. Ordinarily binary constraints like "the primary school should be laid out in the north of the residential area", ternary constraints like "the primary school should be laid out between residential area A and residential area B", or in general n-ary constraints restrict the domain of 2, 3, or n variables. Problem like this is formalized as a constraint satisfaction problem (CSP): given a set of variables R over a domain D and a set A of constraints on the variables R (Renz, 2002). The constraint satisfaction problem is a powerful general framework in which a variety of combinatorial problems can be expressed (Creignou, Khanna, & Sudan, 2001; Marriott & Stuckey, 1998). The aim in a constraint satisfaction problem is to find an assignment of values to the variables subject to specified constraints. In fact, it is the most popular reasoning methods used in qualitative spatial reasoning (Renz, 2002) and a common problem in spatial decision-making process. In this research we restrict to binary CPSs, i.e., CSPs where only binary constraints are used.

To deal with the problem, Ladkin and Maddux (1994) formulated binary CSPs as relation algebras developed by Tarski (1941). This allows treating binary CSPs in uniform way. In fuzzy domain, the relation algebras constitute fuzzy logic reasoning. Fuzzy logic reasoning, one of application domains of fuzzy relationship generalization, is the fundamental of fuzzy spatial reasoning. It implements tasks through logical operations based on usual relation algebra theory. The standard operations of *union*, *intersection*, and *complement* of fuzzy relationship (or fuzzy sets) A_1 and A_2, defined over some domain C, create a new fuzzy relationship (or fuzzy sets) whose membership function is defined as:

$$\text{Union}: \mu_R(z) = \mu_{A1(z)\cup A2(z)}(z) = \max\{\mu_{A1}(z), \mu_{A2}(z)\}, z \in C \quad (2.10)$$

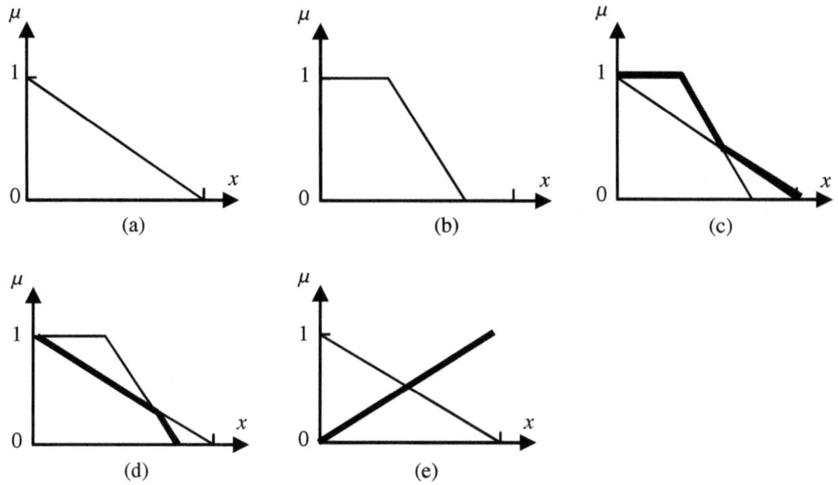

Fig. 2.8 Diagram of fuzzy spatial reasoning operations. (**a**) Fuzzy membership function of A_1. (**b**) Fuzzy membership function of A_2. (**c**) Union of A_1 and A_2. (**d**) Intersection of A_1 and A_2. (**e**) Complement of A_1. *Thick lines* in (**c**), (**d**), and (**e**) stand for the result of operations

$$\text{Intersection}: \mu_R(z) = \mu_{A1(z)\cap A2(z)}(z) = \min\{\mu_{A1}(z), \mu_{A2}(z)\}, z \in C \tag{2.11}$$

$$\text{Complement: } \mu_R(z) = \mu_{A'}(z) = 1 - \mu_A(z), z \in C \tag{2.12}$$

Figure 2.8 illustrates the significance of the operations intuitively. The operations also can be extended to n sets of fuzzy relationships, that is, the operation is applicable to multiple fuzzy sets. Assume that there are n sets of fuzzy relationships. Operations above can be expressed uniformly as:

$$\mu_R(z) = \bigotimes_{i=1}^{n} \mu_{A_i}(z) \quad z \in C \tag{2.13}$$

where, ⊗ denotes operators *union*, *intersection*, and *complement* respectively, and A_i stands for multiple fuzzy relationships. The results of operations stand for the suitableness of satisfying CSPs.

2.5 Application to an Illustrative Case Study

An example is given to illustrate the application of field-based fuzzy spatial reasoning model. In this example, the task is to find a suitable location for a special factory given certain constraining factors as follows:

(a) The factory must be located to the east of the environmental monitoring station.
(b) The factory must not be far from the environmental monitoring station.
(c) The factory must not be situated on land suitable for agriculture.

It is difficult to handle the task just using existing spatial analysis function of traditional GIS software as the constraining factors described above could not be transformed into numeric representation suitable for existing geographical information systems. Alternative way suitable for conventional spatial overlay analysis is to alter the description manner of the constraining factors as:

(a) The azimuth from environmental monitoring station to the factory must be between 45° and 135°.
(b) The distance from the factory to environmental monitoring station must be between 4 and 9 km.
(c) The land must not fall into agricultural category field.

In existing GIS software, this type of reasoning uses the concepts of Boolean map overlay. This is a method for merging different datasets to produce a final output. Various functions are available, including a buffer operator, which increases the size of an object by extending its boundary, and logical operators, such as union, intersection and complement. Figure 2.9 illustrates the process of traditional approach in GIS. But the alternative way is often too rigid and therefore not applicable to scenarios like factory scenario. The reason is that the description like *between 4 and 9 km* may eventually result in an empty map, as they restrict the search space too dramatically by excluding an areas, for example, which are 9.1 km from environmental monitoring station. Such an area, however, might be the best choice available and therefore perfectly acceptable.

It is evident that the result of alternative way in GIS has clear boundary (Fig. 2.9d). If the agriculture area is larger than that of above example, it may be

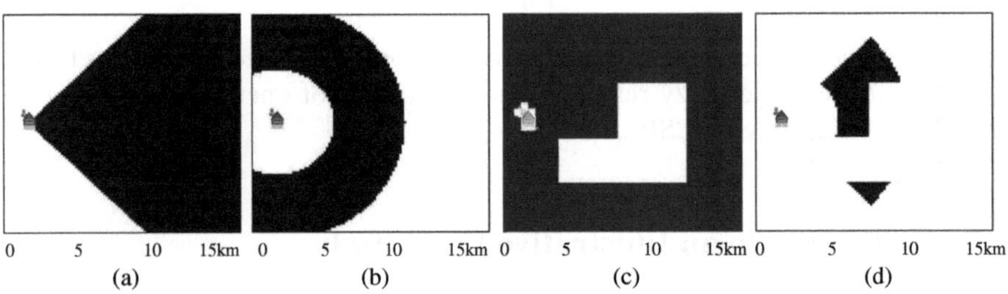

Fig. 2.9 Diagram of traditional approach process in GIS for the example. The *point with a symbol of house* denotes the environmental monitoring station. Satisfaction areas of constraints are expressed as *black*. (**a**) Direction constraint. (**b**) Distance constraint. (**c**) Non agricultural area constraint. (**d**) The result satisfying the constraints

there are no areas satisfying the constraints. Now we solve the problems using field-based fuzzy spatial reasoning approach. A possible solution might be as following:

1. Determine fuzzy membership function for the description *to the east of the environmental monitoring station*, and constitute fuzzy relationship $A_1 = \int_{(x,y)} \frac{\mu_{A_1}(x,y)}{(x,y)}$ on research area by calculating membership value of cell in term of fuzzy membership function. Here we select the membership function discussed in Section 2.3.1 (Fig. 2.10a).
2. Determine fuzzy membership function for *not be so far from the environmental monitoring station*, and constitute fuzzy relationship $A_2 = \int_{(x,y)} \frac{\mu_{A_2}(x,y)}{(x,y)}$. Here we select the following function:

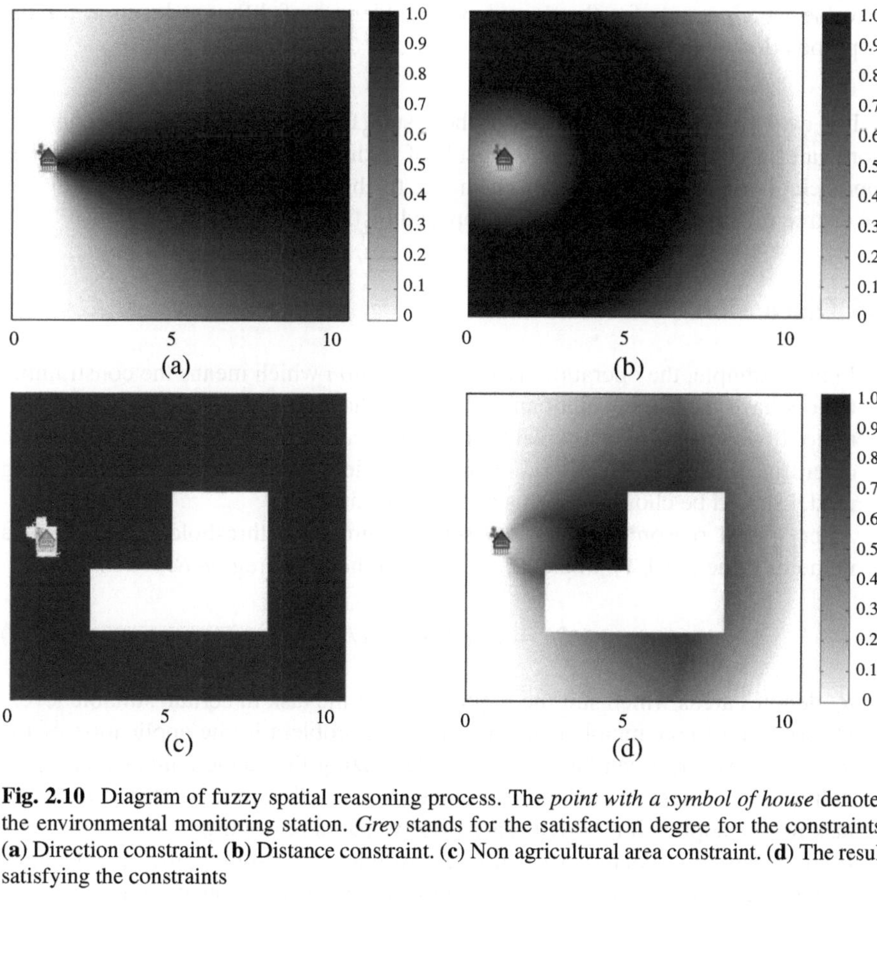

Fig. 2.10 Diagram of fuzzy spatial reasoning process. The *point with a symbol of house* denotes the environmental monitoring station. *Grey* stands for the satisfaction degree for the constraints. **(a)** Direction constraint. **(b)** Distance constraint. **(c)** Non agricultural area constraint. **(d)** The result satisfying the constraints

$$\mu_{notsofar}(x) = \begin{cases} x/4, & x \le 4 \\ 1 & 4 < x \le 9 \\ (15 - x)/6, & 9 < x \le 15 \\ 0 & x \ge 15 \end{cases} \tag{2.14}$$

where, x stands for the distance from the location to the environmental monitoring station with kilometers (Fig. 2.10b).

3. Determine fuzzy membership function for *suitable for agriculture*, and constitute fuzzy relationship $A_3 = \int_{(x,y)} \frac{\mu_{A_3}(x,y)}{(x,y)}$. Here we select classical function to represent the constraint as follows (Fig. 2.10c):

$$\mu_{\text{suit-non-a}} = \begin{cases} 1 & \text{not in agricultural area} \\ 0 & \text{in agricultural area} \end{cases} \tag{2.15}$$

4. Use fuzzy spatial reasoning $R = A_1 \bigwedge A_2 \bigwedge A_3$ to calculate the fuzzy membership value which stands for the degree of suitableness of different location for the proposition of the task (Fig. 2.10d).

It is evident that the information of the result (Fig. 2.10d) is more abundant and more careful compared with that of result of traditional approach (Fig. 2.9d). The approach gives decision-makers more chance to choose suitable result as it provides the degree of suiting the proposition proposed by the users.

2.6 Discussion

1. In this example, the operation used is *intersection* which means the constraining factors have the same weight and all of them have to be satisfied in factory scenario. If we select *union* as operation, it means anyone of the factors satisfied can determine the result. Therefore, operations in field-based fuzzy spatial reasoning model should be chosen in terms of the proposition.
2. After spatial reasoning users can select appropriate threshold α, which is a numeral value in [0, 1], to create α-cut R_α of the fuzzy region R:

$$R_\alpha = \{\mu_R(x, y) \ge \alpha/(x, y)\} \tag{2.16}$$

R_a denotes areas which suit the requirement of the task in certain suitable level.

3. Determining fuzzy membership function is a problem in the application of the model. However, it can be solved by categorizing the vague language and providing every category analogous membership function model. This is a pivotal task for the application of field-based fuzzy spatial reasoning.
4. If the constraining factors need different weights in reasoning, the model can be reconstructed according to weighted multi-criteria evaluation approach.

2.7 Conclusion

Humans' representation about geographic phenomena in natural language is always qualitative rather than quantitative. Qualitative spatial reasoning provides a nice approach which is considered to be closer to the representation. Future GIS commercial software should be equipped with artificial intelligent functions like qualitative spatial reasoning for more and more users, especially spatial decision-makers. This research presents a field-based fuzzy qualitative spatial reasoning framework in the case of constraint satisfaction problems, in which humans' vague language can be represented with field-based model in GIS. Two options available for choosing the membership functions for fuzzy sets are described since the choice of fuzzy membership function is crucial and strongly affects the results derived by a spatial reasoning process. Field-based fuzzy spatial reasoning operations in the case of constraints satisfaction problem are discussed. A simple example compares the approach with traditional spatial reasoning process. It is evident that the approach provides users more chance to choose suitable result as it provides the degree of satisfying the proposition proposed by the users.

Acknowledgements National Natural Science Foundation of China, No. 40901090, 70863014; Foundation of Japan Society for the Promotion of Science (JSPS), No. 19.07003; Talents Introduced into Universities Foundation of Guangdong Province of China, No. 2009–26.

References

Bloch, I., & Ralescu, A. (2003). Directional relative position between objects in image processing: A comparison between fuzzy approaches. *Pattern Recognition*, *36*, 1563–1582.

Bjørke, J. T. (2004). Topological relations between fuzzy regions: Derivation of verbal terms. *Fuzzy Sets and Systems*, *141*, 449–467.

Burrough, P. A. (1996). Natural objects with indeterminate boundaries. In P. A. Burough & A. U. Frank (Eds.), *Geographical objects with indeterminate boundaries* (pp. 3–28). London: Taylor & Francis.

Cheng, T., Molenaar, M., & Lin, H. (2001). Formulizing fuzzy objects from uncertain classification results. *International Journal of Geographical Information Science*, *15*, 27–42.

Claramunt, C., & Thériault, M. (2004). Fuzzy semantics for direction relations between composite regions. *Information Science*, *160*, 73–90.

Clementini, E., Felice, P. D., & Hernández, D. (1997). Qualitative representation of positional information. *Artificial Intelligent*, *95*, 357–408.

Creignou, N., Khanna, S., & Sudan, M. (2001). Complexity classifications of Boolean constraint satisfaction problems. *SIAM Monographs on Discrete Mathematics and Applications*, *7*, 1–96.

Frank, A. U. (1991). Qualitative spatial reasoning about cardinal directions. Proceeding of the 7th Austrian conference on artificial intelligence, pp.157–167.

Frank, A. U. (1996). Qualitative spatial reasoning: Cardinal directions as an example. *International Journal of Geographical Information Systems*, *10*, 269–290.

Goodchild, M. F. (1992). Geographical data modeling. *Computer and Geosciences*, *18*, 401–408.

Guesgen, H. W., & Albrecht, J. (2000). Imprecise reasoning in geographic information systems. *Fuzzy Sets and Systems*, *113*, 121–131.

Hernández, D. (1994). Qualitative representation of spatial knowledge. In *Lecture notes in artificial intelligence* (Vol. 804). Berlin: Springer.

Hernández, D., Clementini, E., & Felice, P. D. (1995). Qualitative distances. *Lecture Notes in Computer Science*, *988*, 45–58.

Kaufman, L., & Rousseuw, P. J. (1990). *Finding groups in data: An introduction to cluster analysis*. Chichester: Wiley.

Ladkin, P. B., & Maddux, R. (1994). On binary constraint problems. *Journal of the Association for Computing Machinery*, *41*, 435–469.

Laurini, R., & Thompson, D. (1992). *Fundamentals of spatial information systems*. London: Academic.

Longley, P. A., Goodchild, M. F., Maguire, D. J., & Rhind, D. W. (2001). *Geographic information systems and science*. England: Wiley.

Marriott, K., & Stuckey, P. (1998). *Programming with constraints: An introduction*. Boston: MIT Press.

Miyajima, K., & Ralescu, A. (1994). Spatial organization in 2D segmented images: Representation and recognition of primitive spatial relations. *Fuzzy Sets Systems*, *65*, 225–236.

Molennar, M., & Cheng, T. (2000). Fuzzy spatial objects and their dynamics. *Photogrammetry & Remote Sensing*, *55*, 164–175.

Pullar, D. V., & Egenhofer, M. J. (1988). Towards the defection and use of topological relations among spatial objects. Proceedings of the 3rd international symposium on spatial data handling, IGU, Colombus, OH, pp. 225–242.

Renz, J. (2002). *Qualitative spatial reasoning with topological information*. Berlin: Springer.

Stefanakis, E., Vazirgiannis, M., & Sellis, T. (1999). Incorporation fuzzy set methodologies in DBMS repository for the application domain of GIS. *International Journal of Geographical Information Science*, *13*, 657–675.

Tarski, A. (1941). On the calculus of relations. *Journal of Symbolic Logic*, *6*, 73–89.

Zadeh, L. A. (1965). Fuzzy sets. *Information Control*, *8*, 338–353.

Zhang, J. X., & Stuart, N. (2001). Fuzzy methods for categorical mapping with image-based land cover data. *International Journal of Geographical Information Science*, *15*, 175–195.

Zhao, Y. L., Zhao, J. S., & Wang, X. Z. (1999). Spatial membership function models of fuzzy geographical phenomenon and their application. Proceedings of the international symposium on spatial data quality, Hong Kong, pp. 585–593.

Chapter 3
Testing Local Spatial Autocorrelation Using *k*-Order Neighbours

Changping Zhang and Yuji Murayama

3.1 Introduction

In a spatial analysis, the spatial objects linked to the study phenomenon must be represented on a map in order to construct their spatial relationships. The scale of a map affects its depiction of above-ground structures (Okuno, 1977). For example, on a 1:2500 scale map, houses and other buildings are shown as two-dimensional objects, polygon, but on 1:25,000 scale maps they are merely points. Polygons or points are also used to show the geographic location of individual buildings, factories, towns, etc., and there are times when various attributes, such as the structure of buildings, production of factories, and municipal population are also displayed. In the former case, only locational information of spatial objects is shown, while the latter involves the depiction of attributes of these objects. Such attributes of spatial objects are usually referred to as "spatial attributes".

If geographical phenomena have relative structural stability over space, the spatial objects, such as points, lines, and polygons, can be linked to the study phenomena and their distribution can be analyzed using spatial statistics. Pattern analysis is one means of understanding distribution of spatial objects, and may help us to verify spatial dependence and discover spatial clusters. Pattern analysis is roughly divided into two different types: (1) patterns based on the locational information about the spatial objects, and (2) patterns concerning the spatial attributes.

Methods for analyzing patterns of spatial attributes include global and local spatial autocorrelation tests. Global spatial autocorrelation tests can verify the relationship between the attribute values and location of spatial objects through the tests of significance of statistics such as Moran's I and Geary's c. While these statistics are effective in measuring the spatial autocorrelation of spatial attributes over an entire region, they are not practical for finding individual local clusters within a

C. Zhang (✉)
Faculty of Regional Development Studies, Toyo University, Tokyo 112-8606, Japan
e-mail: cp-zhang@toyonet.toyo.ac.jp

This chapter is improved from "Changping Zhang and Yuji Murayama (2000), Testing local spatial autocorrelation using *k*-order neighbours, International Journal of Geographical Information Science, 14, 681–692", with permission from Taylor & Francis.

Y. Murayama, R.B. Thapa (eds.), *Spatial Analysis and Modeling in Geographical Transformation Process*, The GeoJournal Library 100, DOI 10.1007/978-94-007-0671-2_3,

region or for searching for heterogeneous regional patterns (Ding & Fotheringham, 1992; Ord & Getis, 1995; Zhang, 1999). To remedy this situation, Getis and Ord (1992) develop a local spatial autocorrelation statistic $G_i(d)$. They also introduce a distance parameter d to a weight coefficient w_{ij} to measure spatial proximity of spatial objects.

The weight coefficient $w_{ij}(d)$ with the distance parameter d, is defined by Getis and Ord (1992) as a binary weight coefficient. When a circle with center i and radius of distance d (hereafter, contiguous subarea of point i) is drawn, the value of $w_{ij}(d)$ is 1 if point j is located in the contiguous subarea of point i, and 0 if it is not.

$$w_{ij}(d) = \begin{cases} 1 \text{ Point } j \text{ is located within the contiguous subarea of point } i\,; \\ 0 \text{ otherwise} \end{cases} \tag{3.1}$$

If there are no points within the spatial weight matrix, the matrix can be expanded until points are included within the subarea, but it is not easy to derive a logical theorem to determine an appropriate distance d. Ordinarily, the spatial interaction and autocorrelation within spatial attributes may be the strongest between the nearest neighbours. As the neighbour order increases, this autocorrelation weakens. This is particularly true for irregularly distributed objects, in which the neighbour order cannot be directly expressed from weight coefficient $w_{ij}(d)$, although it is possible to use $w_{ij}(d)$ to imply the spatial proximity between not only the nearest neighbours, but $k(>1)$-order neighbours. However, in such a scenario it is difficult to define the k-order neighbours and associated measurement for an irregular distribution. It is also difficult to establish a functional relationship between neighbour order of the weight coefficient based k-order neighbours and distance d for an irregular distribution.

Given the above issues, this chapter will propose a new method for analyzing patterns of spatial attributes whose k-order neighbours are measured, and investigate its effectiveness using a case study of Ichikawa City in Japan.

3.2 *k*-Order Neighbours and Their Measurement

3.2.1 Defining k-Order Neighbours Using Delaunay Triangulation

Triangulated irregular networks (TINs) created by Delaunay triangulation will remedy the limitations of defining k-order neighbours when defining the spatial proximity of points. Delaunay triangulation is a method in which points become the vertices of non-overlapping triangles. TINs are constructed so that the circumcircle of each triangle contains no vertices of other adjacent triangles in its interior (Clarke, 1990; Okabe, Boots, & Sugihara, 1992; Tsai, 1993). By defining pairs of connected points within the TIN as neighbour points, the TIN can become a network in which points having a neighbour relationship are spatially connected. For example, Fig. 3.1b shows the result of drawing a TIN based on the point distribution in Fig. 3.1a.

k-order neighbours in the Delaunay triangulation network are defined as follows. Those points directly connected to a point v by the edges of the TIN are called the

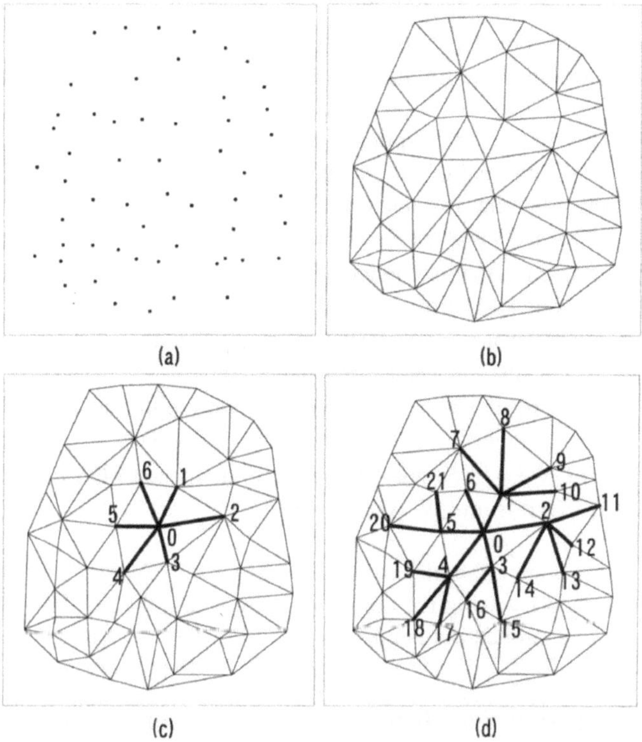

Fig. 3.1 *k*-order neighbours on Delaunay triangulation. (**a**) Point distribution, (**b**) Delaunay triangulation, (**c**) nearest neighbours, (**d**) 2-order neighbours

"nearest neighbours", or "first-order neighbours" of this point. Then, those points which are directly connected to the first-order neighbours, of that are not first-order neighbours themselves, are called "second-order neighbours". By continuing this process we can define *k*-order neighbours for any point, *v*.

3.2.2 Search Algorithms

Now, let us introduce algorithms for searching *k*-order neighbours. We shall start by reviewing the relevant symbols and their meanings in the algorithm.

- DV: the set of points,
- DE: the set of edges in TIN,
- DU: set of *k*-order neighbours,
- SE: set of searched edges,
- QV: queue of points that have already been searched as neighbours and are extendable for searching next order neighbours,
- φ: no element, and
- K: a given neighbour order.

In this chapter, the TIN is considered as an undirected graph G(DV, DE). The following search procedure is continued until either the given neighbour order (K) or the boundary of the region in question is reached. The procedure is as follows:

Step 1: The origin of a point v_o on the TIN is selected, then $v \leftarrow v_o$,DU$\leftarrow\{v_o\}$,SE and QV$\leftarrow\{\varphi\}$,$k \leftarrow 0$ are set.

Step 2: If the edge that is connected to the origin of point v exists in the set of (DE–SE), or if QV$\neq\varphi$ and $k <$ K, the following (I) and (II) are reiterated.

(I) If an edge that is connected with point v exists in (DE–SE), substeps (I-a) and (I-b) are reiterated.

(I-a) One edge in (DE–SE) connected to point v is selected and is called e. Then, SE$\leftarrow$SE $\cup\{e\}$.

(I-b) If point v_i (connected with point v) $\in$ (DV–DU), then DU$\leftarrow$DU $\cup$ $\{v_i\}$, and QV$\leftarrow$QV $\cup$ $\{v_i\}$.

(II) If QV$\neq\varphi$, then $v \leftarrow$ (the point at the head of QV), and that point is removed from QV. Also, $k\leftarrow k$+1.

Figure 3.1c, d is an example, in which 1-order and 2-order neighbours of the point 0 are searched by using the above algorithms and the numbers in the figures represent order of searched out neighbours.

3.2.3 Weight Coefficient Based on k-Order Neighbours

The weight coefficient $w_{ij}(k)$, which measures proximity of k-order neighbours in a point distribution, is determined as follows. $w_{ij}(k)$ is binary coefficient set to 1 if point j is a k-order neighbour of point i; otherwise, it is 0. Hence,

$$w_{ij}(k) = \begin{cases} 1 \text{ Point } j \text{ is a } k \text{ - order neighbour of point } i; \\ 0 \text{ otherwise.} \end{cases} \tag{3.2}$$

The defined weight matrix $\{w_{ij}(k)\}$ is a symmetrical binary matrix. In many cases, point distribution of geographic phenomena is irregular and lopsided and has directional bias. The weight coefficient $w_{ij}(k)$ of the k-order neighbours, which is based in TINs, conforms to spatial distribution having directional bias to measure proximity of points. Furthermore, k-order neighbours are useful because they allow sensible spatial analysis of area feature (polygon) using a point representation (polygon centroid) and are means of defining a contiguity matrix without needing to derive topology for the polygon. In these cases an Euclidean distance metric will poorly represent the spatial interaction between polygons. Thus, the introduction of local spatial autocorrelation statistics based on $w_{ij}(k)$ may likely be able to improve our ability to identify spatial attributes.

Nevertheless, For the Delaunay criterion to create TINs, error may be introduced to the weight coefficient where two points are 1-order neighbours in the TIN, but the polygons are not adjacent. Figure 3.2 shows an example, in which points 2 and 3 are

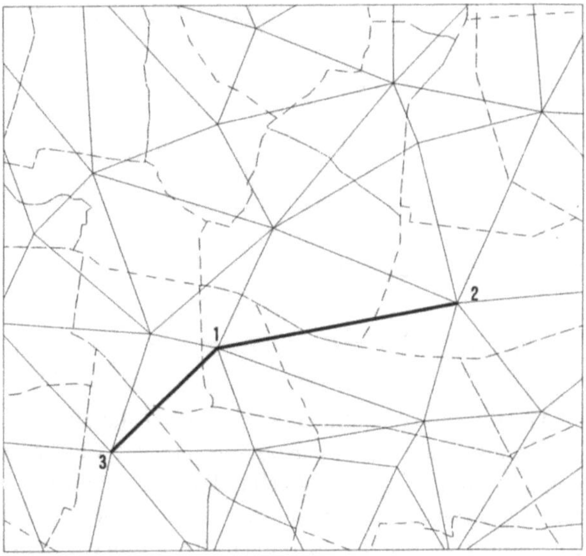

Fig. 3.2 1-order neighbours connected to no adjacent polygon

1-order neighbours of point 1, although polygons using points 2 and 3 as centroids are not adjacent to polygon 1.

3.3 Local Spatial Autocorrelation Statistics with Weight Coefficient of k-Order Neighbours

Getis and Ord's (1992) statistic $G_i(d)$, in which the x value at i is excluded from the summation, is redefined using $w_{ij}(k)$ as follows:

$$G_i(k) = \frac{\sum_j w_{ij}(k)x_j}{\sum_j x_j} \quad j \neq i \tag{3.3}$$

As Eq. (3.3) shows, G_i (k) is derived by $w_{ij}(k)$ which is related to the point locations, as well as x values for j points which are k-order neighbours of point i. The larger the value of attribute of k-order neighbours of i, the higher the proportion of the sum of all x_j values in the entire region (denominator in Eq. (3.3)). Therefore, the value of G_i (k) can be used to measure the concentration of spatial data around a point (i.e., within the realm of the k-order neighbours), and it becomes possible to identify clusters that exist in spatially distributed attribute variable.

Following the principles outlined by Cliff and Ord (1981) and Getis and Ord (1992), if the x_i value at i point is fixed, the set of $(n–1)!$ random permutations of the remaining x values at the j points is considered. Under the null hypothesis of no spatial dependence between the value of x at a point and its neighbours, the expected value and variance of G_i (k) can be obtained.

$$E[G_i(k)] = W_i(n-1) \tag{3.4}$$

Here $W_i = \sum_j w_{ij}(k)$

$$\mathrm{Var}[G_i(k)] = \frac{W_i(n-1-W_i)}{(n-1)^2(n-2)} \frac{S_{i2}}{m_{i1}^2} \tag{3.5}$$

if we set $m_{i1} = \sum_j x_j/(n-1)$

$$S_{i2} = \sum_j x_j^2/(n-1) - m_{i1}^2$$

Then the local spatial autocorrelation statistic $G_i(k)$ can be redefined as a standardized statistic $Z_i(k)$ by using the expected value and variance in Eq. (3.6). A large positive value of $Z_i(k)$ indicates that larger values of x_j than the mean are within realm of k-order neighbours of point i, while a large negative value of $Z_i(k)$ indicates lower values of x_j than mean are within realm of k-order neighbours of point i.

$$Z_i(k) = \frac{(G_i(k) - E[G_i(k)])}{\sqrt{\mathrm{Var}[G_i(k)]}} \tag{3.6}$$

3.4 Example

The validity of the local spatial autocorrelation statistics with the weight coefficient of the k-order neighbours will be examined by using the distribution of the elderly population in the city of Ichikawa. Ichikawa is a medium-sized city of 475,751 people (in 2009) in Chiba Prefecture, Japan. In 1998, just over 10% of the entire population was over 65 years of age, but by 2009, this figure had risen to 16.4%.

As Fig. 3.3 shows, the elderly population tends to be concentrated in the north, and sparse in the south (as demarcated by the JR Sobu Train Line). For example, there are a large number of elderly residents in the northern districts of Ookashiwa, Nakayama, Yawata, and Ichikawa, but in new housing areas in the south such as Gyotoku and Minamigyotoku, as well as in the Shintoku industrial area, the elderly residents are few. However, we cannot divide the city into two spatial clusters (north and south) based merely on this information. That is because if we examine this population distribution in detail, we can see that there are several pockets in the north where there are relatively few elderly residents. At the same time, there are also some older areas in the south such as Hongyotoku (229) and Shiohama 4-chome (199) which have a relatively large elderly population (note: "chome" is a small administrative unit within "cho". For example, Shiohama-cho consists of four chomes). Therefore, it is difficult to determine from distribution maps whether there are clear spatial clusters of elderly residents in Ichikawa.

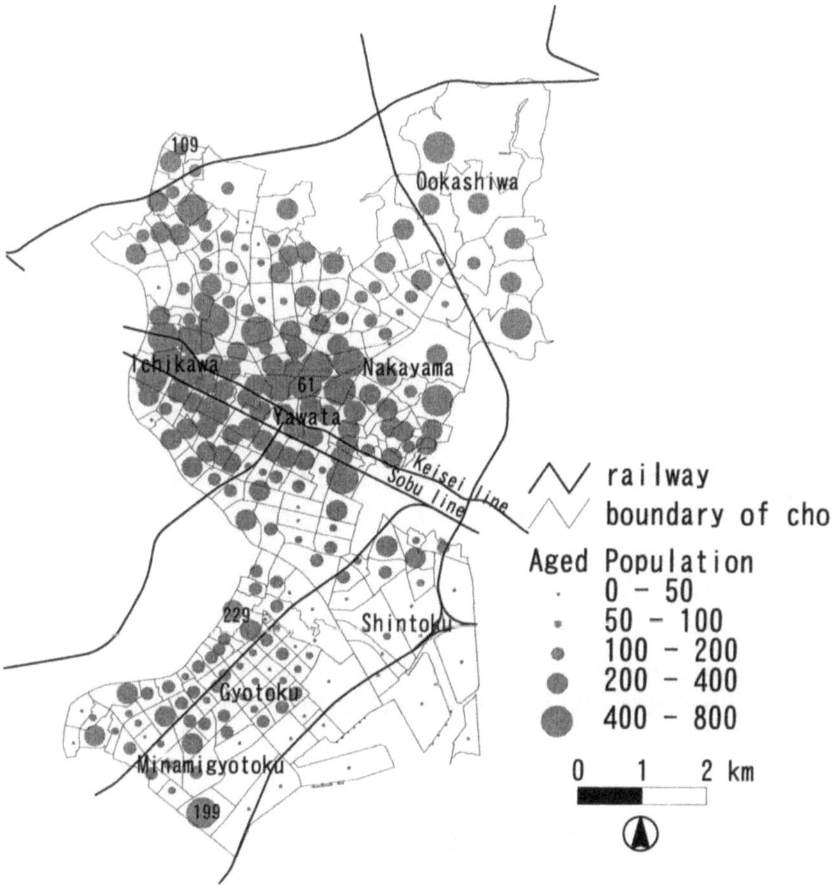

Fig. 3.3 Distribution of the elderly population. 61: Higashisugano 1-chome; 109: Kitakokubun 4-chome; 199: Shiohama 4-chome; 229: Hongyoutoku

3.5 Delaunay Triangulation of Ichikawa

Delaunay triangulation was used to create a triangulated irregular network (TIN) of central points of "cho" districts (using geometrical centroids of "cho" areas) of Ichikawa. Usually, TIN is the size of convex hulls of sets of vertices of triangulation, regardless of the shape of regional boundaries (administrative boundaries in Ichikawa, for example, are highly irregular) (see Fig. 3.4a). However, the pairs of points that straddle the city limits do not necessarily have a relationship with their neighbours, even if they are connected at the TIN edge. Edges that connect unrelated pairs of points, are eliminated using the city limits. The modified TIN of Ichikawa is shown in Fig. 3.4b.

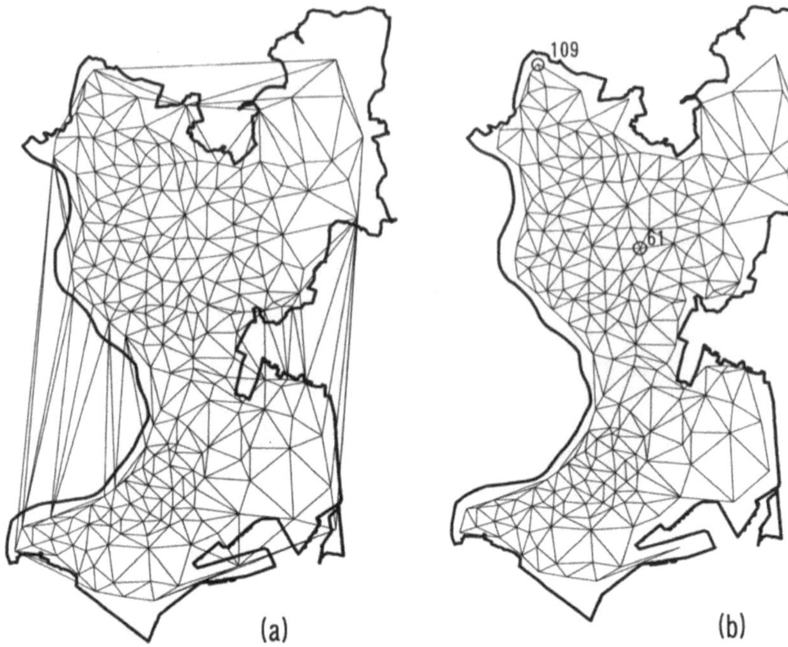

Fig. 3.4 Triangulated irregular network of central point sets of the cho in Ichikawa City. (**a**) Unmodified; (**b**) modified (triangle edges extending outside the city limits are removed); 61: Higashisugano 1-chome; 109: Kitakokubun 4-chome

3.5.1 Identification of Spatial Clusters

Using the above means of spatial autocorrelation statistics, we find that the $G_i(k)$ and $Z_i(k)$ of each point i in the region in question are not independent statistics. This is especially true when point i or its neighbours are located within the realm of the k-order neighbours (or within the contiguous subarea) of a different point j, where a part of k-order neighbours of point i are overlapped with k-order neighbours of point j and we can see correlation between $G_i(k)$ and $G_j(k)$, and $Z_i(k)$ and $Z_j(k)$. Furthermore, as the neighbour order k (or distance d) increases, the number of overlapped neighbours of points i and j increases, and their degree of correlation changes (Ord & Getis, 1995). Furthermore, Ichikawa is long and narrow, extending for 13 km from north to south, and only 6 km from east to west (with the narrowest width a mere 2 km).

Here, as we consider the correlation of local spatial autocorrelation statistics and the shape of the study area (Ichikawa), we can determine that the highest order is 3 and calculate the local spatial autocorrelation statistic $G_i(k)$ and its standardized statistic $Z_i(k)$ with weight coefficients of 1-, 2-, and 3-order neighbours. Figures 3.5a–c show the standardized statistic $Z_i(k)$ for 1-, 2-, and 3-order neighbours, respectively.

Fig. 3.5 Distribution of Z_i value. **(a)** 1-order neighbours; **(b)** 2-order neighbours; **(c)** 3-order neighbours

As can be clearly seen in Fig. 3.5a, regions which show positive value of $Z_i(1)$ are concentrated on the northern side, along the Keisei and JR Sobu train lines, and form the first cluster. In this cluster, districts with a large number of elderly residents are adjacent to one another.

The second cluster is located south of the first cluster, and values of $Z_i(1)$ are large negative (–1.20 to –3.18). This is where the sparsest population of elderly people is found. Between these two clusters, and around them, there are clusters which show small negative values (0 to –1.20) of $Z_i(1)$. These clusters are made up of districts having relatively few elderly residents. In a word, a positive $Z_i(1)$ value indicates a pattern dominated by a clustering of large elderly population around that region; A negative $Z_i(1)$ value indicates a pattern dominated by a clustering of small elderly population around that region.

Now let us look at distribution maps (Figs. 3.5b, c) that use the weight coefficients of 2- and 3-order neighbours. The distribution of Z_i values shows a more distinct spatial cluster division than do the 1-order neighbours. While there are pockets of high Z_i values near the southern boundary of the low Z_i group (i.e., south of the JR Sobu Line), there are also pockets of low Z_i values in the otherwise high Z_i area north of the JR line. However, such pockets do not appear on the distribution maps for the 2- and 3-order. Therefore, the Z_i values with the weight coefficients for 2- and 3-order neighbours appear more effective than 1-order for finding local clusters in the present study area.

Using local spatial autocorrelation test for pattern analysis of spatial attributes proposed by this chapter, we can see that Fig. 3.3, which seems to indicate that the elderly population of Ichikawa is not distributed in orderly patterns, actually implies that there are several clusters in the distribution of orderly residents.

3.5.2 Verifying the Normality of Z_i

This section will examine the propriety of spatial pattern analysis using standardized statistic Z_i. To do that, we use the mean, the standard deviation, skew, and kurtosis of standardized normal distribution to verify the normality of Z_i related to the k-order neighbours and distance parameter based on Monte Carlo simulation. To achieve this goal, 10,000 replications of $\{x_j\}$ were created using Monte Carlo simulation to derive the moments of each distribution for the Z_i values. The results are shown in Table 3.1.

To conduct the simulation, it was first necessary to select the points and set the distance parameters. Ordinarily, the closer points are to boundaries, the greater the possibility that some of their neighbour points will be on the other side of these boundaries. To reduce of this boundary effect, the 1-chome district of Higashisugano (i=61), located in the city center, and the 4-chome district of Kitakokubun (i=109), located near the northeastern city limits, were selected (see Figs. 3.3 and 3.4). Then an effort was made to compare the influence of "boundary effect" on Z_i of the k-order neighbours and the distance parameter. As we can clearly see in the np. column of Table 3.1, the number of k-order neighbours and points within the contiguous

Table 3.1 Moments of Z_i values

	Higashisugano 1-chome ($i = 61$)					Kitakokubun 4-chome ($i = 109$)				
Order/*d*	np.	Mean	s.d.	Skew	Kurtosis	np.	Mean	s.d.	Skew	Kurtosis
1	6	0.0014	1.0069	0.0042	0.2185	3	0.0109	1.0064	0.0208	0.4194
2	21	0.0011	1.0047	0.0080	0.0634	8	0.0032	0.9917	0.0400	0.1532
3	49	0.0001	1.0013	0.0246	0.0123	15	0.0051	0.9939	0.0518	0.1108
771 m	8	0.0024	1.0012	0.0138	0.1424	3	0.0058	1.0018	0.0116	0.3745
1070 m	18	–0.0039	0.9959	0.0284	0.0668	5	–0.0028	1.0033	0.0076	0.1045
1865 m	62	–0.0025	1.0055	–0.0247	0.0102	13	–0.0047	0.9956	–0.0256	0.0108

np.: number of *k*-order neighbors or points in a contiguous subarea of distance *d*; s.d.: standard deviation

subarea of distance differ, depending on the differences in the location of a "cho" and the location, shape, and size of surrounding "chos". It should be noted that to clearly indicate the relation between *k*-order neighbours and distance (*d*), the present study used the mean distance of the *k*-order neighbours. d_k is set to the mean straight line distance between *k*-order neighbours.

$$\text{Mean distance of } k\text{-order neighbours: } d_k = \min_{i\, j \in J} \max \{d_{ij}\} \tag{3.7}$$

Here, *J* is the set of *k*-order neighbours for point *i*. Using Eq. (3.7), values for the mean distance of 1-, 2-, and 3-order neighbours were calculated to be 771, 1070, and 1865 m, respectively. These values were then used as the values of distance parameters corresponding to 1-, 2-, and 3-order neighbours.

The upper section of Table 3.1 shows the moments of each distribution for $Z_i(k)$ values of the *k*-order neighbours, while the lower part shows the moments of each distribution for $Z_i(d)$ values of the distance parameter. As we can clearly see in the table, the distribution for Z_i values of both *k*-order neighbours and distance parameter is normal. However, in the 1-chome district of Higashisugano ($i = 61$), located in the city center, the skew of distribution for Z_i values of the *k*-order neighbours is low. Regarding the mean, the standard deviation, and kurtosis, as the neighbour order increased, all values decreased. This was apparently because the normality of $Z_i(k)$ increased as soon as the neighbour order increased. Furthermore, a comparison of the moments of distance parameter $Z_i(d)$ with the same of *k*-order neighbours showed that there was not a large difference between standard deviation and kurtosis, but generally speaking, the absolute values for mean and skew of the k-order neighbours were lower than those for the distance parameter.

In contrast, in the 4-chome district of Kitakokubun ($i = 109$), near the city limits, the mean, skew and kurtosis of $Z_i(d)$ were generally lower than the same values for $Z_i(k)$. Accordingly, when the standardized statistic Z_i was used to find local clusters, the weight coefficient $w_{ij}(k)$ of the *k*-order neighbours in the city center measured the proximity between points more accurately than did that of the distance parameter, and its standardized statistic $Z_i(k)$ could be used to accurately identify spatial clusters. Near the city limits, however, the opposite held true, and $Z_i(d)$ provided a more accurate means of finding clusters.

3.6 Conclusion

This study verified the validity of local spatial autocorrelation statistics based on k-order neighbours measurement used for pattern analysis of spatial attributes. This new method was applied to the population distribution of elderly residents of Ichikawa City. As a result, local clusters in the distribution could be accurately discovered by use of local spatial autocorrelation statistics with $w_{ij}(k)$. By verifying the normality of Z_i with weight coefficients based on the k-order neighbours and the distance parameter, it was also apparent that there was a difference between the weight coefficients of the k-order neighbours and distance parameter to measure the spatial proximity of districts located in the city center and near the city limits.

References

Clarke, K. C. (1990). *Analytical and computer cartography*. Upper Saddle River, NJ : Prentice Hall.

Cliff, A. D., & Ord, J. K. (1981). *Spatial processes: Models and applications*. London: Pion.

Ding, Y., & Fotheringham, A. S. (1992). The integration of spatial analysis and GIS. *Computers, Environment and Urban System*, *16*, 3–19.

Getis, A., & Ord, J. K. (1992). The analysis of spatial association by use of distance statistics. *Geographical Analysis*, *24*, 189–206.

Okabe, A., Boots, B., & Sugihara, K. (1992). *Spatial tessellations: Concepts and applications of Voronoi diagrams*. New York: Wiley.

Okuno, T. (1977). *Foundmentals of quantitative geography*. Tokyo: Taimeido in Japanese.

Ord, J. K., & Getis, A. (1995). Local spatial autocorrelation statistics: Distributional issues and application. *Geographical Analysis*, *27*, 286–306.

Tsai, V. J. D. (1993). Delaunay triagulations in TIN creation: An overview and a linear-time algorithm. *International Journal of Geographical Information Systems*, *7*, 501–524.

Zhang, C. (1999). Development of a spatial analysis tool for irregular zones using the spatial data framework. *Geographical Review of Japan*, *72*, 166–177 (in Japanese with English abstract).

Chapter 4
Effect of Spatial Scale on Urban Land-Use Pattern Analysis

Yaolong Zhao and Yuji Murayama

4.1 Introduction

Urbanization can be understood as a complex process with their intrinsic characteristics of emergence, self-organizing, self-similarity and non-linear behavior of land-use dynamics (Barredo, Kasanko, McCormick, & Lavalle, 2003). The complex process directly impacts on water quality (Yin et al., 2005), hydrology (Pauleit, Ennos, & Golding, 2005; Carlson & Arthur, 2000), air pollution (Potoglou & Kanaroglou, 2005), microclimate (Solecki & Oliveri, 2004; Stabler, Martin, & Brazel, 2005), and biodiversity (Hostetler & McIntyre, 2001; Pauleit et al., 2005) in urban area. In order to understand the mechanism and spatial process of urban dynamics so as to provide a basis for assessment of the ecological impacts of urban change and spatial decision-making support systems to municipality and civil planners, various spatial dynamic models, especially that based on cellular automata (CA) and geographical information system (GIS), of urban land-use change have been constructed and successfully applied to many cities in the world (Barredo et al., 2003; Batty, 1970; Batty & Xie, 1994; Clarke, Hoppen, & Gaydos, 1997; White & Engelen, 1993; White, Engelen, & Uljee, 1997; Yeh & Xia, 2001).

According to O'Sullivan and Unwin (2002), as spatial pattern in any time is generated from corresponding spatial process, spatial model which aims at simulating the spatial process can be constructed through analyzing dynamic spatial patterns in time-series. Construction of spatial model of urban dynamics also takes the same procedure. That is, spatial model of urban dynamics underlying certain urban change theory needs to be calibrated using urban land-use patterns in time-series (Silva & Clarke, 2002; Straatman, White, & Engelen, 2004). This method

---

Y. Zhao (✉)
School of Geography, South China Normal University, Guangzhou, Guangdong, PR China
e-mail: yaolongzhao@gmail.com

This chapter is improved from "Yaolong Zhao and Yuji Murayama (2006), Effect of spatial scale on urban land-use pattern analysis in different classification systems: An empirical study in the CBD of Tokyo, Theory and Applications of GIS, 14, 29–42", with permission from GIS Association, Japan.

Y. Murayama, R.B. Thapa (eds.), *Spatial Analysis and Modeling in Geographical Transformation Process*, The GeoJournal Library 100, DOI 10.1007/978-94-007-0671-2_4,

also represents the strong need to understand the dynamics of the feedback mechanisms that relate environmental pattern and social process (Nagendra, Munroe, & Southworth, 2004). Therefore, spatial model of urban dynamics is greatly affected by the results of urban land-use pattern analysis.

Generally it is recognized that in the field of landscape ecology, spatial pattern and spatial scale are inseparable in theory and in reality. Spatial pattern occurs on different spatial scales, and spatial scale affects spatial pattern to be observed (Qi & Wu, 1996; Turner, O'Neill, Gardner, & Milne, 1989). Accordingly, the results of urban land-use pattern analysis also show difference in different spatial scales, and it would affect the construction of spatial model of urban land-use changes. Jantz and Goetz (2005) have demonstrated the effect.

Some scholars have paid attentions to the relationship of spatial scale and spatial model of land-use changes. Turner (1987) has studied the difference of results of simulating landscape changes in Georgia by comparing 3 transition models. He argued that it would be useful to incorporate variable scales of land-use transitions into the model of land-use changes. De Koning, Veldkamp, and Fresco (1998) have discussed the spatial scale effects on land-use model. They found different driving factors of land-use changes in Ecuador at different aggregation levels using multiple regression models. Veldkamp et al. (2001) have elaborated on a multi-scale approach as used in CLUE (Conversion of Land-use and its Effects) framework and argued the need for scale sensitive approaches in spatially explicit land-use change modeling.

All the argumentations have confirmed the effect of spatial scale on spatial model of land-use changes. However, the questions regarding "how" and "why" remain largely unanswered, and systematic investigations to address such issue are urgently needed. Moreover, the argumentations above just concentrated on regional or large scale, not on urban area. Urban land-use pattern and dynamics should take on inherent characteristics considering the particularity of urban systems (Barredo et al., 2003). Spatial scale effect on urban land-use pattern analysis may take different characteristics and mechanisms. Zhao and Murayama (2005) have carried out an empirical study on how changing spatial scale affects analysis of urban land-use pattern using spatial autocorrelation index in certain study area. Nevertheless, urban land-use pattern occurs not just at certain scale, but also at certain land-use classification system. Characteristics of spatial scale effect on urban land-use pattern analysis may differ from different land-use classification systems. And the answer to the question of "how the difference is" is not clear yet. In addition, classification system may also influence the construction of spatial model of urban dynamics across a range of scale.

Although scholars have proposed some land-use classification standardizations (Anderson, Hardy, Roach, & Witmer, 1976; Dickinson & Shaw, 1977), most literatures concerning spatial models of urban land-use change just choose their own urban land-use classification system in terms of their own purpose in their research with no theoretical justification (Barredo et al., 2003; White & Engelen, 1993; White et al., 1997). Klosterman (2005) has pointed out that the number of land-use categories which can be projected and the scale at which they can be projected vary substantially for the different types of models. Little systematic investigation has been done as to how the relationship is between urban land-use classification system and spatial pattern of urban land-use across a range of scale. Actually, identification of urban land-use classification systems for spatial model of urban dynamics is also closely linked to spatial scale in the model (see Anderson et al., 1976; Treitz & Rogan, 2004). Study of spatial scale effect on urban land-use pattern analysis for construction of spatial model of urban dynamics entails consideration of urban land-use classification systems. This study concentrates on this issue.

Spatial scale encompasses both grain and extent. Grain refers to the resolution of the data (spatial resolution), i.e., the area represented by each data unit. Extent refers to the overall size of the study area (Turner et al., 1989). For certain urban area, the extent is fixed. Therefore, this research focuses on the scale effects from a grain size point of view at certain extent of study area.

Objective of this research is to investigate the effect of changing spatial scale on the result of urban land-use pattern analysis for different classification systems in terms of spatial autocorrelation and fuzzy mathematics by an empirical study in the CBD (Central Business District) of Tokyo so as to provide useful information for construction of multi-scale or hierarchical spatial model of urban land-use changes.

4.2 Methods

4.2.1 Data and Study Area

The data set used for this study is "Detailed Digital Information (10 m grid land-use) Metropolitan Area" of Tokyo, which was released in 1998 (Geographical Survey Institute, 1998). The data set was produced by the Geographical Survey Institute, the Ministry of Construction in Japan, and it has the data on the category of land-use of each 10 m square cell, which we call "basic cell unit", surveyed with the visual interpretation of aerial photography in 1979, 1984, 1989 and 1994. It provides the possibility of changing spatial resolution of urban land-use map from high resolution (10 m) to low resolution (200 m, even less than 200 m) in analyzing the effect of spatial resolution on urban land-use pattern analysis.

In "Detailed Digital Information (10 m grid land-use) Metropolitan Area" of Tokyo, the area is divided into regular grid map with 3 km $\times$4 km, and land-use information is stored in TDU format for every grid map (Geographical Survey Institute, 1998). 120,000 base units exist in one grid map. Considering the speed of calculation in computer, we select one of the grids map located in the CBD of Tokyo as study area (Fig. 4.1), and use the digital data of 1994.

4.2.2 Land-Use Classification Systems

Classifying land-use is one of manners to understand environment so as to provide important information to nation level or city level plans for overcoming the problems of haphazard, uncontrolled development, deteriorating environmental quality,

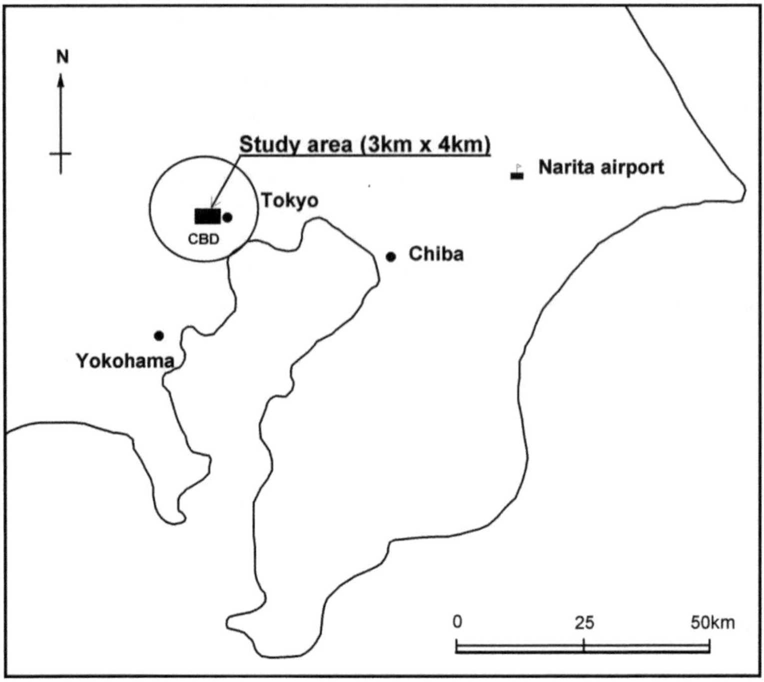

Fig. 4.1 Study area

loss of prime agricultural lands, destruction of important wetlands, and loss of fish and wildlife habitat (Anderson et al., 1976). Although land-use relates to physical form, land-use classifications mostly come from social purposes. Bibby and Shepherd (2000) have discussed the complexity of social purposes in terms of possibility of specifying purposes at many levels of generality, interaction between form and function as well as multi-networks of purposes. Classifying land-use, therefore, can not be deemed as a straightforward process. In this chapter, as we aim to investigate the effect characteristics of spatial scale on the result of urban land-use pattern analysis for different classification systems, we do not identify how to classify urban land-use. We designed two types of urban land-use classification system, systems A and B, based on the land-use classification system of "Detailed Digital Information (10 m Grid Land-use) Metropolitan Area" of Tokyo for study area (Table 4.1). We assumed that this design would not affect our understanding of the characteristics of spatial scale effect on urban land-use pattern analysis for different classification systems in this study.

There are 9 categories of land-use in system A and 5 in system B. In order to systematically compare the change of urban land-use pattern across a range of scale for these two systems, the categories of vacant, Industrial and commercial in system B are the same as that in system A; two categories of low-storey residential and high-storey residential in system A were grouped into category of residential

Table 4.1 Two types of urban land-use classification system

Categories in system A	Categories in system B
1. Vacant	1. Vacant
2. Industrial	2. Industrial
3. Commercial	3. Commercial
4. Low-storey residential	4. Residential
5. High-storey residential	
6. Public	5. Public
7. Park and woods	
8. Road	
9. Water	

in system B, and four categories of public, park and woods, road and water into category of public in system B.

4.2.3 Data Processing

In order to allow a systematic analysis of spatial scale effect, the original grid cells (basic cell unit, BCU, here 10 m $\times$10 m) were aggregated into larger grid cells for both land-use classification systems in the following way. Each BCU was treated as one basic unit, and therefore the grain size at this scale was expressed as 1 by 1. A 2 $\times$2 areal unit, then, corresponded to the grain size that contained four BCUs (two on each side). This was accomplished by aggregating four adjacent basic cell units into one larger grid cell. This procedure was repeated until the entire region of the data sets was covered. In total, 20 different grain sizes (spatial resolution) were created, ranging from 1 $\times$1 through 20 $\times$20 BCUs (i.e., 1, 2^2, 3^2..., 20^2). That is, a series of urban land-use map with spatial resolution from 10 m $\times$10 m through 200 m $\times$200 m were created for this study.

In the aggregation process, sometimes the original data set had to be modified (edge rows or columns were omitted) to obtain integer numbers of rows and columns. While this kind of modification was necessary only for technical convenience, we assumed that this modification would not affect the results of analysis greatly because of the relatively large size of the data at small spatial resolution level.

Every cell at certain level of spatial resolution is always assigned just one main state of land-use at stated period in traditional method, especially in CA approach (Batty & Xie, 1994; Clarke et al., 1997; White & Engelen, 1993; White et al., 1997). The state of the cell which is aggregated from some BCUs is identified usually under majority rule (e.g. Turner et al., 1989). Figure 4.2a, b indicate the schematic process of aggregation of four BCUs into one cell with 20 m $\times$20 m. In Fig. 4.2a, as the sate of all the BCUs is park and woods, the aggregated cell is assigned land-use type of park and woods. In Fig. 4.2b, the proportion of park and woods, road, commercial to the area of four BCUs is 50, 25 and 25% respectively. As the proportion of park and woods is more than that of others, the aggregated cell is assigned land-use category of park and woods.

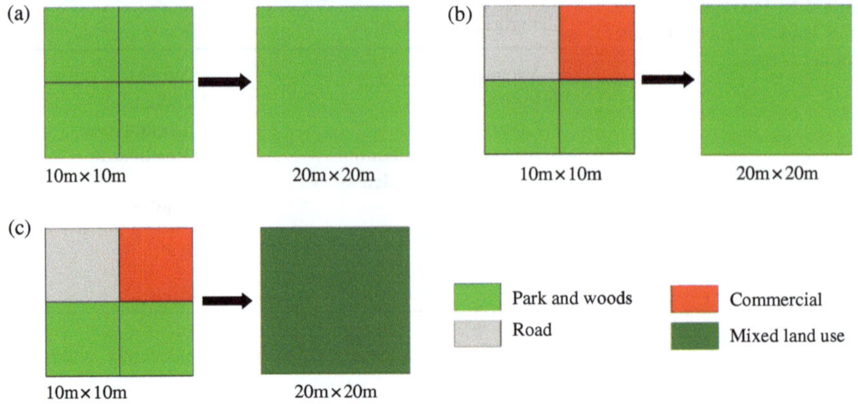

Fig. 4.2 Schematic process of aggregation of four BCUs into one cell. (**a**) Four BCUs with same category of land-use in single-state structure; (**b**) Four BCUs with different categories of land-use in single-state structure; (**c**) Four BCUs with different categories of land-use in multi-states structure

Here, every cell is considered as homogeneous rather than heterogeneous. We call the result of the simplification process of land-use as representation of single-state structure. However, the process of simplification may generate area errors of land-use (e.g. Moody & Woodcock, 1994). The errors inevitably influence the result of urban land-use pattern analysis (Zhao & Murayama, 2005). In order to compare the difference of urban land–use pattern at different classification systems across the range of scale between two situations of existence and no existence of area errors, we adopt the concept of multi-state structure (Zhao & Murayama, 2005). At multi-state structure, the concept of grade of fuzzy membership is introduced to represent the attribute information of the cell at every level of spatial resolution. In fact, because of the limitation of corresponding spatial resolution, land-use attribute of every cell in certain spatial scale takes characteristics of fuzzy uncertainty (Zhang & Stuart, 2001). That is, as the land-use of the cell at the scale is heterogeneous, the cell is assigned different land-use type at corresponding grade of fuzzy membership. Multi-state structure can be represented as following:

$$R_{ij} = \{\mu_{R_{ij}}(1), \mu_{R_{ij}}(2), \ldots, \mu_{R_{ij}}(k)\} \tag{4.1}$$

where, R_{ij} stands for one cell in ith row and jth column; k land-use category; $\mu_{R_{ij}}(k)$ the extent to which the cell belongs to land-use category k, i.e., grade of fuzzy membership in [0, 1]. $\sum_k \mu_{R_{ij}}(k) = 1$ for one cell is required so as to ensure the integrality of attribute information of cells. Figure 4.2c, for example, indicates the schematic process of aggregation of four BCUs into one cell with 20 m ×20 m using multi-state structure. The state of cell with 20 m ×20 m can be represented as follows:

$$R_{ij} = \left\{ \begin{array}{l} \mu_{ij}(public), \mu_{ij}(commercial), \mu_{ij}(park\ \&\ woods), \\ \mu_{ij}(vacant), \mu_{ij}(low - s - resi), \mu_{ij}(high - s - resi), \\ \mu_{ij}(road), \mu_{ij}(water), \mu_{ij}(industry) \end{array} \right\}$$

where

$$\mu_{ij}(public) = 0.25, \mu_{ij}(commercial) = 0.25,$$
$$\mu_{ij}(park\ \&\ woods) = 0.5, \mu_{ij}(others) = 0$$

For the cell in Fig. 4.2a, $\mu_{ij}(park\ \&\ woods) = 1, \mu_{ij}(others) = 0$. This means that single-state structure can be considered a special case of multi-state structure.

As we aggregate BCUs (10 m ×10 m) into larger grid cells systematically in data processing, at all the levels of spatial resolution we can identify attribute information of land-use for every cell in both structures of single-state and multi-state. The results of urban land-use pattern analysis in both structures for both classification systems are used to compare so as to analyze the effect of simplification of land-use with changing spatial resolution in single-state structure under majority rule for different land-use classification systems.

It should be noted that although the multi-state structure based on the concept of fuzzy mathematics looks similar formally with the proportion structure of land use types (e.g., Verburg, de Koning, Kok, Veldkamp, & Bouma, 1999) for each grid cell, it derives from fuzzy classification method (Benz, Hofmann, Willhauck, Lingenfelder, & Heynen, 2004; Zhang & Foody, 1998; Zhang & Stuart, 2001) which grasps the intrinsic characteristics of fuzzy uncertainty of the classification for the grid cell at certain resolution. The presentation of multi-state structure based on fuzzy mathematics may provide more helps to the extension research of this study for the construction of spatial model of urban dynamics using fuzzy set approaches (Liu & Phinn, 2003; Wu, 1998).

4.2.4 Detection of Spatial Autocorrelation

Spatial autocorrelation is a general geographical phenomenon in nature, which indicates spatial association and spatial dependence of geographic phenomenon (Anselin, 1988; O′Sullivan & Unwin, 2002). The existence of spatial autocorrelation is reflected in the proposition which Tobler (1970) has referred to as the "first law of geography: everything is related to everything else, but near things are more related than distant things." Although spatial autocorrelation could be seen as a methodological disadvantage, but on the other hand it is exactly what gives us information on spatial pattern, structure and processes and fundamental to much geographical work (Gould, 1970). Here, we use spatial autocorrelation index to represent the general pattern of urban land-use.

Measures of spatial autocorrelation work by examining how objects at one location are similar to objects located nearby. If features situated close together have similar attribute information, then the pattern in the data can be described as exhibiting positive autocorrelation. When features close together are more dissimilar in attribute value than features further away, pattern in the data is negatively auto correlated. Zero autocorrelation exists when attributes or their values are independent of location (Goodchild, 1986). Moran's *I* and Geary's c are two common indices for detection of spatial autocorrelation (Cliff & Ord, 1981; Goodchild, 1986; Moran, 1950). Qi and Wu (1996) have used both the indices to analyze the effect of changing spatial resolution on the results of topography and biomass pattern in 1972 of Peninsular Malaysia. They did not found appreciable difference among them with regularly grid data sets. Therefore, we select Moran's *I* as analysis index in this research. Moran's *I* is defined as follows:

$$I = \frac{n \sum_{i=1}^{n} \sum_{j=1}^{n} W_{ij}(X_i - \overline{X})(X_j - \overline{X})}{\sum_{i=1}^{n} \sum_{j=1}^{n} W_{ij} \sum_{i=1}^{n} (X_i - \overline{X})^2} \tag{4.2}$$

where, X_i *and* X_j stand for feature value of two cells nearby; $\overline{X}$ average value of feature of all the study area; n total number of cells; W weight (connectivity) matrix. When cell i and cell j are neighboring, $W_{ij}=1$, otherwise $W_{ij}=0$. By convention, $W_{ii}=0$. The value of Moran's *I* generally vary between 1 and -1, although values lower than -1 or higher than $+1$ may occasionally by obtained. Positive autocorrelation in the data translates into positive values of *I*; negative autocorrelation produces negative values. No autocorrelation results in a value close to zero (Goodchild, 1986).

Definition above shows that the value of Moran's *I* is mostly determined by two factors: the value of cell and weight matrix. Here, the value of cell adopts the grade of fuzzy membership to which the cell belongs to one or more categories of land-use. The weight matrix is constructed in queen case (i.e., a grid cell is adjacent to the neighboring cells in eight directions: left, upper-left, upper, upper-right, right, lower-right, lower, lower-left), similar to Moore neighborhood in cellular-based model of urban dynamics (Overmars, De Koning, & Veldkamp, 2003).

4.3 Results and Analysis

Firstly, we calculated the area of all types of land-use in both classification systems and their proportion to the study area at BCU level (Fig. 4.3). Then the value of Moran's *I* (VMI) of the land-use categories for both classification systems at all the levels of spatial resolution was calculated using Geoda software. Figure 4.4 shows the variogram of VMI of all the land-use categories with different classification system in the range of scale using single-state structure.

Figure 4.4 illustrates that the spatial autocorrelations of all the categories of urban land-use in both classification systems are scale-dependent as VMI of all the categories decreases with increasing grain size. In the extent from 10 m $\times$10 m to 50 m $\times$50 m, VMI of all the categories of urban land-use in both classification systems decreases rapidly, indicating that in both classification systems, spatial pattern of all

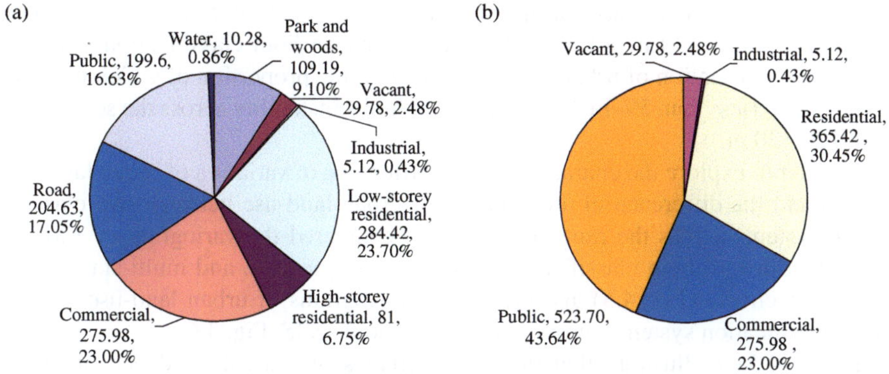

Fig. 4.3 Land-use structure of study area for both land-use classification systems. (**a**) System A; (**b**) System B (Note: The data for one category stands for category, area in ha. and proportion to the study area orderly)

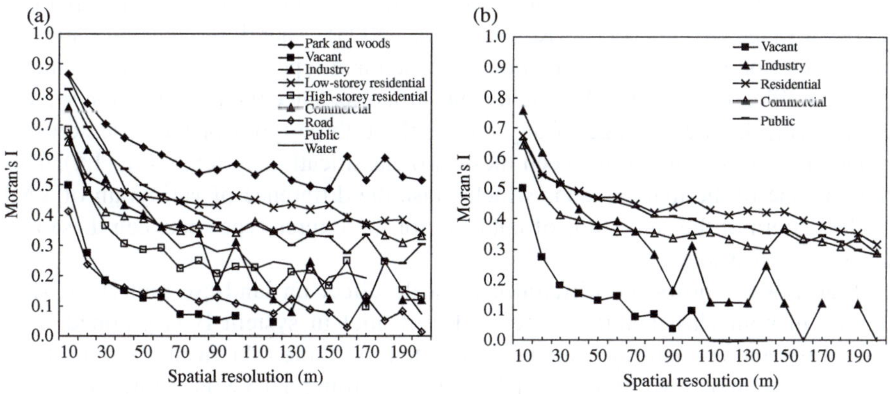

Fig. 4.4 Variogram of VMI of urban land-use in both classification systems across the range of scale. (**a**) System A; (**b**) System B

the categories of urban land-use shows strong scale-dependence in this extent. In the scale range of more than 50 m ×50 m, while any one of all the land-use categories is different from the other in the variation of VMI, all the variogram of VMI decreases slowly. That is, scale-dependence in this extent is not as obvious as that in extent from 10 m ×10 m to 50 m ×50 m. It also means that 50 m ×50 m may be a critical point of scale-dependence of urban land-use pattern in both classification systems.

Variogram of VMI of land-use vacant, water, Industrial in system A and vacant, water in system B, all the proportions of which to the study area are less than 5% (Fig. 4.3), shows the characteristics of bouncing or breaking off (no VMI) in the extent of more than 50 m ×50 m scale. In particular, VMI of vacant in both systems vanishes in the range of more than 120 m ×120 m. We examined area change of vacant across the range of spatial resolution and found that area of vacant in both classification systems nearly disappear and VMI can not be worked out in some scales more than 120 m $\times$120 m. It indicates that in both classification systems, spatial autocorrelation of urban land-use category, the proportion of which to study area is small (less than 5% in this research), exhibits instability across the scale more than 120$\times$20 m.

In order to explore the characteristics of difference of variation of urban land-use pattern and the difference of area change of urban land-use between two classification systems across the range of scale, we compared the variogram of VMI of urban land-use of both classification systems in single-state and multi-state structures respectively (Fig. 4.5) and the change histograms of urban land-use area in both classification systems across a range of spatial scale (Fig. 4.6).

Figure 4.5a–c illustrate that the characteristics of scale-dependence of urban land-use pattern of vacant, industrial and commercial does not change from system A to system B in multi-state structure. However, the variogram of VMI of the pattern of the same category of land-use in the three types differs a little between two classification systems in single-state structure. In multi-state structure, any one of the three types of land-use holds same content in both systems and no area lost across a range of scale (Zhao & Murayama, 2005). Therefore, characteristics of scale effect on spatial autocorrelation of these three types of land-use are same in both classification systems. In single-state structure, since the classification system is different, the area change of any one of these three types of land-use across a range of scale is different between these two classification systems (Fig. 4.6). This may be one of the main reasons which cause the difference of variogram of VMI of the pattern of the same type of land-use for these three types between these two classification systems.

Figure 4.5d shows the difference of scale effect on land-use pattern of residential in both classification systems. Residential in system B is grouped from low-storey and high-storey residential in system A (Table 4.1). In single-state structure, variogram of VMI of residential is different from both VMIs of low-storey and high-storey residential. Variogram of VMI of residential differs greatly from that of high-storey residential but is close to that of low-storey residential. In Fig. 4.3a, the proportion of low-storey residential to study area (23.7%) is more than that of high-storey (6.75%). This difference may indicate that the characteristics of scale effect of one type M (e.g. type residential) of urban land-use in one classification system (e.g. system B) is mainly influenced by the type of urban land-use with higher proportion to study area (here low-storey residential) in all the types of land-use, which compose type M of land-use, in another classification system (e.g. system A). In multi-state structure, characteristics of difference of scale effect on residential, low-storey and high-storey residential show the same as that in single-state structure. That is, scale effect on residential in system B is mainly influenced by low-storey residential in system A. But here the variogram of VMI of residential is flatter than that of low-storey residential. This may means that in the process of transformation from one classification system to another system, with the combination from other types of urban land-use and increase of area, scale effect on one type of urban land-use decreases.

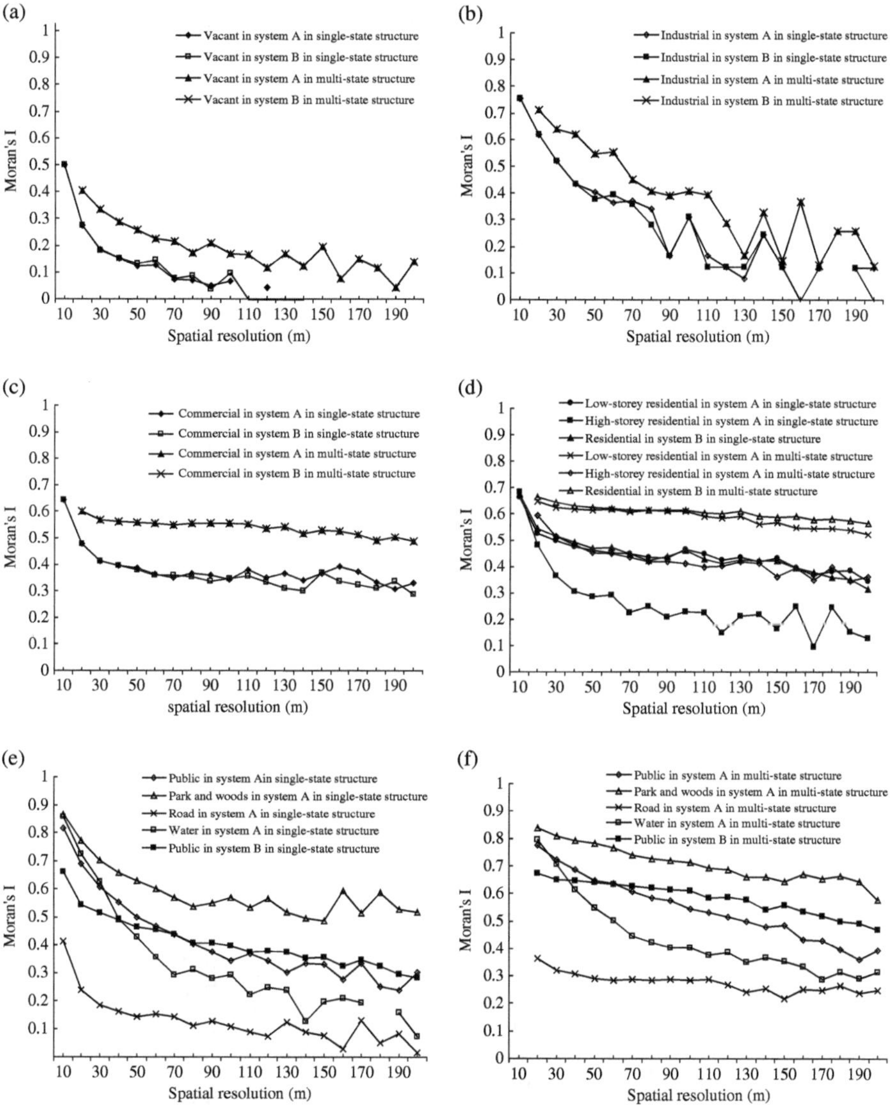

Fig. 4.5 Comparison of VMI of urban land-use with different classification systems across a range of spatial resolution. (**a**) Vacant; (**b**) industrial; (**c**) commercial; (**d**) residential; (**e**) public in single-state structure; (**f**) public in multi-state structure

Land-use of public in system B is grouped from land-use of public, water, road, park and woods in system A. Figure 4.5e, f illustrate the difference of spatial pattern of urban land-use of public between these two classification systems. In single-state structure, VMI of land-use of public of system B at basic spatial unit intervenes that of system A which compose public of system B. In Fig. 4.3, we find that the

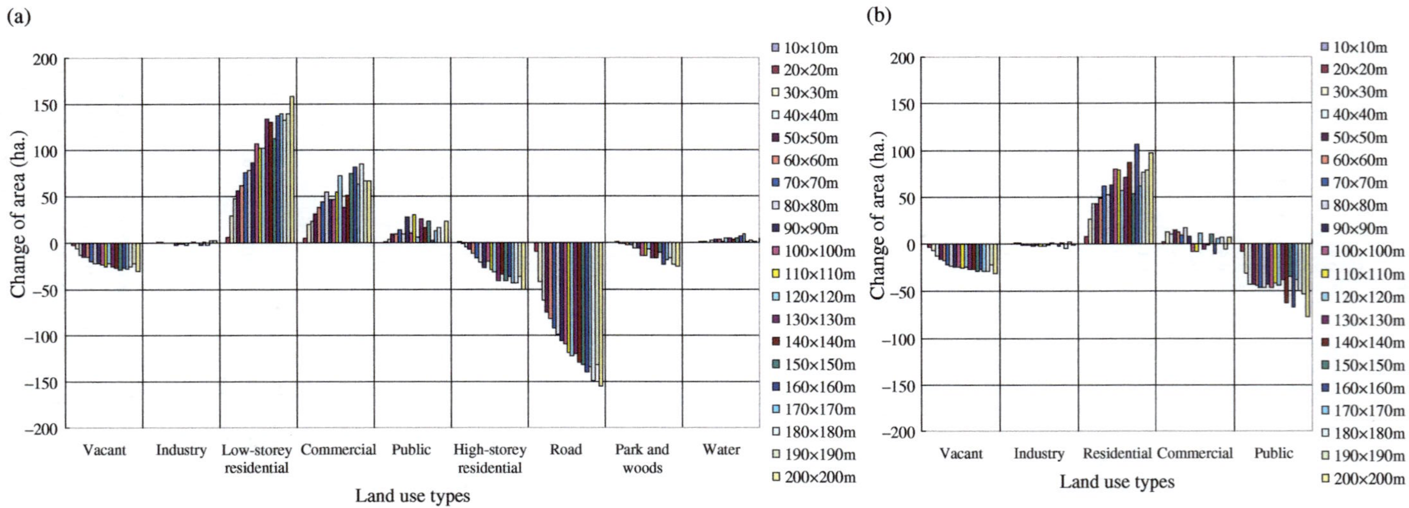

Fig. 4.6 Change histogram of urban land-use area in single-state structure across a range of spatial resolution. (**a**) System A; (**b**) System B (Note: for each type of urban land-use in both classification systems, the pillars from left to right stand for change of area of land-use in 10 ×10 m, 20 ×20 m,..., 200 ×200 m respectively)

proportion of public to study area (16.63%) in system A nearly equal that of road (17.05%). All VMI of road across the range of scale are lower very much than that of others. So add of road to public may influence VMI of land-use of public of system B at basic spatial unit. But scale effect of public of system B is not as notable as that of others. Since the proportion of land-use of water is small (0.86%) compared with that of other categories, it may not influence the public greatly. However, land-use of park and woods, the proportion of which is bigger (9.1%), may give more influence than water to the public as their variograms parallel each others. In multi-state structure, characteristics of difference of scale effect on category of public in system B and other categories in system A show the same as that in single-state structure. But all variograms of VMI have no sharp decline in the extent less than the critical point of 50 m ×50 m.

Figure 4.6 illustrates the histograms of urban land-use change in both classification systems across the range of spatial scale. We find that the amount of area change of land-use in different classification systems across the range of scale is different. Change of land-use area across the range of spatial scale may be one of main reasons of generating spatial resolution effect on urban land-use pattern analysis (Zhao & Murayama, 2005). Because of the difference of land-use area change in two classification systems across the range of scale, therefore, Fig. 4.6 indicates that change of land-use area across the range of scale may be one of the reasons of generating the difference of scale effect on urban land-use between two classification systems. Moreover, the amount of area change of land-use in system B is less than that in system A across the range of scale. It means that reducing number of urban land-use categories may diminish the loss of information of land-use area across the range of scale and the effect of spatial scale on urban land-use pattern analysis.

4.4 Discussion

(1) The detected spatial autocorrelation can be used to describe and compare the spatial pattern of the data. The general change trends of spatial autocorrelation of urban land-use for different classification systems show similar across a range of scale in this study. For most categories of urban land-use in both classification systems, the scale of 50 m ×50 m is a critical point of scale-dependence of spatial pattern. Because urban land-use pattern changes rapidly in the extent from 10 m ×10 m to 50 m ×50 m, considering the effect of spatial pattern of land-use on the identification of driving forces (see De Koning et al., 1998) and the relationship between land-use pattern and process (see Nagendra et al., 2004), driving force, mechanism accordingly, of urban dynamics in this extent may differ greatly between different levels of spatial scale. Therefore, construction of spatial model of urban dynamics in this extent needs to take carefully into account the problem of spatial scale. For instance, in this extent, intervals of spatial resolution in construction of hierarchical or multi-scale spatial model of urban dynamics can be set finer in order to

grasp more clearly the spatial process of urban dynamics across the scales in both classification systems.

While the value of Moran's I of all the urban land-use categories in these different classification systems changes across the range of scale more than 50 m ×50 m, the difference of it at different scale in this extent is not so obvious as that in extent from 10 m ×10 m to 50 m ×50 m. This means that the extent of scale more than 50 m ×50 m seems to take not so sensitive impact on construction of spatial model of urban dynamics. Therefore, intervals of spatial resolution in this extent for construction of multi-scale spatial model of urban dynamics may be set wider than that in extent from 10 m ×10 m to 50 m ×50 m in both classification systems.

However, the characteristics of the spatial scale effect on spatial autocorrelation of urban land-use categories in both classification systems exhibit just in this certain study area, whether it take the same in other urban areas, whether 50 m ×50 m is a critical point and whether we can set the intervals of the scale according to the characteristics for the construction of spatial model of urban dynamics or not needs to be demonstrated further.

(2) Area is one type of important information in urban land use pattern analysis. Although Fig. 4.6 indicates the fact that the process of reducing number of urban land-use categories can diminish the loss of information of land-use area across the range of scale in certain extent, the improvement of area accuracy seems to be the result of the tradeoff of diminishing the cognition of urban land use category information as some categories in system A are aggregated into higher level category in this process and not taken into account in system B anymore. Low-storey and high-storey residential, for example, in system A were aggregated into one category of residential in system B. The information of the difference between low-storey residential and high-storey residential disappeared in system B.

(3) The weight matrix in the definition of Moran's I depicts the relation between an element and its surrounding elements. Weight can be based, for example, on contiguity relations or distance. In this study, the weight matrix is directly based on contiguity and queen case. Where, a value unequal to zero in the matrix represents pairs of elements with a certain contiguity relation and a zero represents pairs without contiguity relation. Obviously, construction of the weight matrix affects the detection of spatial autocorrelation. Whether does it influence the analysis of spatial scale effect? In order to answer this question, we compared the VMI of commercial and low-storey in system A across the range of scale between queen case and the other general case: rook case (i.e., a grid cell is adjacent to the neighboring cells in four directions: left, upper, right and lower) (Fig. 4.7).

Figure 4.7 shows that while the value of Moran's I of land-use in rook case is bigger than that in queen case, the characteristics of spatial scale effect is similar. It means that the construction of weight matrix in Moran's I seems not to influence the analysis of spatial scale effect obviously. However, as weight matrix is one of key factors which determine the value of Moran's I in detecting spatial autocorrelation, changing weight matrix in study of urban land-use pattern analysis would retain to be a value extension of the current study.

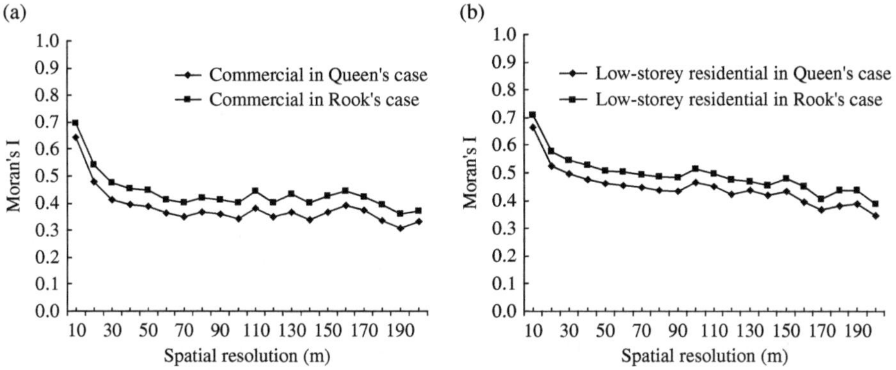

Fig. 4.7 Variogram of VMI of commercial (**a**) and low-storey (**b**) across the range of scale in queen case and rook case

4.5 Conclusions

Construction of spatial model of urban dynamics entails the linkage of urban land-use pattern and process. Different pattern of urban land-use at different scale and different land-use classification systems affects the understanding of mechanism and spatial process of urban dynamics. This chapter aims to explore the difference of the effect characteristics of spatial scale on urban land-use pattern analysis in different classification systems using spatial autocorrelation index and fuzzy mathematics by an empirical study in the CBD of Tokyo.

Due to the different purpose of using land-use information, criterion of land-use classification is different as well. Therefore, in different perspectives in the classification process, even same term of land-use category may contain different features. The difference of features contained in land-use categories generate different spatial pattern of urban land-use and effect characteristics of spatial scale on different classification systems. The characteristics of spatial scale effect on urban land-use pattern analysis in different land-use classification systems is demonstrated using one of the spatial autocorrelation indices – Moran's I in this study. Result shows that the spatial autocorrelation of urban land-use categories even at same scale between both classification systems take different values. This indicates the effect of land-use classification system on urban land-use pattern analysis.

Since the classification systems affect the result of urban land-use pattern analysis, result also shows that the effect of spatial scale on urban land-use pattern analysis in different classification systems take different characteristics in certain range of scale. Nevertheless, the general change trends of spatial autocorrelation of urban land-use for both classification systems show similar across the range of scale.

Multi-state structure based on the concept of fuzzy mathematics was used to compare the difference of spatial scale effect on urban land-use pattern analysis between both classification systems with that using single-state structure in order to explore some reasons. Spatial autocorrelation indices show less effect of spatial scale on urban land-use pattern analysis using multi-state structure than that using single-state structure in both classification systems. It means that change of urban land-use area across the range of spatial scale may be one of main reasons of generating spatial scale effect on urban land-use pattern analysis even in different classification systems. Furthermore, by comparing the change of land-use area in both classification systems across the range of scale we found that reducing number of urban land-use types may diminish the loss of information of land-use area across the range of scale and the effect of spatial scale on urban land-use pattern analysis in certain extent.

Acknowledgements National Natural Science Foundation of China, No. 40901090, 70863014; Foundation of Japan Society for the Promotion of Science (JSPS), No. 19.07003; Talents Introduced into Universities Foundation of Guangdong Province of China, No. 2009–26.

References

Anderson, J. R., Hardy, E. E., Roach, J. T., & Witmer, R. E. (1976). A land-use and land cover classification system for use with remote sensor data. *Geological Survey Professional Paper 964*. http://landcover.usgs.gov/pdf/anderson.pdf

Anselin, L. (1988). *Spatial econometrics: Methods and models*. Dordrecht: Kluwer.

Barredo, J. I., Kasanko, M., McCormick, N., & Lavalle, C. (2003). Modeling dynamic spatial processes: Simulation of urban future scenarios through cellular automata. *Landscape and Urban Plannning, 64*, 145–160.

Batty, M. (1970). Modeling cities as dynamic systems. *Nature, 231*, 426–428.

Batty, M., & Xie, Y. (1994). From cells to cities. *Environment and Planning B, 21*, 31–48.

Benz, U. C., Hofmann, P., Willhauck, G., Lingenfelder, I., & Heynen, M. (2004). Multi-resolution, object-oriented fuzzy analysis of remote sensing data for GIS-ready information. *ISPRS Journal of Photogrammetry & Remote Sensing, 58*, 239–258.

Bibby, P., & Shepherd, J. (2000). GIS, land-use, and representation. *Environment and Planning B: Planning and Design, 27*, 583–598.

Carlson, T. N., & Arthur, S. T. (2000). The impact of land-use - land cover changes due to urbanization on surface microclimate and hydrology: A satellite perspective. *Global and Planetary Change, 25*, 49–65.

Clarke, K. C., Hoppen, S., & Gaydos, L. (1997). A self-modifying cellular automaton model of historical urbanization in the San Francisco Bay area. *Environment and Planning B: Planning and Design, 24*, 247–261.

Cliff, A. D., & Ord, J. K. (1981). *Spatial processes: Models and applications*. London: Pion.

De Koning, G. H. J., Veldkamp, A., & Fresco, L. O. (1998). Land-use in Ecuador: A statistical analysis at different aggregation levels. *Agriculture, Ecosystems and Environment, 70*, 231–247.

Dickinson, G. C., & Shaw, M. A. (1977). What is land-use? *Area, 9*, 38–42.

Geographical Survey Institute (1998). *Detailed digital information (10 m grid land-use) metropolitan area 1974, 1979, 1984, 1989 and 1994*. Tokyo: Japan Map Center (in CD-ROM).

Goodchild, M. F. (1986). *Spatial autocorrelation*, Catmog 47. Norwich: GeoBook.

Gould, P. R. (1970). Is statistics inferens the geographical name for a wild goose? *Economic Geography, 46*, 439–448.

Hostetler, N. E., & McIntyre, M. E. (2001). Effects of urban land-use on pollinator (Hymenoptera: Apoidea) communities in a desert metropolis. *Basic and Applied Ecology, 2*, 209–217.

Jantz, C. A., & Goetz, S. J. (2005). Analysis of scale dependencies in an urban land-use-change model. *International Journal of Geographical Information Science, 19*, 217–241.

Klosterman, R. E. (2005). An update on planning support systems. *Environment and Planning B, 32*, 477–484.

Liu, Y., & Phinn, S. R. (2003). Modelling urban development with cellular automata incorporating fuzzy-set approaches. *Computers, Environment and Urban Systems, 27*, 637–658.

Moody, A., & Woodcock, C. E. (1994). Scale dependent errors in the estimation of land cover proportions: Implications for global land cover data sets. *Photogrammetric Engineering & Remote Sensing, 60*, 585–594.

Moran, P. A. P. (1950). Notes on continuous stochastic phenomena. *Biometrika, 37*, 17–23.

Nagendra, H., Munroe, D. K., & Southworth, J. (2004). From pattern to process: Landscape fragmentation and the analysis of land-use/land cover change. *Agriculture, Ecosystems and Environment, 101*, 111–115.

Overmars, K. P., De Koning, G. H. J., & Veldkamp, A. (2003). Spatial autocorrelation in multi-scale land-use models. *Ecological Modelling, 164*, 257–270.

O'Sullivan, D., & Unwin, D. (2002). *Geographic information analysis*. Hoboken, NJ: Wiley.

Pauleit, S., Ennos, R., & Golding, Y. (2005). Modeling the environmental impacts of urban land-use and land cover change-a study in Merseyside, UK. *Landscape and Urban Planning, 71*, 295–310.

Potoglou, D., & Kanaroglou, P. S. (2005). Carbon monoxide emissions from passenger vehicles: Predictive mapping with an application to Hamilton, Canada. *Transportation Research Part D, 10*, 97–109.

Qi, Y., & Wu, J. G. (1996). Effects of changing spatial resolution on the results of landscape pattern analysis using spatial autocorrelation indices. *Landscape Ecology, 11*, 39–49.

Silva, E. A., & Clarke, K. C. (2002). Calibration of the SLEUTH urban growth model for Lisbon and Porto, Portugal. *Computers, Environmental and Urban Systems, 26*, 525–552.

Solecki, W. D., & Oliveri, C. (2004). Downscaling climate change scenarios in an urban land-use change model. *Journal of Environmental Management, 72*, 105–115.

Stabler, L. B., Martin, C. A., & Brazel, A. J. (2005). Microclimates in a desert city were related to land-use and vegetation index. *Urban Forestry & Urban Greening, 3*, 137–147.

Straatman, B., White, R., & Engelen, G. (2004). Towards an automatic calibration procedure for constrained cellular automata. *Computers, Environmental and Urban Systems, 28*, 149–170.

Tobler, W. R. (1970). A computer movie simulating urban growth in the Detroit region. *Economic Geography, 46*, 234–240.

Treitz, P., & Rogan, J. (2004). Remote sensing for mapping and monitoring land-cover and land-use change: An introduction. *Progress in Planning, 61*, 349–363.

Turner, M. G. (1987). Spatial simulation of landscape changes in Georgia: A comparison of 3 transition models. *Landscape Ecology, 1*, 29–36.

Turner, M. G., O'Neill, R. V., Gardner, R. H., & Milne, B. T. (1989). Effects of changing spatial scale on the analysis of landscape pattern. *Landscape Ecology, 3*, 153–162.

Veldkamp, A., Verburg, P. H., Kok, K., de Koning, G. H. J., Priess, J., & Bergsma, A. R. (2001). The need for scale sensitive approaches in spatially explicit land-use change modeling. *Environmental Modeling and Assessment, 6*, 111–121.

Verburg, P. H., de Koning, G. H. J., Kok, K., Veldkamp, A., & Bouma, J. (1999). A spatial explicit allocation procedure for modelling the pattern of land use change based upon actual land use. *Ecological Modelling, 116*, 45–61.

White, R., & Engelen, G. (1993). Cellular automata and fractal urban form: A cellular modeling approach to the evolution of urban land-use patterns. *Environment and Planning A, 25*, 1175–1199.

White, R., Engelen, G., & Uljee, I. (1997). The use of constrained cellular automata for high-resolution modeling of urban land-use dynamics. *Environment and Planning B, 24*, 323–343.

Wu, F. (1998). Simulating urban encroachment on rural land with fuzzy-logic-controlled cellular automata in a geographical information system. *Journal of Environmental Management, 53*, 293–308.

Yeh, A. G. O., & Xia, L. (2001). A constrained CA model for the simulation and planning of sustainable urban forms by using GIS. *Environment and Planning B, 28*, 733–753.

Yin, Z., Walcott, S., Kaplan, B., Cao, J., Lin, W., Chen, M., et al. (2005). An analysis of the relationship between spatial pattern of water quality and urban development in Shanghai, China. *Computers, Environment and Urban Systems, 29*, 197–221.

Zhang, J., & Foody, G. M. (1998). A fuzzy classification of sub-urban land cover from remotely sensed imagery. *International Journal of Remote Sensing, 19*, 2721–2738.

Zhang, J., & Stuart, N. (2001). Fuzzy methods for categorical mapping with image-based land cover data. *International Journal of Geographical Information Science, 15*, 175–195.

Zhao, Y., & Murayama, Y. (2005). Effect characteristics of spatial resolution on the analysis of urban land-use pattern: A case study of CBD in Tokyo using spatial autocorrelation index. In Y. Murayama & G. Du (Eds.), *Cities in global perspective: Diversity and transition* (pp. 585–594). Tokyo: IGU Urban Commission.

Chapter 5
Modeling Neighborhood Interaction in Cellular Automata-Based Urban Geosimulation

Yaolong Zhao and Yuji Murayama

5.1 Introduction

One of the most important developments in Geographic Information Science (GIScience) is the expansion of theories, models, and technologies to effectively discern and interpret spatiotemporal patterns, relationships, and interactions among features, activities, processes, and events in geographic domains. In current era, as rapid changes of urban land-use all over the world have greatly impacted on local (Lin & Ho, 2003; McKinney, 2006; Paul & Meyer, 2001) and global environmental changes (Grimm, Grove, Pickett, & Redman, 2000; Lambin et al., 2001), the issue of modeling the spatial process of urban growth to better understand the mechanism and consequences of urbanization and explore the extent of future urban land-use change has attracted sweeping attentions of scientists with background in different disciplines ranging from anthropology to mathematical programming. This issue also has enriched the theory and technology of simulation model of geographic phenomena.

An important component in Cellular Automata (CA)-based urban geosimulation models is the local spatial interaction between neighborhood land-use types. The neighborhood interaction is often addressed based on the notion that urban development can be conceived as a self-organizing system in which natural constraints and institutional controls (land-use policies) temper the way in which local decision-making processes produce macroscopic urban form. Different processes can explain the importance of neighborhood interaction. At large scale, simple mechanisms for economic interaction between locations were provided by the central place theory (Christaller, 1933) that describes the uniform pattern of towns and cities in space as a function of the distance that consumers in the surrounding region travel to the nearest facilities. Spatial interaction between the location of facilities, residential

Y. Zhao (✉)
School of Geography, South China Normal University, Guangzhou, Guangdong, PR China
e-mail: yaolongzhao@gmail.com

This chapter is improved from "Yaolong Zhao and Yuji Murayama (2007), A new method to model neighborhood interaction in Cellular Automata-based urban geosimulation, Lecture Notes in Computer Science, 4488, 550–557", with permission from Springer.

Y. Murayama, R.B. Thapa (eds.), *Spatial Analysis and Modeling in Geographical Transformation Process*, The GeoJournal Library 100, DOI 10.1007/978-94-007-0671-2_5,

areas and industries has been given more attention in the work of Krugman (Fujita, Krugman, & Mori, 1999; Krugman, 1999). The spatial interaction is explained by a number of factors that either cause concentration of urban functions (centripetal forces: economies of scale, localized knowledge spill-over, thick labor markets) and others that lead to a spatial spread of urban functions (centrifugal forces: congestion, land rents, factor immobility etc.).

In keeping with the spirit of simplicity, neighborhood interaction in the application of CA on urban geosimulation models most often adopt either the Von Neumann 3 ×3 (or 5 ×5) or the Moore 3 ×3 neighborhood (Batty, 1998; Wu, 1998; Yeh & Li, 2001). For most physical systems, these are clearly the most appropriate definitions since such systems typically have only local causation. However, in the case of human systems like cities, the idea of locality may be much larger, since people and institutions are aware of their surroundings in a wider space (White & Engelen, 2000). Thus it is desirable to define a neighborhood large enough (i.e. extended neighborhood) to capture the operational range of the local processes being modeled by CA. White and Engelen (1993) firstly proposed this kind of configuration of neighborhood for exploring the relationship of CA-based model of urban form evolution (White & Engelen, 1993). In 1997, White et al. calibrated the neighborhood effect by means of a trial and error approach for geosimulation of Cincinnati city (White, Engelen, & Uljee, 1997). In 2004, this research group proposed automatic calibration procedure for this kind of neighborhood effect (Straatman, White, & Engelen, 2004).

Objective of this research is to improve the methodology of existing CA models by proposing a theoretical framework to model and calibrate the neighborhood effect. This framework aims to assist modelers in the implementation and quantification of neighborhood interaction in urban geosimulation models.

5.2 Modelling Neighborhood Interaction

Tobler's First Law of Geography (FLG), "Everything is related to everything else, but near things are more related than distant things", is the fundamental theory in this framework. This law, firstly proposed in 1970 (Tobler, 1970), has brought strong controversy in geography domain. In 2003, a panel on this law was organized in AAG meeting in New Orleans. Five famous geographers presented their comments in this panel and these comments were published in a forum of Annals of the Association of American Geographers in 2004. Some professors agreed with Tobler, others not. However, all of them accepted the actual geography phenomena illustrated by the FLG. The divarication existed on the word "law". Goodchild especially discussed the validity and usefulness of the FLG in GIScience and geography (Goodchild, 2004). Here, the controversy of whether phenomena can be expressed as "law" was discarded, and the local knowledge expressed in Tobler's FLG was accepted. It is assumed that the effect of cell states in the neighborhood area of developable cell accord with the rule of distance decay described in the FLG.

The expression of Tobler's FLG is very qualitative, and a distance decay function is needed for representing the law. Herein the idea of Reilly's law of retail gravitation

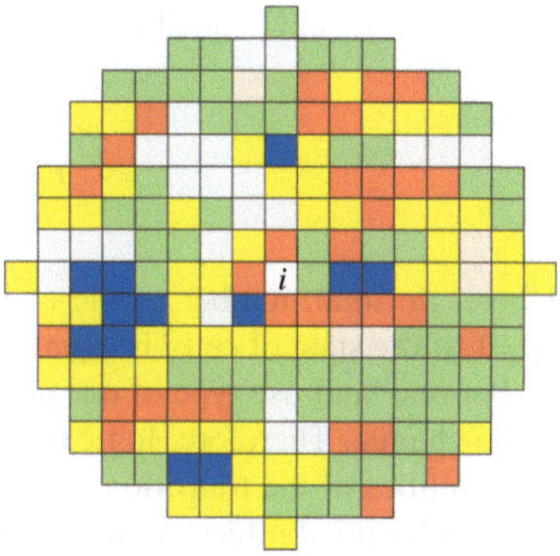

Fig. 5.1 An extended neighborhood pattern

(Reilly, 1931) was adopted, which states that "A city will attract retail trade from a town in its surrounding territory, in direct proportion to the population size of the city and in inverse proportion to the square of the distance from the city". Figure 5.1 shows one of the extended neighborhood patterns of one developable cell *i*, which is defined as all cells within a radius of eight cells, an area containing 196 cells. It is assumed that in cellular environment all the cells in the neighborhood contribute to the conversion of developable cell *i*. The contribution of one cell is associated with the state of itself and the distance to the developable cell *i*. It can be expressed as follows:

$$f_{kh} = G_{kh}\frac{A_j}{d_{ji}^2} \tag{5.1}$$

where, f_{kh}: constribution of one cell j with land-use k in the neighborhood to the conversion of the developable cell i to land-use h for next stage; A_j: area of the cell j; d_{ji}: the Euclidean distance between the cell j in the neighborhood area and the developable cell i, and G_{kh}: constant of the effect of land-use k on the transition to land-use h. + stands for positive, – repulsive.

Figure 5.2 indicates the scheme of the impact gradient using this function. Note that this is a modificatory Reilly's function and no unit problem exists inthis func-

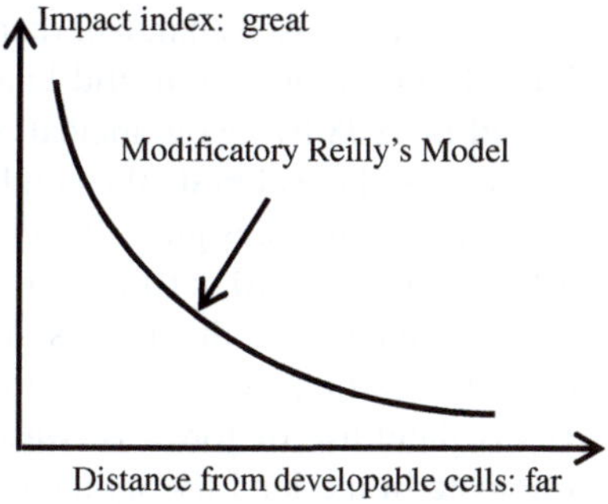

Fig. 5.2 Scheme of the impact gradient

tion. Then the aggregated effect of the cells in the neighborhood can be expressed as:

$$F_{kh} = G_{kh} \sum_{j=1}^{m} \frac{A_j}{d_{ji}^2} I_{kj} \tag{5.2}$$

where, m: total number of the cells in the neighborhood, and I_{kj}: index of cells. I_{kj}=1, if the state of cell j is equal to k; I_{kj}=0, otherwise.

For one land-use type of one cell, there are just two results of transition: change or no change. Therefore, logical regression approach was selected to calculate the probabilities of the transition of cell i under neighborhood effect. The contribution of neighborhood effect to the probability of conversion to land-use h of a cell (N_i) is described as a function of a set of aggregated effect of different land-use types:

$$Log(\frac{N_{ih}}{1-N_{ih}}) = \beta_{oi} + \sum_k \beta_{ikh} F_{ikh} = \beta_{oi} + \sum_k \beta_{ikh} G_{kh} \sum_m \frac{A_m}{d_{mi}^2} I_{mk} \tag{5.3}$$

As G_{kh} is a constant, let:

$$\beta'_{0i} = \beta_{0i}, \quad \beta'_{ikh} = \beta_{ikh} G_{kh}$$

Then:

$$Log(\frac{N_{ih}}{1-N_{ih}}) = \beta'_{0i} + \sum_k \beta'_{ikh} \sum_m \frac{A_m}{d_{mi}^2} I_{mk} \tag{5.4}$$

where, β'_{oi} and β'_{ikh} are the coefficients which should be calibrated.

5.3 Calibration of Neighborhood Interaction

5.3.1 Study Area and Data Set

The Tokyo metropolitan area was identified as study area to confirm the approach as this area possesses abundant dataset of land-use. The dataset is named Detailed Digital Information (10 m grid land-use) Metropolitan Area of Tokyo which was released in 1998 by Geographical Survey Institute of Japan. The data in 1984 and 1989 was used to calibrate the neighborhood effect.

The dataset was designed 10 cell states of land-use and was aggregated into 100 ×100 m grid from 10 ×10 m in majority rule. Land-use of water represents fixed features in the model, that is, this feature is not assumed to change and which therefore do not participate in the dynamics in order to protect the life environment. Two types of land-use of forest & wasteland, and cropland take passive features that participate in the land-use dynamics. However, the dynamics are not driven by an exogenous demand for land. They appear or disappear in response to land being taken or abandoned by the active functions. The active functions are four land-use categories which are forced by exogenously generated demands for land to CA in response to the growth of the urbanized area: vacant, industrial, residential, and commercial land. Other three types of land-use of road, public land, and special land show passively active features which are dynamics in the model. Active and passively active types of land-use are assumed to participate in the neighborhood interaction.

5.3.2 Calibration of Neighborhood Interaction

As one of statistical analysis techniques, logistical regression has to consider the problem of spatial statistics like spatial dependence and spatial sampling (Irwin & Geoghegan, 2001; Cheng & Masser, 2003) in the calibration procedure. The integration of both systematic and random sampling method was adopted to eliminate spatial dependence effect. Firstly, land-use changes were detected from the data set in 1984 and 1989. Systematic sampling was implemented and approximately half cells of the changes of every one of four active land-use types were remained. Then random selection of unchanged cells was carried out to create nearly 1:1 ratio for changed cells and unchanged cells. Its total size was 27,070 cells (Table 5.1).

Table 5.1 illustrates that all the values of PCP of four active land-use types are more than 80% and all of the values of ROC more than 0.9, thus showing goodness of fit of this approach. The results of the test also indicate the existence of neighborhood interaction in urban land-use changes.

Table 5.1 Result of calibration of neighborhood interaction

	Active land-use types			
Factors and test	Vacant land	Industrial land	Residential land	Commercial land
Total size (cells) of sampling	11,034	1732	11,596	2708
$\beta'_{\text{Vacant land}, h}$	1.147	0.091*	0.190	0.158
$\beta'_{\text{Industrial land}, h}$	0.334	1.446	0.262	0.457
$\beta'_{\text{Residential land}, h}$	0.103	**	0.562	0.209
$\beta'_{\text{Commercial land}, h}$	0.348	0.727	0.181	1.821
$\beta'_{\text{Road}, h}$	0.199	**	0.421	0.561
$\beta'_{\text{Public land}, h}$	0.198	**	0.199	0.224
Constant β'_0	–2.428	–1.988	–2.830	–2.763
Test				
PCP (%)	84.3	87.6	83.6	86.3
ROC	0.924	0.937	0.905	0.937

PCP: Percentage Correctly Predicted (0–100%); ROC: Relative Operating Characteristic (0–1).
*Significant at $p<0.05$; **non-statistically significant; others significant at $p<0.001$.

5.4 Simulation and Results

Constrained CA model was put forward to confirm this method. In this model, transition potentials for each cell are calculated as follows:

$$^{t}P_{ik} = (1 + {}^{t-1}N_{ik})(1 + {}^{t-1}S_{ik})(1 + {}^{t-1}Z_{ik})(1 + {}^{t-1}A_{ik})^{t-1}v \quad (5.5)$$

where, $^{t}P_{ik}$ = the potential of the cell i for land use k in time t; $^{t-1}N_{ik}$ = the neighborhood effect on the cell i for land-use k at time t–1, which equals the value of N_{ih} in Eq. (5.4); $^{t-1}S_{ik}$ = the intrinsic suitability of the cell i for land-use k at time t–1; $^{t-1}Z_{ik}$ = the zoning status of the cell i for land-use k at time t–1; $^{t-1}A_{ik}$ the accessibility of the cell i to transportation for land-use k at time t–1; ^{t-1}v is the scalable random perturbation term at time t–1. Four active land-use types were changed by an intervention that is exogenous to the CA model from regional systems.

In this schema the neighborhood factor N makes city works like a non-linear system as suitability, accessibility and land-use zoning status are relatively stable in certain period.

In a learning stage, $^{t-1}S_{ik}$, $^{t-1}Z_{ik}$, and $^{t-1}A_{ik}$ also were calibrated through the dataset in 1984 and 1989. The calibrated model was used to simulate the spatial process of urban growth of the Tokyo metropolitan area from 1989 to 1994. Land-use maps in simulation and reality are shown in Fig. 5.3. Note that in order to make the maps clearer, land-use types of forest & wasteland and cropland have been grouped into non-urbanized area and other land-use types except water into urbanized area.

A good CA-based model produces results which have all the patterned complexity of the real system (White & Engelen, 2000). Comparison of the simulated result with the actual data in terms of fractal dimension and spatial metrics was carried out towards the test of this proposed approach. These two indices are excellent for presenting the pattern of complex system like city (Barredo & Demicheli, 2003).

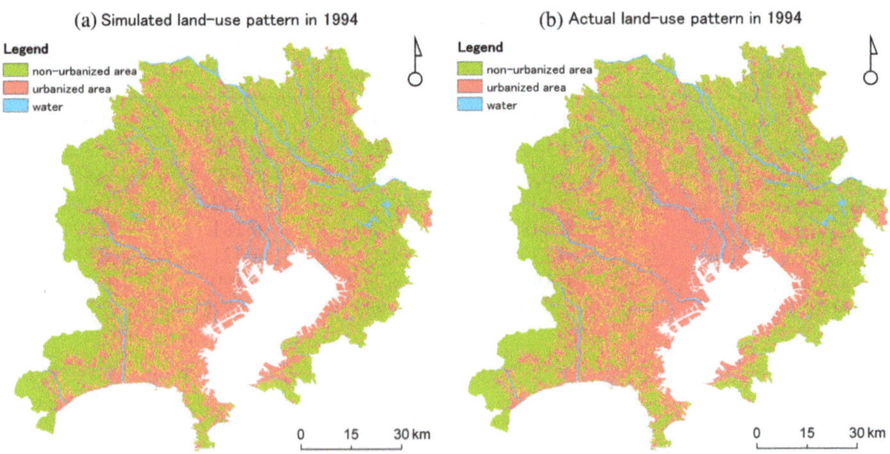

Fig. 5.3 Land-use maps in the study area: (**a**) simulation in 1994; (**b**) reality in 1994

Table 5.2 Assessment of the simulated urbanized area in terms of fractal dimension

Fractal dimension in different radius zones	Reality in 1989	Reality in 1994	Simulation in 1994
In 0–16 km radius	1.94	1.95	1.95
In 16–50 km radius	1.45	1.48	1.48

Table 5.2 shows the assessment result of the simulated urbanized area in terms of fractal dimension. Areas, the distance from which to Tokyo station is more than 50 km, were omitted considering the effect of boundary in this table. In order to understand the change of urbanized area structure and confirm the ability of this model in capturing the change of structure, fractal dimension of urbanized area in 1989 also is shown. Table 5.2 indicates that urbanized area shows bifractal structure in the study area. Urbanized area had grown more greatly in the second radius zone with 16–50 km than in the first one with 0–16 km. Model well captures this characteristic.

Figure 5.4 shows the assessment result of simulation in terms of spatial metrics. Two metrics were used for the assessment: NP (number of patches) and PD (patch density). Values of NP and PD had declined from 1989 to 1994, thus indicating the characteristics of compact growth or conglomeration of the existing urbanized area in the study area. Simulated urbanized area presents the same characteristics. However, if we take out the component of neighborhood interaction from the model,

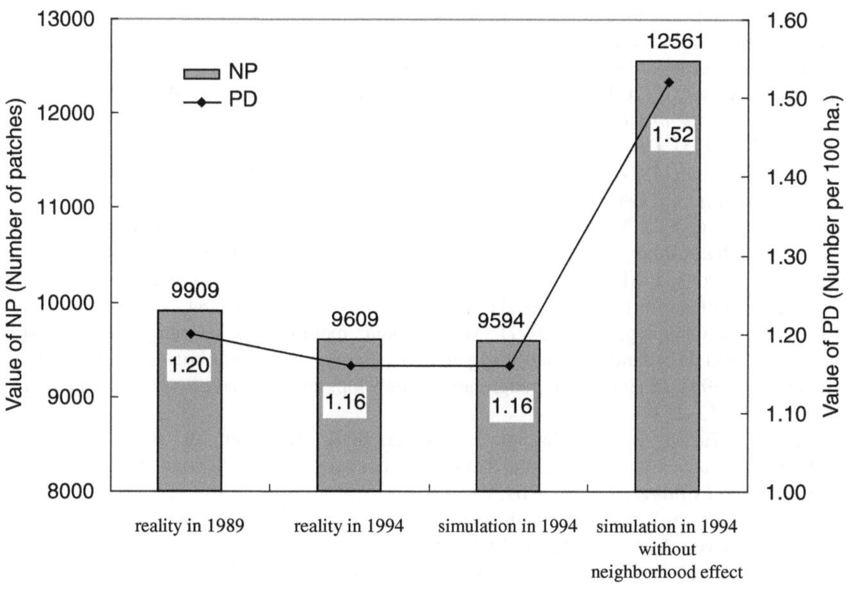

Fig. 5.4 Comparison of the significance of spatial metrics of urbanized area between reality and simulation in the Tokyo metropolitan area

NP and PD turgidly increased. This confirms the utility of the proposed approach in modeling neighborhood interaction.

5.5 Concluding Remarks

This study developed a novel approach for modeling and calibrating neighborhood interaction in CA-based urban geosimulation. The proposed method provides a theoretical framework for presenting neighborhood effect of CA. The results of simulation using the Tokyo metropolitan area as a case study indicates that urban geosimulation model which embeds this method well captures the main characteristics of spatial process of urban growth. The results also confirmed the utility of this method for presenting dynamics of complex system. This approach can be used not only in regular grid cells, but also in irregular cells, like in vector structure, as it considers the area of cells and the distance decay. To discuss this issue would be a value extension to the current study.

Acknowledgements National Natural Science Foundation of China, No. 40901090, 70863014; Foundation of Japan Society for the Promotion of Science (JSPS), No. 19.07003; Talents Introduced into Universities Foundation of Guangdong Province of China, No. 2009–26.

References

Barredo, J. I., & Demicheli, L. (2003). Urban sustainability in developing countries' megacities: Modelling and predicting future urban growth in Lagos. *Cities, 20*, 297–310.

Batty, M. (1998). Urban evolution on the desktop: Simulation with the use of extended cellular automata. *Environment and Planning A, 30*, 1943–1967.

Cheng, J., & Masser, I. (2003). Urban growth pattern modeling: A case study of Wuhan city, PR China. *Landscape and Urban Planning, 62*, 199–217.

Christaller, W. (1933). *Central places of southern Germany*. London: Prentice Hall.

Fujita, M., Krugman, P., & Mori, T. (1999). On an evolution of hierarchical urban systems. *European Economic Review, 43*, 209–251.

Goodchild, M. F. (2004). The validity and usefulness of laws in geographic information science and geography. *Annals of the Association of American Geographers, 94*, 300–303.

Grimm, N. B., Grove, J. M., Pickett, S. T. A., & Redman, C. L. (2000). Integrated approaches to long-term studies of urban ecological systems. *Bioscience, 50*, 571–584.

Irwin, E. G., & Geoghegan, J. (2001). Theory, data, methods: Developing spatially explicit economic models of land use change. *Agriculture, Ecosystems & Environment, 85*, 7–23.

Krugman, P. (1999). The role of geography in development. *International Regional Science Review, 22*, 142–161.

Lambin, E. F., Turner, B. L., Geist, H. J., Agbola, S. B., Angelsen, A., Bruce, J. W., et al. (2001). The causes of land-use and land-cover change: Moving beyond the myths. *Global Environmental Change, 11*, 261–269.

Lin, G. C. S., & Ho, S. P. S. (2003). China's land resources and land-use change: Insights from the 1996 land survey. *Land Use Policy, 20*, 87–107.

McKinney, M. L. (2006). Urbanization as a major cause of biotic homogenization. *Biological Conservation, 127*, 247–260.

Paul, M. J., & Meyer, J. L. (2001). Streams in the urban landscape. *Annual Review of Ecology Systematics, 32*, 333–365.

Reilly, W. J. (1931). *The law of retail gravitation.* New York: Knicherbocker.

Straatman, B., White, R., & Engelen, G. (2004). Towards an automatic calibration procedure for constrained cellular automata. *Computers, Environment and Urban Systems, 28*, 149–170.

Tobler, W. (1970). A computer movie simulating urban growth in the Detroit region. *Geographical Analysis, 46*, 234–240.

White, R., & Engelen, G. (1993). Cellular automata and fractal urban form: A cellular modelling approach to the evolution of urban land-use patters. *Environment and Planning A, 25*, 1175–1199.

White, R., & Engelen, G. (2000). High-resolution integrated modelling of the spatial dynamics of urban and regional systems. *Computers, Environment and Urban Systems, 24*, 383–400.

White, R., Engelen, G., & Uljee, I. (1997). The use of constrained cellular automata for high-resolution modeling of urban land-use dynamics. *Environment and Planning B, 24*, 323–343.

Wu, F. (1998). SimLand: A prototype to simulate land conversion through the integrated GIS and CA with AHP-derived transition rules. *International Journal of Geographical Information Science, 12*, 63–82.

Yeh, A. G. O., & Li, X. (2001). A constrained CA model for the simulation and planning of sustainable urban forms by using GIS. *Environment and Planning B, 28*, 733–753.

Part II
Urban Analysis: Zonation and Population Structure

Chapter 6
Estimation of Building Population from LIDAR Derived Digital Volume Model

Ko Ko Lwin and Yuji Murayama

6.1 Introduction

Many methods to map population distribution have been practiced in geographic information systems (GIS) and remote sensing fields. Common cartographic forms of population mapping are the choropleth map and the dasymetric map. Choropleth maps provide an easy way to visualize how a measurement varies across a geographic area. However, they have limited utility for detailed spatial analysis of population data, especially where the population is concentrated in a relatively small number of villages, towns and cities. Moreover, choropleth maps cannot express statistical variation within the administrative areal units, for example changing population density. One way to avoid this limitation is by transforming the administrative units into smaller and more relevant map units through a process known as dasymetric mapping (Bielecka, 2005). Voss, Long and Hammer (1999) noticed that most census boundaries do not coincide with the boundaries of geographic features such as land use/land cover, soil type, geological units, and floodplain and watershed boundaries; this is known as "spatial incongruity" and it arises when spatially aggregated data are available for one set of geographic areal units but not the areal units of primary interest. Spatial incongruity presents a major obstacle to the integration of social and natural science data and consequently places limitations on interdisciplinary research efforts.

The segregation of population data method was firstly utilized and termed dasymetric by the Russian cartographer Tian-Shansky, who developed the multi-sheet population density map of European Russia, scale 1:420,000, published in the 1920s (Preobrazenski, 1954). As populated territory, Tian-Shansky mapped areas within

K.K. Lwin (✉)
Division of Spatial Information Science, Graduate School of Life and Environmental Sciences, University of Tsukuba, Tsukuba, Ibaraki, Japan
e-mail: kokolwin2002@yahoo.com

This chapter is improved from "Ko Ko Lwin and Yuji Murayama (2010), Development of GIS tool for dasymetric mapping. International Journal of Geoinformatics, 6, 11–18", with permission from Association for Geoinformation Technology.

Y. Murayama, R.B. Thapa (eds.), *Spatial Analysis and Modeling in Geographical Transformation Process*, The GeoJournal Library 100, DOI 10.1007/978-94-007-0671-2_6,

the equidistant of one verst (1067 m) from built-up terrains. The first cartographer who popularized dasymetric mapping was Wright (1936). He set forth a new method of presenting population density based upon the division of a given administrative unit into smaller areas complying with different types of geographical environments. Although dasymetric mapping has been in use since at least the early 1800s, it has never achieved the ubiquity of other types of thematic mapping, and thus the means of producing dasymetric maps have never been standardized and codified in the way other types of thematic mapping techniques have been (Eicher & Brewer, 2001; Slocum, 1999). Therefore, dasymetric methods remain highly subjective, with inconsistent criteria. The reason for this relative lack of popularity and the paucity of standard methodology surely lies at least partially in the difficulty inherent in constructing dasymetric maps, and until recently, the difficulties in obtaining the necessary data, as well as access to the computer power required to generate them (Maantay, Maroko, & Herrmann, 2007). Transferring data from one set of geographic zones or districts to another set of non-coincident zones is often necessary in spatial analysis. For instance, we might have data on the number of people living within a certain census tract but need to estimate the number of people in a smaller area within the tract, or an area that includes only part of that tract and part of other tracts. We may be interested in population or other data at a watershed level and only have population data available at the census enumeration units.

Several methods have been developed to generate smaller geographical units of population distribution based on aggregated values with ancillary datasets, commonly known as "Dasymetric Mapping" by using GI Science theory and practice. Maantay et al. (2007) extensively reviews on existing dasymetric methods and techniques. The following are some developed methods and approaches in dasymetric mapping: areal interpolation, filtered areal weighting (binary method) (Eicher & Brewer, 2001), filtering with land use/land cover data (Sleeter, 2004), and cadastral-based expert dasymetric system (CEDS) (Maantay et al., 2007). Recent years, research into micro-spatial analysis has increased due to remote sensing data available at finer spatial resolution with more diverse geo-information sources (IKONOS, QuickBird, LIDAR, etc.) and the availability of fine-scale GIS data with enhanced attribute information (e.g. building footprints with the number of floors, building use type and building name). In this chapter, we estimate building population from building footprints and census tracts by integrating LIDAR derived Digital Volume Model, DVM.

6.2 Development of GIS Tool

6.2.1 Methodology

Lwin and Murayama (2009) introduce two building population estimation methods: (1) Areametric (which does not require information on the number of building floors); and (2) Volumetric (which does require information on the number of floors). For improved accuracy, the two methods allow filtering by other categories such as: minimum footprint area and building use types, e.g., commercial,

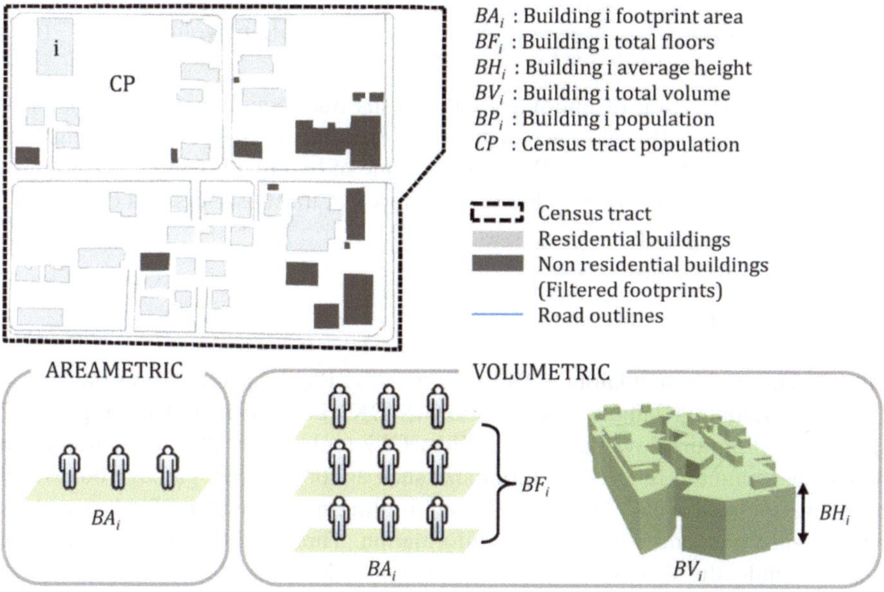

Fig. 6.1 Graphical illustration of equations (modified from Lwin & Murayama, 2009)

industrial, educational, and other building use types that are not occupied by residents. Figure 6.1 shows the abstract idea of estimations and the calculation is demonstrated by the following mathematical expressions:

Areametric method:

$$\mathrm{BP}_i = \left(\frac{\mathrm{CP}}{\sum_{k=1}^{n} \mathrm{BA}_k}\right) \mathrm{BA}_i \quad \text{Using building footprint surface area} \tag{6.1}$$

Volumetric method:

$$\mathrm{BP}_i = \left(\frac{\mathrm{CP}}{\sum_{k=1}^{n} \mathrm{BA}_k.\mathrm{BF}_k}\right) \mathrm{BA}_i.\mathrm{BF}_i \quad \text{Using number of floors information} \tag{6.2}$$

The two algorithms have been tested with actual building population data acquired from City Office for study purpose by applying various filtering footprint sizes. The beset result was achieved in volumetric method, filtered by 20 m^2 footprints size. Moreover, advances in remote sensing data acquisition technologies such as LIDAR can be used for the extraction of building footprints, building height (Digital Height Model (DHM)) and building volume (Digital Volume Model (DVM)). Equations (6.3) and (6.4) can be used for LIDAR data:

$$\mathrm{BP}_i = \left(\frac{\mathrm{CP}}{\sum_{k=1}^{n} \mathrm{BA}_k.\mathrm{BH}_k}\right) \mathrm{BA}_i.\mathrm{BH}_i \quad \text{Using average building height} \tag{6.3}$$

$$BP_i = \left(\frac{CP}{\sum_{k=1}^{n} BV_k}\right) BV_i \quad \text{Using total building volume} \tag{6.4}$$

where: BP_i, Population of building *i;*CP, Census tract population; BA_i, Footprint area of building *i;* BF_i, Number of floors of building *i;* BH_i, Average height of building *i* (from LIDAR data); BV_i, Total volume of building *i* (from LIDAR data); *i*, *k*, Summation indices; *n*, Number of buildings that meet user-defined criteria and fall inside the CP polygon.

6.2.2 A GIS Tool

We implemented a standalone GIS tool named as PopShapeGIS using the Visual Basic programming language and TatukGIS DK (Development Kit). Figure 6.2 shows the program flowchart of PopShapeGIS tool. Under this tool, users can define the minimum ignored footprint size such as for porticos, garbage boxes and other unpopulated areas. They can also apply filtering by attribute field(s) such as building-use type and other attribute information. Three additional approaches are available under the Volumetric method, namely Use Number of Floors, Use Average Building Height and Use Total Building Volume (see in Fig. 6.3). After processing, the estimated building population attribute field, "EST_POP", appears in a new ESRI Shape file. A map viewer is also provided for viewing the processed results by performing common GIS functions such as add map layer, zoom in, zoom out, get attribute information, label by attribute field and change map layer properties.

More details about this tool and program can be downloaded from the following URL: http://giswin.geo.tsukuba.ac.jp/sis/en/gis_software.html

6.2.3 Data Requirement

We need two GIS dataset, census tracts and building footprints dataset. Areametric method requires only building footprints and volumetric method requires either number of floor or building height or building volume. Areametric method is suitable for rural area and volumetric method is suitable for urban area. Building use type is also required for estimation of residential building population. Nowadays, building footprints data with number of floors, tenant information and other attribute information can be purchased from commercial GIS data vendors. In addition, by utilizing modern remote sensing data acquisition system like LIDAR, we can measure building height (known as Digital Height Model DHM) and building volume (known as Digital Volume Model DVM) that can be used in volumetric method.

LIDAR techniques have been studied and utilized since the early 1960s, but appear to have become more prominent in the past few years. LIDAR has found applications in a wide variety of fields of study, including atmospheric science, bathymetric data collection, law enforcement, telecommunications, and even steel production (Maune, Daniel, & Damron, 2000). Advantages of using LIDAR for terrain and urban applications include the following: LIDAR allows rapid generation

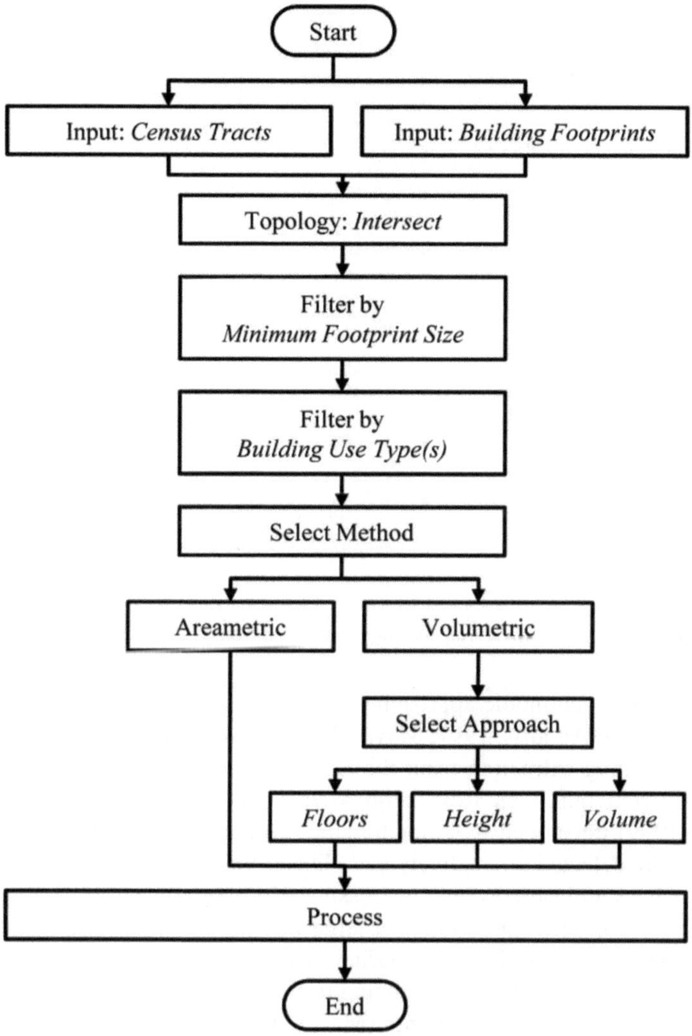

Fig. 6.2 PopShapeGIS program flowchart (Lwin & Murayama, 2009)

of a large-scale DTM (digital terrain model); LIDAR is daylight independent, is relatively weather independent, and is extremely precise. In addition, because LIDAR operates at much shorter wavelengths, it has higher accuracy and resolution than microwave radar (Jelalian, 1992).

6.2.4 Data Acquisitions

During the past two decades many researchers in Photogrammetry, Remote Sensing and computer vision communities have been trying to study and develop the

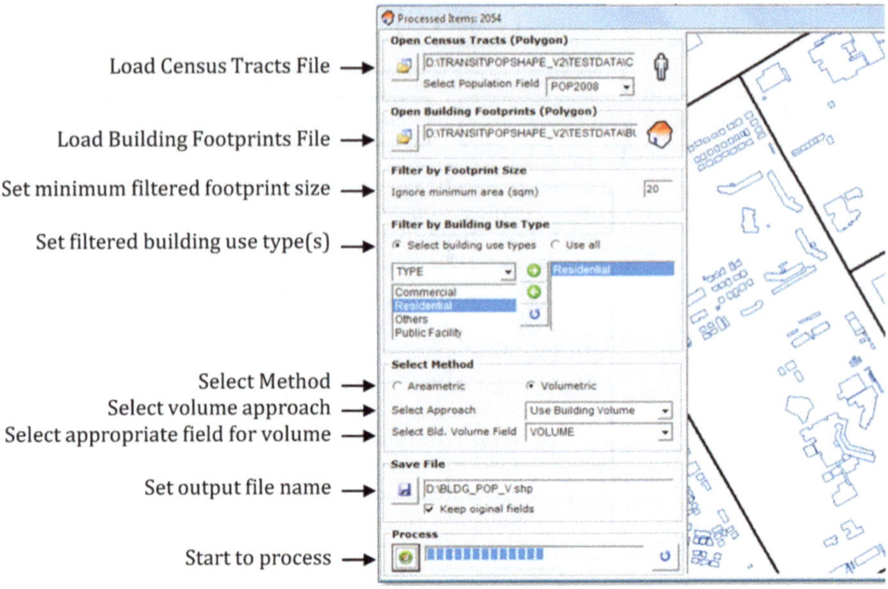

Fig. 6.3 PopShapeGIS graphical user interface (modified from Lwin & Murayama, 2009)

automatic or semi-automatic approaches for building extraction and reconstruction (Gruen, Kubler, & Agouris, 1997; Mayer, 1999). Several approaches have been presented for building extraction from the laser altimeter data. Maas and Vosselman (1999) extracted buildings from original laser altimeter point data. Sahar and Krupnik (1999) developed a semiautomatic building extraction approach, which buildings were detected interactively and 3D building outlines were extracted using shadow analysis and stereoscopic processing. However, the aerial photographs are typically very complex and contain a large number of objects in the scene. The automatic building extraction from aerial photograph has proven to be quite difficult. Those approaches are far from being useful in practice for images of different characteristics and complex contents (Mayer, 1999). There are alternate ways to acquire building footprints data. In some countries, building footprints data can be purchased from commercial map vendors. For example, in Japan, building footprints and other fine scale GIS dataset can be purchased from Zenrin Map vendor which product is known as ZmapTOWNII. Moreover, availability of commercial high resolution satellite data such as QuickBird (0.61 m at Nadir) or orthorectified aerial images can be used as base map for onscreen building footprints digitizing.

6.3 Dasymetric Mapping

We have produced Tsukuba City dasymetric map using building footprints (ZmapTOWNII product) and census tracts population from city office. LIDAR data provided by PASCO Corp. was utilized for measurement of building heights and volume in order to use in volumetric method. Additional iTownpage (Internet town page) from Nippon Telegraph & Telephone Corp. (NTT) which includes address of business centers, government organizations, public facility centers and other business activities information was converted into point features in ESRI Shape format using geocoding and address matching software in which geocoding accuracy is building level. This iTownpage data was used to separate residential, non-residential and mixed building use types. Figure 6.4 shows the building height and volume extraction from LIDAR point cloud data.

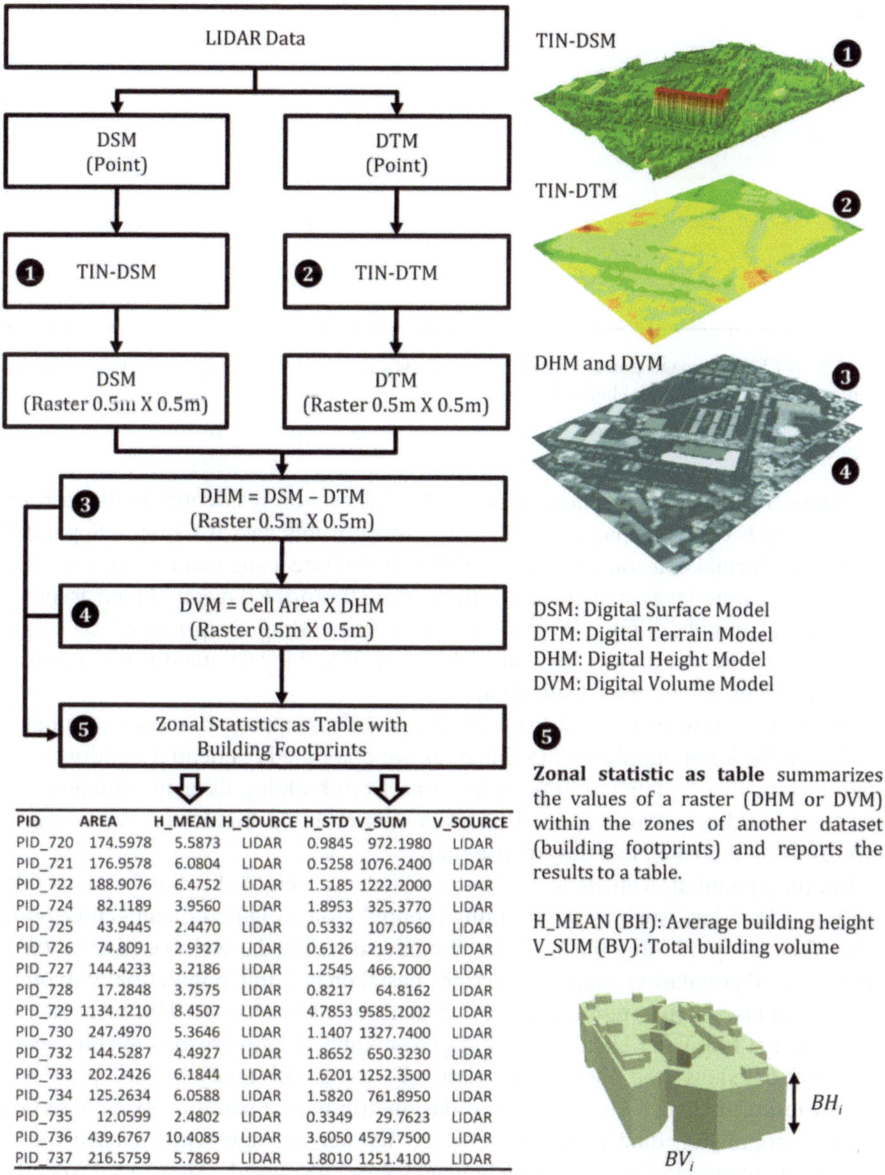

PID	AREA	H_MEAN	H_SOURCE	H_STD	V_SUM	V_SOURCE
PID_720	174.5978	5.5873	LIDAR	0.9845	972.1980	LIDAR
PID_721	176.9578	6.0804	LIDAR	0.5258	1076.2400	LIDAR
PID_722	188.9076	6.4752	LIDAR	1.5185	1222.2000	LIDAR
PID_724	82.1189	3.9560	LIDAR	1.8953	325.3770	LIDAR
PID_725	43.9445	2.4470	LIDAR	0.5332	107.0560	LIDAR
PID_726	74.8091	2.9327	LIDAR	0.6126	219.2170	LIDAR
PID_727	144.4233	3.2186	LIDAR	1.2545	466.7000	LIDAR
PID_728	17.2848	3.7575	LIDAR	0.8217	64.8162	LIDAR
PID_729	1134.1210	8.4507	LIDAR	4.7853	9585.2002	LIDAR
PID_730	247.4970	5.3646	LIDAR	1.1407	1327.7400	LIDAR
PID_732	144.5287	4.4927	LIDAR	1.8652	650.3230	LIDAR
PID_733	202.2426	6.1844	LIDAR	1.6201	1252.3500	LIDAR
PID_734	125.2634	6.0588	LIDAR	1.5820	761.8950	LIDAR
PID_735	12.0599	2.4802	LIDAR	0.3349	29.7623	LIDAR
PID_736	439.6767	10.4085	LIDAR	3.6050	4579.7500	LIDAR
PID_737	216.5759	5.7869	LIDAR	1.8010	1251.4100	LIDAR

Fig. 6.4 The work process of building height and volume extraction from LIDAR point cloud data

Fig. 6.5 NTT iTownpage point layer overlays on building footprints polygon layer to separate residential vs. non-residential buildings

Figure 6.5 shows the separation of residential vs. non-residential building footprints using NTT iTownpage data. Original form of this data is Comma Separated Value CSV format and converted into ESRI Point feature using geo-coding software. This point feature layer is intersects with building footprints layer and then assigned to non-residential buildings. We also assigned the building footprints which height is less than 2 m and area is less than 20 m^2 such as bicycle stand roofs, garbage boxes, porticos, etc. to non-residential.

Figure 6.6 shows the Tsukuba City dasymetric mapping based on estimated residential building population. In this map, we computed residential building population from two volumetric approaches, one from building floor information and one form building volume applied to residential building footprints. We averaged two results and created this dasymetric map.

Building population attribute field can be used as a weighted factor for many GIS analytical functions such as determining a population weighted mean center point rather than using building mean center. We have developed a web-based interactive micro-spatial population analysis functions based on building population with other ancillary dataset such as public facility and transportation network. This web site can be reached at following URL http://land.geo.tsukuba.ac.jp/microspa/welcome.aspx

Figure 6.7 shows the population results of various GIS analyses (i.e., point and line buffering) between choropleth and dasymetric map. Spatial analysis based on building population data is key benefit for disaster management teams in order to prepare humanitarian assistance when disaster occurred. They need specific

Fig. 6.6 Dasymetric mapping of Tsukuba city central area based on GIS estimated building population

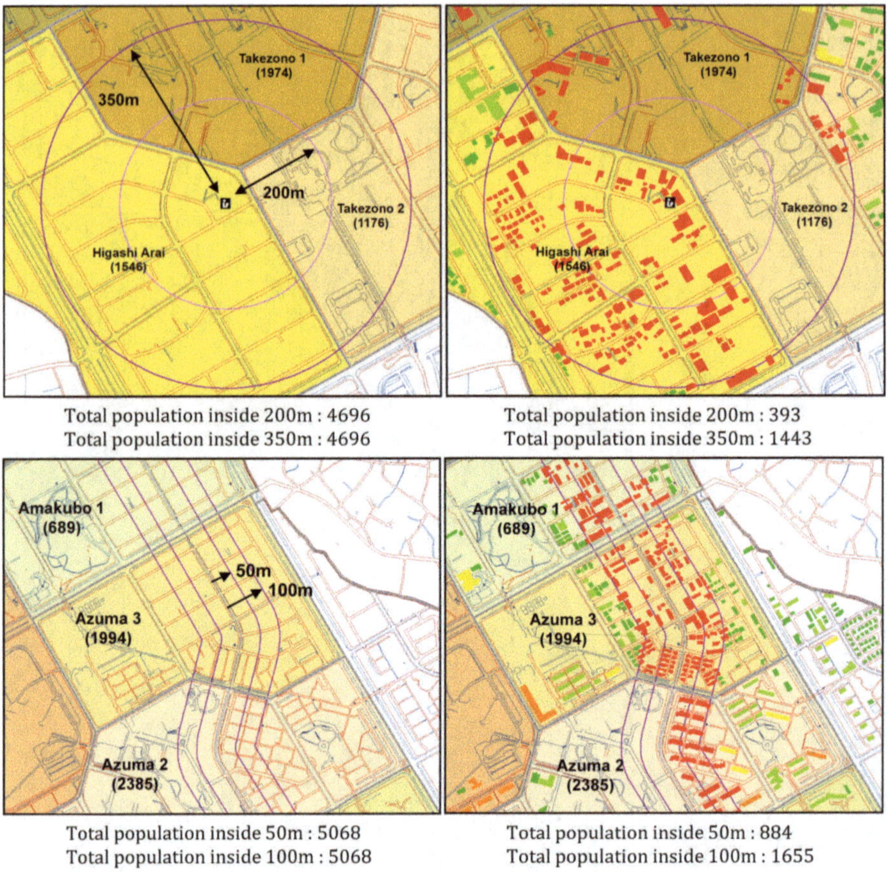

Fig. 6.7 Comparison of various GIS analyses between choropleth and dasymetric map example in point and line buffering

quantitative amount of population with certain geographical unit such as 500 m away from coastal lines in the case of tsunami strike or 5 km distance from earthquake's epicenter.

6.4 Conclusion

The estimated or quantitative mapping of building population is essential for micro-spatial analysis especially in terms of emergency management. Effective disaster preparedness requires quantitative spatial distribution patterns of population in order to position emergency response centers and prepare food and shelter in the event of disaster. Building population data is also required for improved accuracy in cost estimation of food and shelter for emergency preparedness and other humanitarian assistance. City and urban planners need to know how many local residents will benefit from newly constructed public facilities such as bus centers, railway stations and hospitals. Hydrologists require an estimate on the number of people living on a floodplain. Potential business owners can define their business location and perform consumer analysis. Quantitative building population data can be used as a weighted factor in spatial statistical analysis such as for determining population mean center and standard distance. This is important for decision making related to population such as in selecting a voting site or constructing a new public facility. Specific sub-population distribution for urban areas in order to develop an improved "denominator," which would enable the calculation of more correct rates in GIS analyses involving public health, crime, and urban environmental planning. Additionally, the knowledge of accurate population distribution can be extremely valuable in the sphere of urban planning. The understanding of the locational characteristics of target populations would allow for more equitable resource allocation in areas such as community infrastructure development, provision of open space and recreational opportunities, transportation access, and necessary environmental facilities (Maantay et al., 2007).

In this dasymetric mapping, we considered population data as a 3D space and developed a GIS tool for generating smaller geographical unit of population (i.e., building population) based on building footprints and census tracts. Although LIDAR data is expensive and require additional computer resources to handle, still we can use building surface area for rural area and building floor information for urban area. This approach is suitable for micro-scale population analysis such as market competition analysis, public health, traffic noise impact studies, public facility planning and disaster management.

References

Bielecka, E. (2005). A dasymetric population density map of Poland. *Proceedings of the International Cartographic Conference, July*, 9–15.

Eicher, C. L., & Brewer, C. A. (2001). Dasymetric mapping and areal interpolation: Implementation and evaluation. *Cartography and Geographic Information Science*, *28*, 125–138.

Gruen, A., Kubler, O., & Agouris, P. (1997). *Automatic extraction of man-made objects from aerial and space images (II)*. Basel-Boston-Berlin: Birkhauser Verlag.

Jelalian, A. (1992). *Laser radar systems*. Boston: Artech House.

Lwin, K. K., & Murayama, Y. (2009). A GIS Approach to estimation of building population for micro-spatial analysis. *Transactions in GIS*, *13*, 401–414.

Maantay, J. A., Maroko, A. R., & Herrmann, C. (2007). Mapping population distribution in the urban environment: The cadastral-based expert dasymetric system (CEDS). *Cartography and Geographic Information Science*, *34*, 77–102.

Maas, H. G., & Vosselman, G. (1999). Two algorithms for extracting building models from raw laser altimetry data. *ISPRS Journal of Photogrammetry and Remote Sensing*, *54*, 153–163.

Maune, D., Daniel, C., & Damron, J. (2000). *LIDAR and IFSAR: Pitfalls and opportunities for our future*. Proceedings of the American society for photogrammetry and remote sensing conference. Washington, DC.

Mayer, H. (1999). Automatic object extraction from aerial imagery-a survey focusing on buildings. *Computer Vision and Image Understanding*, *74*, 138–149.

Preobrazenski, A. (1954). Dorewolucjonnyje i sovietskije karty razmieszczenija nasielenia. *Woprosy Geografii Kartographia, 34*, 134–149.

Sahar, L., & Krupnik, A. (1999). Semiautomatic extraction of building outlines from large-scale aerial images. *Photogrammetric Engineering and Remote Sensing, 65*, 459–466.

Sleeter, R. (2004). *Dasymetric mapping techniques for the San Francisco Bay region, California.* Urban and regional information systems association annual conference proceedings (pp. 7-10), Reno: NV.

Slocum, T. A. (1999). *Thematic cartography and visualization.* Upper Saddle River, NJ: Prentice Hall.

Voss, P., Long, D., & Hammer, R. (1999). *When census geography doesn't work: Using ancillary information to improve the spatial interpolation of demographic data.* Center for Demography and Ecology, University of Wisconsin, Madison.

Wright, J. K. (1936). A method of mapping densities of population: With Cape Cod as an example. *Geographical Review, 26*, 103–110.

Chapter 7
Accuracy Assessment of GIS Based Building Population Estimation Algorithm

Ko Ko Lwin and Yuji Murayama

7.1 Introduction

Research into micro-spatial analysis has increased due to the emergence of high resolution satellite images for urban areas and the availability of fine-scale GIS data with enhanced attribute information (e.g., building footprints with the number of floors, building use type and building name). In view of the advances in modern geospatial information technologies, this is a good time for studying the world at a micro level. Population count is a key anchor for much of the official statistical system and the benchmark for many commercial and research surveys and analyses (Cook, 2004). GIS plays a critical role in population studies and analyses by means of mapping the spatial extent and analyzing it along with other GIS datasets. However, estimating and mapping the population are not an easy task due to the nature of human activities which change over space and time. Normally, population can be estimated using statistical and spatial (remote sensing and GIS) approaches. For example, night-time city lights imagery has been shown to demonstrate a reasonable correlation with population (Sutton, Roberts, Elvidge & Baugh, 2001).

Developments in computer hardware and mapping software have already encouraged many statistical and census offices to move from traditional cartographic methods to digital mapping and geographic information systems (GIS) (Rhind, 1991; United Nations, 1997). The benefits of geographic data automation in statistics are shared by the users of census and survey data. The data integration functions provided by GIS, which allow linking of information from many different subject areas, have led to much wider use of statistical information. This, in turn, has increased the pressure on statistics agencies to produce high-quality spatially

---

K.K. Lwin (✉)
Division of Spatial Information Science, Graduate School of Life and Environmental Sciences, University of Tsukuba, Tsukuba, Ibaraki, Japan
e-mail: kokolwin2002@yahoo.com

This chapter is improved from "Ko Ko Lwin and Yuji Murayama (2009), A GIS Approach to estimation of building population for micro-spatial analysis, Transactions in GIS, 13, 401–414", with permission from John Wiley and Sons.

Y. Murayama, R.B. Thapa (eds.), *Spatial Analysis and Modeling in Geographical Transformation Process*, The GeoJournal Library 100, DOI 10.1007/978-94-007-0671-2_7,

referenced information for small geographic units. The types of applications for such data are almost limitless. Examples include planning of social and educational services, poverty analysis, utility service planning, labor force analysis, marketing analysis, voting district delineation, emergency planning, epidemiological analysis, floodplain modeling and agricultural planning (United Nations, 2000). The main objective of this study was to apply GIScience theory and practice to produce a smaller geographic unit of population data (i.e., at the building level) for improved accuracy in the decision-making process at the micro level.

7.2 Problems in Micro-Spatial Analysis

Openshaw (1989) identified the following sources of error in micro-spatial analysis: errors in the positioning of objects; errors in the attributes associated with objects; and errors in modeling spatial variation (e.g., by assuming spatial homogeneity between objects). Population data exhibits spatial variation, especially in areas with a mix of high and low-rise buildings such as Tokyo, and residential areas mixed with unpopulated spaces (paddy fields, parks, playgrounds or government institutions) such as in Tsukuba City. Moreover, the population data used in GIS analyses is generally at the level of census tracts, townships or prefectures, since building population data is not available for public use due to privacy concerns. For spatial information users, population data has generally only been available in township polygons or city point features with aggregated population data. For non-spatial information users, population data (text and tables) can be obtained from the National Census Bureau or local government offices. All of these datasets are suitable for local and regional analysis, but not for micro-spatial analysis and decision-making processes.

In general, population mapping has two purposes: firstly, to cartographically portray the extent and density of population across an area of interest; and secondly, to derive a quantitative estimation of population density for use in subsequent spatial analytical modeling tasks (Bielecka, 2005). Common cartographic forms of population mapping are the choropleth map and the dasymetric map. Choropleth maps provide an easy way to visualize how a measurement varies across a geographic area. However, choropleth maps have limited utility for detailed spatial analysis of population data, especially where the population is concentrated in a relatively small number of villages, towns and cities. Moreover, choropleth maps cannot express statistical variation within the administrative areal units, such as changing population density. One way to avoid this limitation is by transforming the administrative units into smaller and more relevant map units through a process known as dasymetric mapping (Bielecka, 2005). There are some developed methods and approaches in dasymetric mapping: areal interpolation, filtered areal weighting (binary method) (Eicher & Brewer, 2001), filtering with land use/land cover data (Sleeter, 2004), and cadastral-based expert dasymetric system (CEDS) (Maantay, Maroko, & Herrmann, 2007). Dasymetric maps use ancillary information to help delineate new zones that better reflect the changing patterns over space. Recent research suggests that dasymetric mapping can provide small-area population estimates that are more accurate than many areal interpolation techniques that do not use ancillary data (Mrozinski & Cromley, 1999; Gregory, 2002). The U.S. Geological Survey (USGS) has refined and extended automated processes for improving spatial accuracy and visualization in mapping population distribution using dasymetric mapping (Sleeter, 2004). This technique aims to refine the spatial accuracy of aggregated data by using ancillary information to partition space into zones that better reflect the statistical variation in population. However, most census boundaries do not coincide with or intersect the boundaries of geographic features such as land use/land cover, soil type, geological unit, and floodplain and watershed boundaries - an issue that is known as spatial incongruity. This may introduce inaccurate population results for environmental assessment and emergency preparedness. Moreover, under the GIS domain, spatial analysis functions performed within the census tract do not acquire any significant changes in population. To overcome these problems, this chapter introduces a GIS approach for estimating building population for micro-spatial analysis.

7.3 Methodology

7.3.1 Applied Method

Here, we introduce two estimation methods: (1) Areametric (which does not require information on the number of building floors); and (2) Volumetric (which does require information on the number of floors). For improved accuracy, we also allow filtering by other categories into the computation, such as filtering by minimum footprint area and building use types, e.g., commercial, industrial, educational, and other building use types that are not occupied by residents. The calculation is demonstrated by the following mathematical expressions:

Areametric method:

$$\mathrm{BP}_i = \left(\frac{\mathrm{CP}}{\sum_{k=1}^{n} \mathrm{BA}_k}\right) \mathrm{BA}_i \quad \text{Using building footprint surface area} \tag{7.1}$$

Volumetric method:

$$\mathrm{BP}_i = \left(\frac{\mathrm{CP}}{\sum_{k=1}^{n} \mathrm{BA}_k.\mathrm{BF}_k}\right) \mathrm{BA}_i.\mathrm{BF}_i \quad \text{Using number of floors information} \tag{7.2}$$

Moreover, advances in remote sensing data acquisition technologies such as LIDAR can be used for extraction of building footprints, building height (Digital Height Model (DHM)) and building volume (Digital Volume Model (DVM)). Equations (7.3) and (7.4) can be used for LIDAR data:

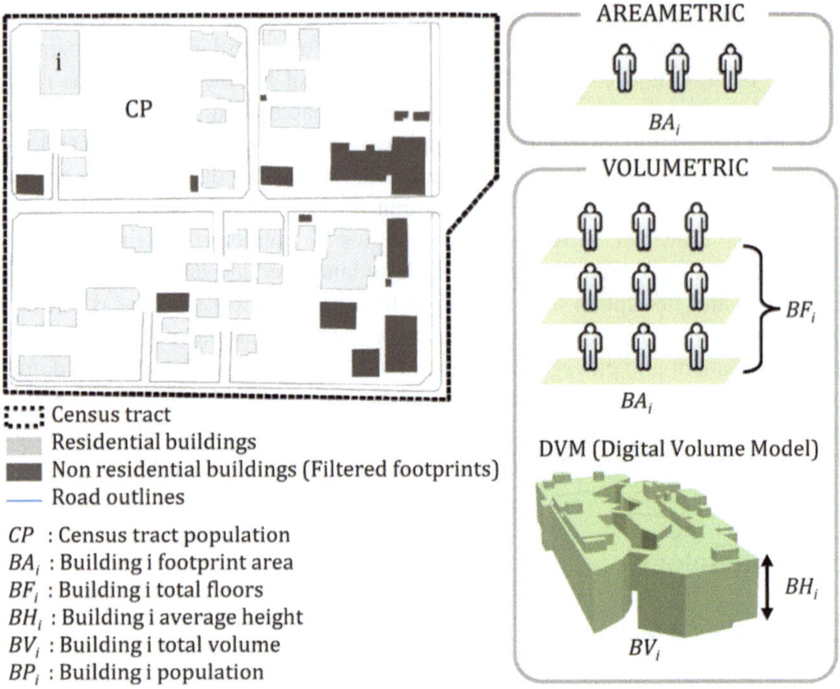

Fig. 7.1 Graphical illustration of equations

$$\mathrm{BP}_i = \left(\frac{\mathrm{CP}}{\sum_{k=1}^{n} \mathrm{BA}_k.\mathrm{BH}_k}\right)\mathrm{BA}_i.\mathrm{BH}_i \quad \text{Using average building height} \tag{7.3}$$

$$\mathrm{BP}_i = \left(\frac{\mathrm{CP}}{\sum_{k=1}^{n} \mathrm{BV}_k}\right)\mathrm{BV}_i \quad \text{Using total building volume} \tag{7.4}$$

where BP_i = the population of building i, CP = the census tract population, BA_i = the footprint area of building i, BF_i = the number of floors of building $_i$, BH_i = the average height of building i (from LIDAR data), BV_i = the total volume of building i (from LIDAR data), i, k = summation indices, and n = the number of buildings that meet user-defined criteria and fall inside the CP polygon (Fig. 7.1). The Areametric method is suitable for low-rise buildings especially in rural areas while the Volumetric method is suitable for high-rise buildings, especially in downtown areas.

7.3.2 Test Data

The two methods were evaluated using actual building population data acquired from the city administration office for study purposes. These data include detailed information about each building such as age, construction material, building type (detached, non-detached, semi-detached, flat or apartment), building use type (residential, commercial or educational, etc.), number of floors, number of households and total number of people, which is intended for use in disaster management. The test data was converted into the ESRI Shape file format and after conversion the building footprint polygons totaled 9,913, the census tracts totaled 28 and the total population was 28,000.

7.3.3 Test Method

In order to identify the best results based on data availability, we employed both methods (Areametric and Volumetric) with filtering by various footprint areas such as 0, 5, 10, 15, 20, 25, 30, 35, and 40 m^2 applied to the residential building use type.

7.3.4 Implementation of GIS Tool

To achieve the goal, we implemented a standalone GIS tool using the Visual Basic programming language and TatukGIS DK (Development Kit). This latter tool is partially embedded in the SIOSSS system (Spatially Integrated Online Social Survey System), which makes it possible to collect, store, share and visualize spatially distributed public survey datasets for local government decision-making processes such as for public facility management, emergency response and disaster preparedness, and market analysis. We are testing this system in Tsukuba City, a planned city located in Ibaraki Prefecture, Japan. Under the SIOSSS system, the PopShape tool (Fig. 7.2) will automatically populate survey items into building footprints

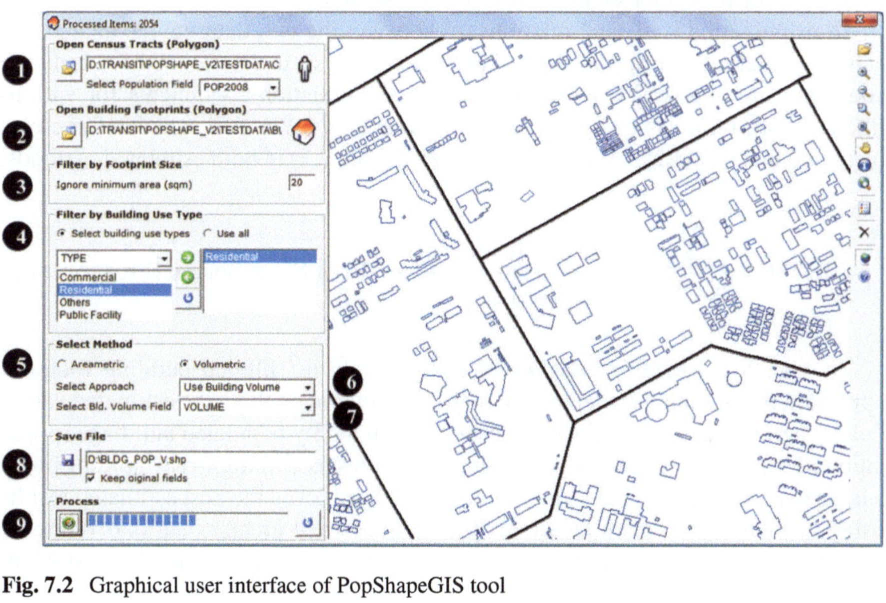

Fig. 7.2 Graphical user interface of PopShapeGIS tool

for micro-spatial analysis users, since SIOSSS does not collect particulars from respondents such as room number, building number and street name due to privacy concerns. Users can define the minimum ignored footprint size such as for porticos, garbage boxes and other unpopulated areas. They can also apply filtering by attribute field(s) such as building use type and other attribute information. Three additional approaches are available under the Volumetric method, namely Use Number of Floors, Use Average Building Height and Use Total Building Volume. After processing, the estimated building population attribute field, EST_POP, appears in a new ESRI Shape file. A map viewer is also provided for viewing the processed results by performing common GIS functions such as add map layer, zoom in, zoom out, get attribute information, label by attribute field and change map layer properties. The operational steps employed (with numbers corresponding to those shown in Fig. 7.2) were as follows: (1) Open census tracts file (Shape polygon); (2) open building footprints file (Shape polygon); (3) filter by footprint size; (4) filter by building use type(s); (5) select method (Areametric or Volumetric); (6) select approach (use Number of Floors or Building Height or Building Volume); (7) select appropriate field (Floor or Height or Volume attribute field); (8) assign output file name; and (9) start to process (see http://giswin.geo.tsukuba.ac.jp/sis/en/gis_software.html for additional details).

7.4 Results and Accuracy Assessment

All estimated values were evaluated using actual building population data by means of visual, statistical and spatial approaches. We obtained the best results using the Volumetric method filtered at the 25 m^2 footprint category. Figures 7.3 and 7.4 show the results for the preferred method visually for example.

To examine the statistical relationship between the two datasets (estimated and actual population), we applied linear regression analysis to determine the correlation coefficient, R^2. Table 7.1 shows the results of correlation coefficients for various filtered footprint areas using two estimation methods. We also calculated the root mean square error (RMSE) for each category (Table 7.2) in both estimated methods:

$$\mathrm{RMSE} = \sqrt{\frac{\sum_{i=1}^{n}(\mathrm{Actual} - \mathrm{Estimated})^2}{n}} \quad \text{Root Mean Square Error} \qquad (7.5)$$

The smallest RMSE was achieved using the 25 m^2 filtered building footprint applied in the Volumetric method. With the Areametric method, the estimated values did not agree with the actual values, especially in highly populated buildings (population >50). This may have occurred due to the presence of high-rise buildings with a large number of floors but a small footprint. We achieved the best estimated results using the Volumetric method where all R^2 values were greater than 0.9. While all R^2 values are acceptable in the Volumetric method, the best value ($R^2 = 0.9488$)

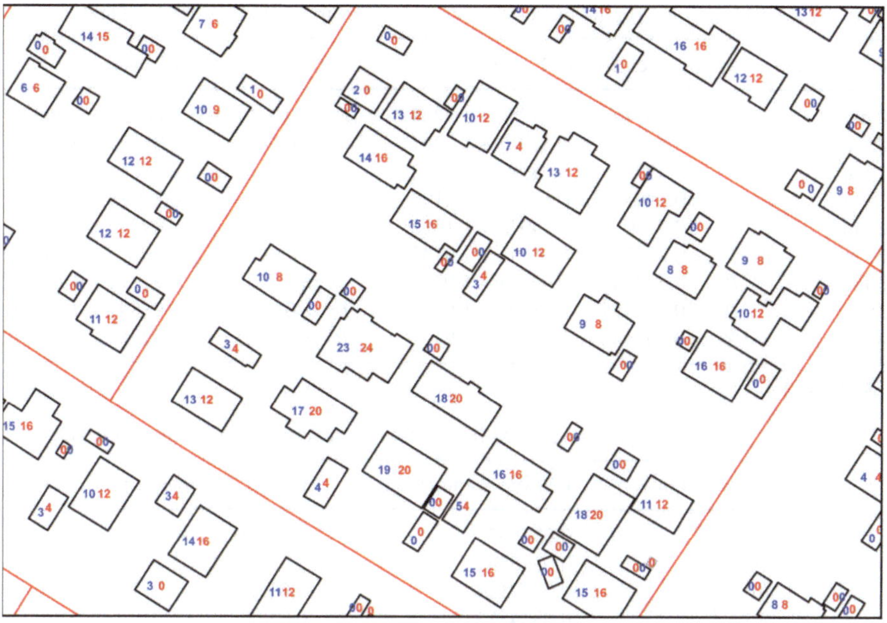

Fig. 7.3 Estimated population (*left value*) vs. actual population (*right value*) in low-rise building area (filtered area 25 m^2, Volumetric method applied to residential buildings)

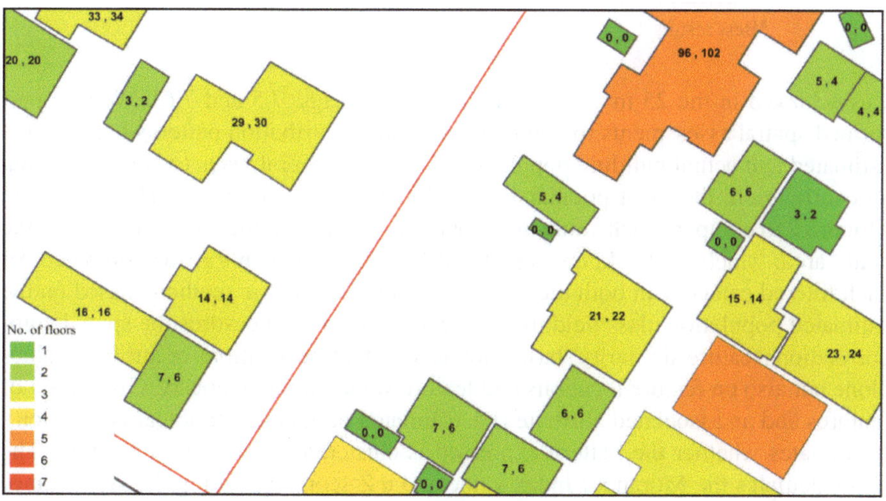

Fig. 7.4 Estimated population (*left value*) vs. actual population (*right value*) in highrise-building area (filtered area 25 m^2, Volumetric method applied to residential buildings)

Table 7.1 Various correlation coefficients for Areametric and Volumetric methods

Filtered area	Areametric method R^2 and Y	Volumetric method R^2 and Y
0 m^2	$R^2 = 0.8004$ ($y = 0.5785x + 1.3214$)[a]	$R^2 = 0.9461$ ($y = 0.8625x + 0.3699$)
05 m^2	$R^2 = 0.8003$ ($y = 0.5795x + 1.3241$)	$R^2 = 0.9461$ ($y = 0.8628x + 0.3708$)
10 m^2	$R^2 = 0.7995$ ($y = 0.5898x + 1.3022$)	$R^2 = 0.9467$ ($y = 0.8688x + 0.3765$)
15 m^2	$R^2 = 0.7995$ ($y = 0.6160x + 1.2197$)	$R^2 = 0.9468$ ($y = 0.8794x + 0.3748$)
20 m^2	$R^2 = 0.7990$ ($y = 0.6431x + 1.1348$)	$R^2 = 0.9479$ ($y = 0.8934x + 0.3402$)
25 m^2	$R^2 = 0.8002$ ($y = 0.6631x + 1.0703$)	$R^2 = 0.9488$ ($y = 0.9037x + 0.3088$)[a]
30 m^2	$R^2 = 0.7971$ ($y = 0.6739x + 1.0312$)	$R^2 = 0.9458$ ($y = 0.9117x + 0.2773$)
35 m^2	$R^2 = 0.7954$ ($y = 0.6823x + 1.0078$)	$R^2 = 0.9439$ ($y = 0.9189x + 0.2571$)
40 m^2	$R^2 = 0.7944$ ($y = 0.6926x + 0.9751$)	$R^2 = 0.9425$ ($y = 0.9275x + 0.2306$)

[a]Best result

Table 7.2 Root mean square error (RMSE) for both Areametric and Volumetric methods

Filtered area	RMSE (Areametric)	RMSE (Volumetric)
0 m^2	0.031552	0.015302
05 m^2	0.0315193	0.0152933
10 m^2	0.0312114	0.0150973
15 m^2	0.0304115	0.014867
20 m^2	0.0296994	0.014507
25 m^2	0.0291742	0.0142609[a]
30 m^2	0.0290846	0.0145403
35 m^2	0.0290064	0.0147061
40 m^2	0.0288724[a]	0.0148335

[a]Best results

was achieved in the 25 m^2 filtered area category (Figs. 7.5 and 7.6). We also performed spatial assessments to compare the spatial distribution patterns between the estimated and actual building population. There are several ways to perform spatial autocorrelation, the most popular being Moran's *I* and Geary's *C*. Here, we used Moran's *I* to compare each value in the pairs to the mean value for all features in the study area, which is also known as global Moran's *I*. Moran's *I* was computed for each filtered category in both the Areametric and Volumetric methods based on the estimated population class field using ArcGIS. This tool measures the spatial autocorrelation (feature similarity) based not only on feature locations or attribute values alone but also on feature locations and feature values simultaneously. Given a set of features and an associated attribute (i.e. estimated population or actual population), it evaluates whether the pattern expressed is clustered, dispersed, or random. The tool calculates the Moran's *I* index value and a *Z* score evaluating the significance of the index value. In general, a Moran's *I* index value near +1.0 indicates clustering while an index value near -1.0 indicates dispersion. A high positive *Z* score for a feature indicates that the surrounding features have similar values. A low negative *Z* score indicates that the feature is surrounded by dissimilar values. Moreover, Moran's *I* for actual population is also computed for comparison. Although Moran's

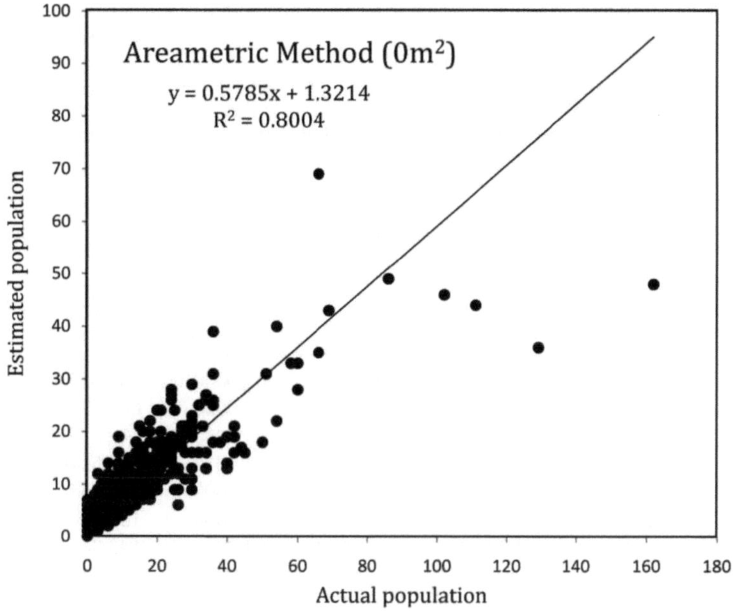

Fig. 7.5 Scatter plot for 0 m^2 filtered area in Areametric method (total sample size = 8,854)

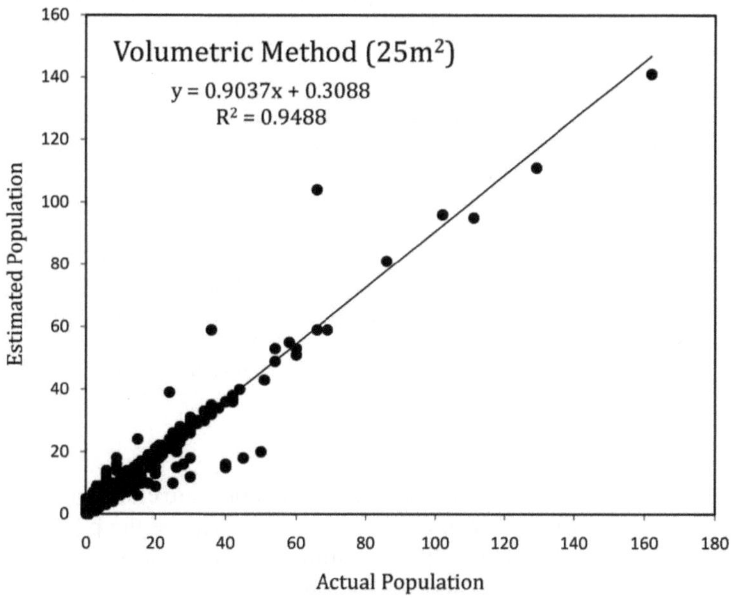

Fig. 7.6 Scatter plot for 25 m^2 filtered area in Volumetric method (total sample size = 8,854)

Table 7.3 Moran's *I* and *Z* score for actual building population

Actual building population				
Filtered area	Index	Expected	Variance	*Z* score
None	0.03271	–0.00021	0.00000	62.05492

Table 7.4 Moran's *I* and *Z* score for estimated building population in Areametric method

Areametric method				
Filtered area	Index	Expected	Variance	*Z* score
None	0.02508	–0.00014	0.00000	65.48493
05 m^2	0.02529	–0.00014	0.00000	66.06920
10 m^2	0.03178[a]	–0.00014	0.00000	81.18271[a]
15 m^2	0.04734	–0.00016	0.00000	111.56468
20 m^2	0.06082	–0.00018	0.00000	130.03565
25 m^2	0.06845	–0.00020	0.00000	137.12042
30 m^2	0.07450	–0.00021	0.00000	143.36652
35 m^2	0.07615	–0.00021	0.00000	142.61562
40 m^2	0.07727	–0.00022	0.00000	141.69034

[a]Best results

Table 7.5 Moran's *I* and *Z* score for estimated building population in Volumetric method

Volumetric method				
Filtered area	Index	Expected	Variance	*Z* score
None	0.01851	-0.00018	0.00000	39.79721
05 m^2	0.01726	-0.00011	0.00000	54.03807
10 m^2	0.01867	-0.00018	0.00000	40.15002
15 m^2	0.02128	-0.00018	0.00000	45.31663
20 m^2	0.02845	-0.00019	0.00000	58.47552
25 m^2	0.03451[a]	-0.00020	0.00000	68.89462[a]
30 m^2	0.03799	-0.00021	0.00000	73.47143
35 m^2	0.03892	-0.00022	0.00000	73.43831
40 m^2	0.04002	-0.00022	0.00000	74.06918

[a]Best results

I measures the patterns to determine whether the features are clustered or dispersed, the purpose of using Moran's *I* in this study was to measure the patterns for each category and then compare them with the feature patterns of the actual building population. Tables 7.3, 7.4 and 7.5 show the Moran's *I* indexes in each filtered category for both the Areametric and Volumetric methods.

Although both indices intersect at certain filtered areas, one of the *Z* scores of the Volumetric method intersected at a point between the 20 and 25 m^2 footprint areas (Fig. 7.7). This is probably the average single-unit living space in the study area.

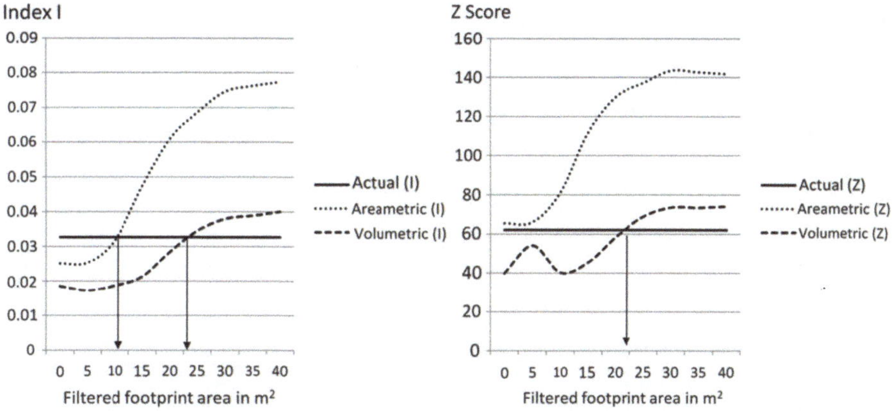

Fig. 7.7 Moran's *I* (*left*) and *Z* score (*right*) for actual population vs. estimated population using Areametric and Volumetric methods for various filtered footprint areas

7.5 Potential Applications

The estimated or quantitative mapping of building population is essential for microspatial analysis especially in terms of emergency management. Effective disaster preparedness requires quantitative spatial distribution patterns of population in order to position emergency response centers and prepare food and shelter in

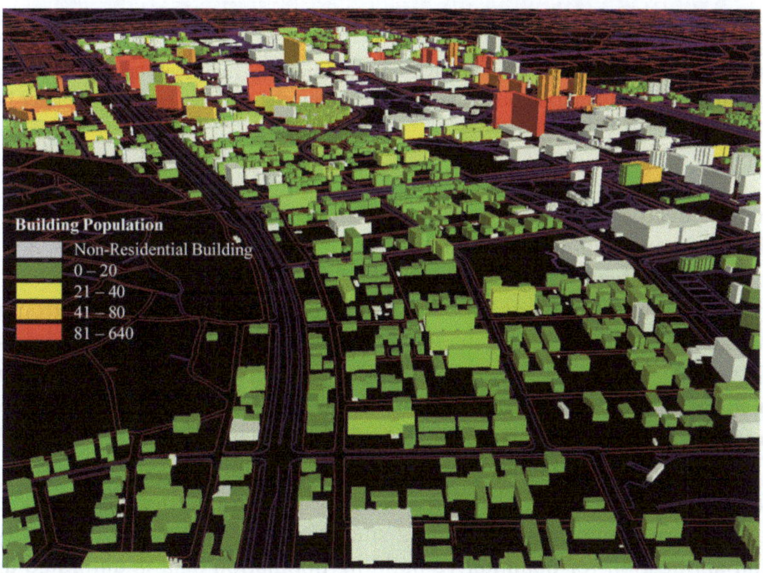

Fig. 7.8 Example of 3D quantitative building population mapping (3D visualization is one of the techniques used for effective public facility and utility planning)

Fig. 7.9 Example of web-based interactive decision-making tool for planning local community bus route based on building population (determining the shortest route with larger building population within a specific buffer zone)

the event of disaster. Building population data is also required for improved accuracy in cost estimation of food and shelter for emergency preparedness and other humanitarian assistance (Fig. 7.8).

Urban planners need to know how many local residents will benefit from newly constructed public facilities such as bus centers, railway stations and hospitals (Fig. 7.9). Hydrologists require an estimate on the number of people living on a floodplain. Potential business owners can define their business location and perform consumer analysis. Quantitative building population data can be used as a weighted factor in spatial statistical analysis such as for determining population mean center and standard distance (Fig. 7.10). This is important for decision making related to population such as in selecting a voting site or building a new public facility. The potential application for quantitative building population data is unlimited and we hope to increase the accuracy in various spatial decision-making processes at the micro level.

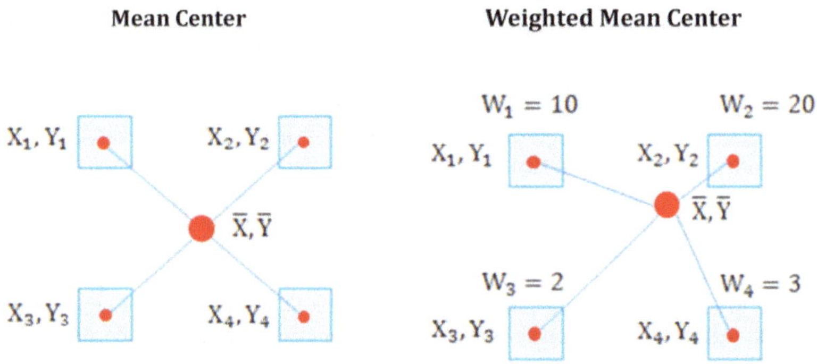

Fig. 7.10 Example of determining the mean center: (**a**) mean center without weighted factor (spatially oriented); and (**b**) mean center using building population as a weighted factor (population oriented)

7.6 Conclusion and Future Work

In this chapter, we have discussed the application of two building population estimation methods based on census tract and building footprint data which may be applied when building population data is not publicly available for privacy reasons. We have also developed a tool to generate a GIS-ready dataset with quantitative building population attribute information. The estimated results were evaluated using visual, statistical and spatial approaches. We have achieved reasonable results, confirming model suitability for use in micro-spatial analysis. However, further development is required to improve the accuracy by incorporating other factors such as an adjustment factor for mixed building use type (i.e. residential buildings mixed with commercial) and building status (i.e. newly constructed or abandoned buildings). Through integration with modern spatial data acquisition technologies such as LIDAR and other high-resolution satellite imagery, we hope to achieve more accurate estimation of building population.

References

Bielecka, E. (2005). A dasymetric population density map of Poland. *Proceedings of the International Cartographic Conference, July*, 9–15.

Cook, L. (2004). The quality and qualities of population statistics, and the place of the census. *Area, 36*, 111–123.

Eicher, C. L., & Brewer, C. A. (2001). Dasymetric mapping and areal interpolation: Implementation and evaluation. *Cartography and Geographic Information Science, 28*, 125–138.

Gregory, I. N. (2002). The accuracy of areal interpolation techniques: Standardising 19th and 20th century census data to allow long-term comparisons. *Computers, Environment and Urban Systems, 26*, 293–314.

Maantay, J. A., Maroko, A. R., & Herrmann, C. (2007). Mapping population distribution in the urban environment: The cadastral-based expert dasymetric system (CEDS). *Cartography and Geographic Information Science, 34*, 77–102.

Mrozinski, R. D., & Cromley, R. G. (1999). Singly-and doubly-constrained methods of areal interpolation for vector-based GIS. *Transactions in GIS*, *3*, 285–301.

Nations, U. (1997). *Geographical information systems for population statistics*. New York: United Nations Studies in Methods Series F, No 68.

Nations, U. (2000). *Handbook on geographic information systems and digital mapping*. New York: United Nations Department of Economic and Social Affairs Statistics Division.

Openshaw, S. (1989). Learning to live with errors in spatial databases. In M. Goodchild and S. Gopal (Eds.), *Accuracy of Spatial Databases* (pp. 263–276). London: Taylor & Francis.

Rhind, D. W. (1991). Counting the people: The role of GIS. In P.A. Longley et al. (Eds.), *Geographical Information Systems*, *2* (pp. 127–137). London: John Wiley & Sons.

Sleeter, R. (2004). Dasymetric mapping techniques for the San Francisco Bay region, California. Urban and regional information systems association annual conference proceedings (pp. 7–10) Reno: NV.

Sutton, P., Roberts, D., Elvidge, C., & Baugh, K. (2001). Census from Heaven: An estimate of the global human population using night-time satellite imagery. *International Journal of Remote Sensing*, *22*, 3061–3076.

Chapter 8
The Application of GIS in Education Administration

Fatemeh Ahmadi Nejad Masouleh, Yuji Murayama, and Todd Wendell Rho'Dess

8.1 Introduction

While most developed countries have delineated school attendance boundaries, they have never been officially created for the Iranian schools. Though enrolment by proximity is an official educational policy, the lack of defined school attendance areas has resulted in an informal open enrolment system where parents can choose public schools outside their residential areas. Two major consequences of parental choice are longer commutes to schools and increased use of motor vehicle transport in a country where the rate of car accidents is dangerously high. A study of the travel patterns of students to the public (government) female junior high schools of Rasht city in Northern Iran showed that almost 60% of students surveyed travelled to school by motorized vehicles (Ahmadi Nejad Masouleh, 2006).

The results also indicate that travel on taxies and buses to schools mean exposure to highly inconvenient and in numerous cases life-threatening rides. For shared rental mini-buses, for example, the complaints are that the rides can be uncomfortable due to factors such as poor seating, damaged roofs and severe overcrowding. Rides in city taxies are potentially dangerous for students since informal sector drivers may drive recklessly without regard for traffic safety. Reports of kidnapping and rape of female passengers on rides in these vehicles are also serious concerns given that informal taxi drivers are self-employed and lack government oversight. Other problems encountered by students include exposure to pollution in the form of heavy vehicle fumes, stressful journeys from loud clanking sounds sound, and a high incidence of breakdowns during commutes. In addition to the added expense

F.A.N. Masouleh (✉)
Department of Geography, Sinclair Community College, Dayton, OH, USA
e-mail: f_ahmadinejad2000@yahoo.com

This chapter is improved from "Fatemah Ahmadi Nejad Masouleh, Yuji Murayama, and Todd Wendell Rho'Dess (2009), The application of GIS in education administration: Protecting students from hazardous roads, Transactions in GIS, 13, 105–123", with permission from John Wiley and Sons.

Y. Murayama, R.B. Thapa (eds.), *Spatial Analysis and Modeling in Geographical Transformation Process*, The GeoJournal Library 100, DOI 10.1007/978-94-007-0671-2_8,

of daily bus or taxi fees, in every case, commutes in such vehicles also carry the extra cost of lateness to school and loss of energy after exhausting rides.

Aside from these vehicle hazards, serious traffic and environmental dangers in the case of walking to the schools of Rasht makes the reduction of commutes on foot an equally pressing priority. In Iran, as well as in Rasht, not only is there widespread disregard for pedestrian rights at intersections, but in many cases vehicles often pass within inches of pedestrians at high speeds. These risks are compounded by the fact that numerous traffic lights at intersections are not in working order. By analyzing the travel patterns and the dangers involved under informal school choice in Rasht, the research concluded that shorter safer commutes on foot and the elimination of motorized commutes through the construction of school attendance areas for the city are needed (Ahmadi Nejad Masouleh, 2006).

To increase the convenience and safety of journeys to schools, this research focuses on construction of more effective school attendance areas in Rasht city that can yield the greatest possible reduction in current commuting time and distances for school students.

8.2 Methods

8.2.1 Study Site

The public female junior high (PFJH) schools of Rasht city in Gilan province were chosen for the purpose of constructing new attendance areas. Given that Iranian male and female students are segregated, students in Rasht city may attend any of the 40 male and 45 female junior government high schools. However, since the environment around schools is often more unsafe for female students and public schools outnumber private ones, the focus of the research here is on PFJH schools of Rasht city.

Although overcrowding can be a primary cause of parents choosing schools outside local areas, analysis of local population statistics reveals that this is not true in the case of Rasht. Recent data show that Rasht currently has just over 519,000 residents (SCI, 2005). While the city covers an area of just 41 km^2, population growth over time has resulted in a relatively high population density of 9,555 per km^2 – when compared with the average density of just 3,003 per km^2 for all of Gilan province in 1994 (Iran Management and Planning Organization, 1996). Despite this fact, the city has been experiencing a population decline since 1996 which translates into decreases in the number of school students. Rasht city has been losing students at its female junior high schools (both public and private) and in the seven year period from 1997 to 2004 the number of female junior high school students dropped by 3,000 (Fig. 8.1). It is also notable that during these years, two female junior high schools with a total capacity of 700 had to be closed due to insufficient student numbers (Iran Ministry of Education, 2004). Although one expected benefit of a drop of population would have been increased accessibility to local schools,

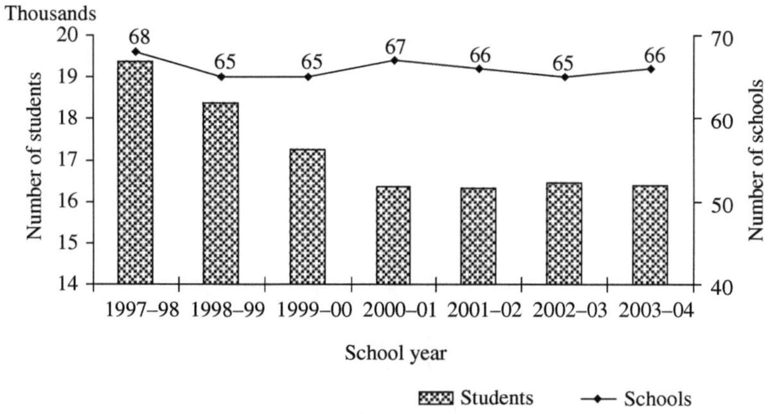

Fig. 8.1 Total number of students at public and private female junior high schools in Rasht City from 1997 to 2004 by number of schools
Source: Iran Ministry of Education

the lack of defined school attendance areas has led to the opposite; students have to commute outside their local areas.

Most schools in Iran also have to overcome the obstacle of a limited number of constructed school facilities by alternating the use of school buildings in morning and afternoon shifts to maximize the use of available resources. While different schools might occupy the same building and use the same resources, they remain completely autonomous each with their own principals, teachers and administrative staff. The arrangement of shared educational facilities also characterizes the 45 Rasht PFJH schools chosen for the purpose of delineation, and these schools utilize 26 educational buildings.

Figure 8.2 shows the location of PFJH school buildings in the city. While there are seven schools that own buildings for their exclusive use, the remaining 38 schools share 19 educational buildings in morning and afternoon shifts that alternate weekly. These schools are referred to as either Shift 1 or Shift 2 for ease of identification (see Table 8.1). School names correspond to the shift they occupy, Amupur 1 and Amupur 2, for example. In doing so, seven morning shift schools were excluded from the Shift 2 data since these schools are unavailable to students for the second afternoon shift time slot. In conducting spatial analysis for the purpose of research, two special schools of Fajr and Shahed were excluded since their populations included enrolled students from outside the study area.

In order to focus on student commuting distances and examine whether constructed boundaries can minimize school travel times and distances for Rasht PFJH schools, five schools (Aban, Anvar, Azadi, Felestin and Rahzahra) were chosen from Shift 1 for more detailed examination. These five schools have diverse features that make them good representatives of the whole to fit the goal of generalizing any results. The five are located on both main roads and secondary routes, have differing reputations, and enroll students from both the center of the city and its suburbs.

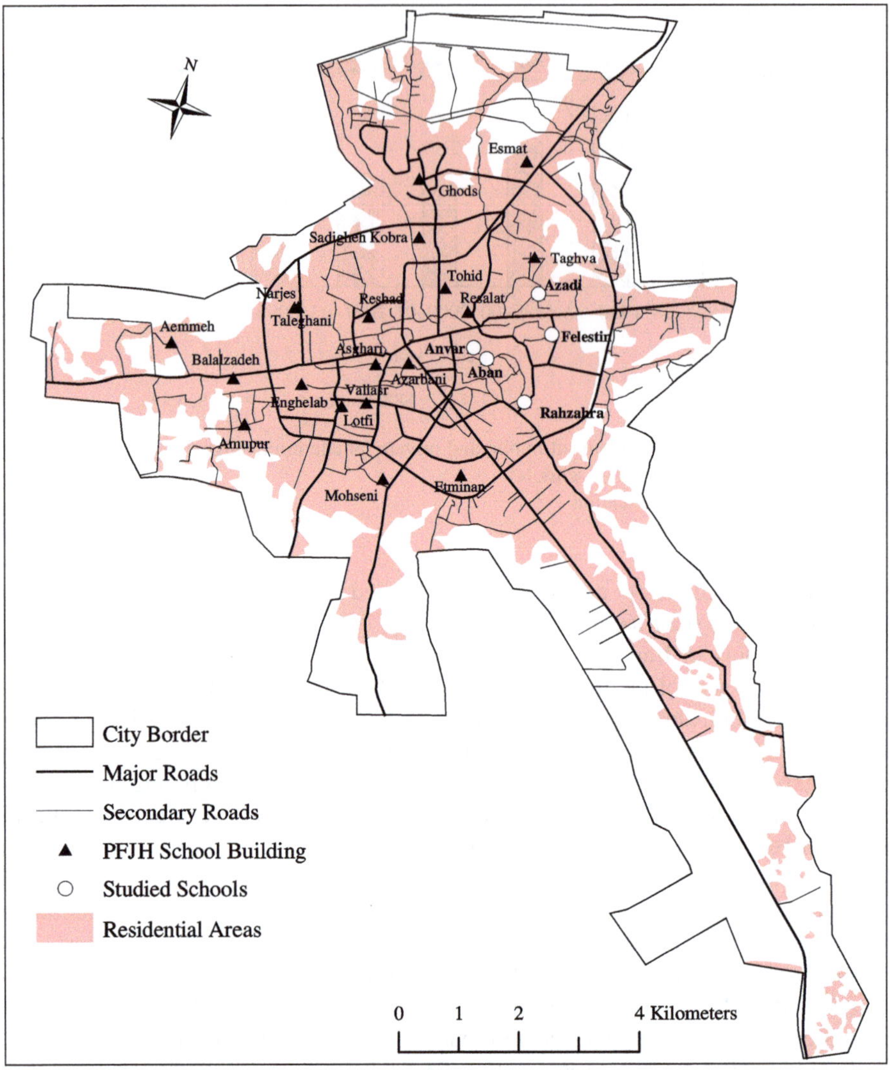

Fig. 8.2 Location of all Rasht PFJH school buildings by name including studied schools and city residential area. *Note*: Two especial schools of Fajr and Shahed were excluded from the map

While Felestin and Rahzahra are located on main routes, Azadi, Aban and Anvar are located on secondary roads. Aban and Anvar are the two most well-known schools in the study area. For those who live far from well-known schools this often means a longer commute to school. Though students have access to schools closer to their residence, parents choose to enroll their children outside their local areas and rent vehicles for the commute. Since Shift 1 and 2 student populations encounter the same environment and hazardous difficulties in commuting, and generalization of the results of the research from either shift was possible, the five schools were chosen randomly from Shift 1 PFJH schools.

Table 8.1 Number of students in Rasht city PFJH schools according to school shifts for the 2003–2004 school year

Shift 1		Shift 2	
School name	Number of students	School name	Number of students
Amupur 1	451	Amupur 2	552
Etminan 1	423	Etminan 2	433
Mohseni 1	433	Mohseni 2	423
Asghari 1	400	Asghari 2	403
Narjes 1	363	Narjes 2	386
Azarbani 1	363	Azarbani 2	357
Enghelab 1	286	Enghelab 2	407
Lotfi 1	297	Lotfi 2	327
Valiasr 1	207	Valiasr 2	238
Balalzadeh 1	203	Balalzadeh 2	125
Reshad	326	Ghods 2	442
Taleghani	247	Azadi 2	348
Aemmeh	211	Felestin 2	336
Ghods 1	420	Aban 2	210
Azadi 1	328 (241)	Anvar 2	206
Felestin 1	336 (327)	Esmat 2	206
Fajr	547	Sadigheh Kobra 2	169
Aban 1	326 (307)	Tohid 2	189
Anvar 1	236 (175)	Rahzahra 2	211
Esmat 1	219		
Sadigheh Kobra 1	243		
Tohid 1	201		
Taghva	332		
Shahed	300		
Rahzahra 1	249 (221)		
Resalat	211		
Total	8,158	Total	5,968

Source: Education Office of Rasht

The numbers in parentheses are the school population in 2004–2005 school year for the studied schools and were collected by conducting an interview with the school principles. The data on the school population in 2004–2005 was not published for the remaining schools

8.2.2 Research Method

To provide essential data on student residential locations, commute distance, and modes of transportation, 1,271 questionnaires were distributed to all students attending the five PFJH schools in the 2004–2005 school year which had the student populations of 307, 175, 241, 327, and 221 for Aban, Anvar, Azadi, Felestin and Rahzahra, respectively. Of the total, students returned 919 questionnaires (73%). Once the authors acquired the home addresses of students from the returned questionnaires, it was then possible to construct a map showing the current allocation of students to the five PFJH schools. In order to delimit school attendance areas the research used the multiplicatively weighted Voronoi diagram (MWVD) method. Following a review of studies related to bounding techniques

(Clarke & Surkis, 1968; Maxfield, 1972; McDaniel, 1975; Jennergren & Obel, 1980; Woodall, Cromley, Semple, & Green, 1980; Ghosh & Rushton, 1987; Okabe, Boots, Sugihara, & Chiu, 2000; Mu, 2004) the MWVD method was chosen for a number of reasons.

First, as a widely applied bounding technique, it has been utilized successfully in numerous studies seeking to delimit areas such as geographical delivery systems (Huff & Lutz, 1979) and market areas (Hubbard, 1970; Wood, 1974; Von Hohenbalken & West, 1984). Though it has been tested in other fields, the application of weighted Voronoi has never been used to delineate school areas, at least to our knowledge. Second, a MWVD-based delineated area is a non-empty set with joined areas and all regions are divided into a set of enclosed polygons. The properties of MWVDs make it the most appropriate method for the purpose of the research – the delineation of residential areas of the city into regions around the PFJH schools. Another benefit is that by use of this method any non-assigned area will also be kept within the city. The weighted Voronoi approach is also useful for the purpose of weighting school areas. By constructing MWVD polygons to represent proposed school attendance areas, all locations enclosed within a single area will be closest to designated PFJH schools, ensuring the closest commutes and optimum capacities.

The research applied a specialist GIS-based package known as WVD18 (Mu, 2004) for the construction of polygons that would represent the attendance areas for each school. Each polygon was then constructed in relation to the number of students at each school to represent the weighted distance. Due to lack of data, the authors could not consider other factors such as the school-aged population density around each school in weighting the school attendance areas. Given the approach, larger schools tended to have larger attendance areas and conversely for smaller schools. Once the boundaries representing each school attendance area were completed, they were analyzed using ArcInfo 9.1.

8.3 Results and Discussion

8.3.1 Characteristic Features of the Constructed School Attendance Areas

Figure 8.3 shows the proposed attendance areas for the Rasht PFJH schools in each shift. To give a fuller picture of the nature and effectiveness of the attendance areas in achieving the goal of reduced commutes a brief overview of four general characteristics is needed.

The first characteristic is that the attendance areas are defined for both morning and afternoon shifts for each school building. Since the number of schools, school size and distance between them differ for each shift, the constructed areas differ for each shift in size and shape.

The second characteristic of the proposed attendance areas is that with the exception of some on the outskirts of town, most are small in size to allow for the shortest

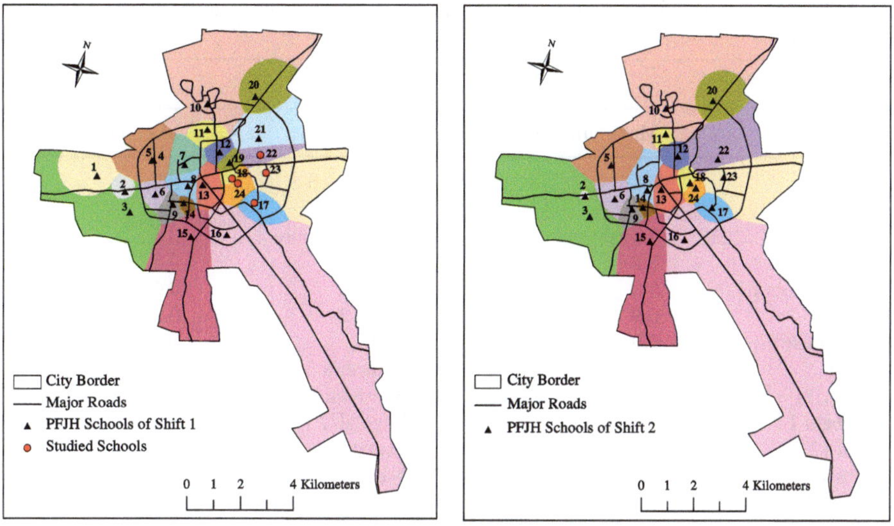

Fig. 8.3 Proposed attendance areas for Rasht PFJH schools of Shift 1 and Shift 2. 1: Aemmeh, 2: Balalzadeh, 3: Amupur, 4: Taleghani, 5: Narjes, 6: Enghelab, 7: Reshad, 8: Asghari, 9: Lotfi, 10: Ghods, 11: Sadigheh Kobra, 12: Tohid, 13: Azarbani, 14: Valiasr, 15: Mohseni, 16: Etminan, 17: Rahzahra, 18: Anvar, 19: Resalat, 20: Esmat, 21: Taghva, 22: Azadi, 23: Felestin, 24: Aban

journeys possible for the majority of students. There are 43 PFJH schools (minus the two excluded special schools) in Rasht for both the morning and afternoon shifts. Of this total, 28 PFJH schools, or more than half, have small proposed attendance areas with a maximum of a 1.7 km projected walking distance (Table 8.2). The number of students that can travel this reduced distance is 7,521. The number is even greater if we add those students residing in larger school attendance areas who will also benefit from a shorter commute. As shown by a previous study, the parents of PFJH students in Rasht city send their children to school by vehicle when the commute exceeds 1.7 km (Ahmadi Nejad Masouleh, 2006). This means that the reduced commute distance within the proposed attendance to a maximum 1.7 km can reduce exposure to the dangers of commuting – thus meeting parental safety concerns.

As the third characteristic, the constructed polygons were also contiguous with no gaps between the schools. Hence, all locations within the city are allocated to a local PFJH school and parents living in any location within the city can be guided to enroll their children in the nearest school. The final feature towards this end was that all areas are closed.

8.3.2 Spatial Allocation of Students

The current spatial allocation of students to the five studied schools with respect to the proposed attendance areas provides a clear illustration of how school populations are currently interspersed with students from outside local areas requiring

Table 8.2 Maximum commuting distance within proposed attendance areas for both shifts of PFJH schools

Shift 1		Shift 2	
School name	Maximum distance	School name	Maximum distance
Anvar 1	0.6	Balalzadeh 2	0.2
Engelab 1	0.7	Aban 2	0.4
Valiasr 1	0.8	Sadigeh Kobra 2	0.5
Resalat	0.8	Valiasr 2	0.5
Sadigeh Kobra 1	0.8	Anvar 2	0.6
Tohid 1	0.9	Tohid 2	0.7
Balalzadeh 1	0.9	Lotfi 2	0.7
Lotfi 1	1.0	Rahzahra 2	1.0
Asghari11	1.0	Azarbani 2	1.0
Azarbani 1	1.1	Engelab 2	1.0
Aban 1	1.2	Esmat 2	1.4
Aemmeh	1.4	Narjes 2	1.6
Reshad	1.5	Asghari 2	1.7
Rahzahra 1	1.7	Azadi 2	2.2
Azadi 1	1.7	Mohseni 2	2.7
Taleghani	1.9	Felestin 2	2.9
Narjes 1	1.9	Amupur 2	3.8
Esmat 1	2.3	Ghods 2	4.3
Taghva	2.5	Etminan 2	11.6
Mohseni 1	3.2		
Felestin 1	3.5		
Ghods 1	3.6		
Amupur 1	4.5		
Etminan 1	14.3		

Note: Two schools of Fajr and Shahed were excluded from the study
Maximum distance has been calculated from school location to farthest residential location in a straight line within the proposed school attendance areas

longer commutes than are necessary. The result is a large volume of students attending the studied schools from different areas which are outside the designated local areas designed to allow for the most convenient commutes (Fig. 8.4). The number of students attending the most reputable of the five studied schools is a case in point. As the map shows, a sizable number commuting to Aban includes students whose homes are located in a neighboring attendance area where the Etminan School would in fact have been the closest commute. The high occurrence of students travelling to schools further away from their homes is even clearer when we take a closer look at the number of attendance areas from which students attend Aban and the other four studied schools.

Analysis of questionnaire responses revealed that students who attend the more reputable schools of Aban and Anvar, for example, travel to school from 11 different neighboring attendance areas. This phenomenon is equally pronounced for the remaining three PFJH schools of Azadi, Felestin, and Rahzahra. Despite having less appeal relative to the two most popular ones, these schools also continue

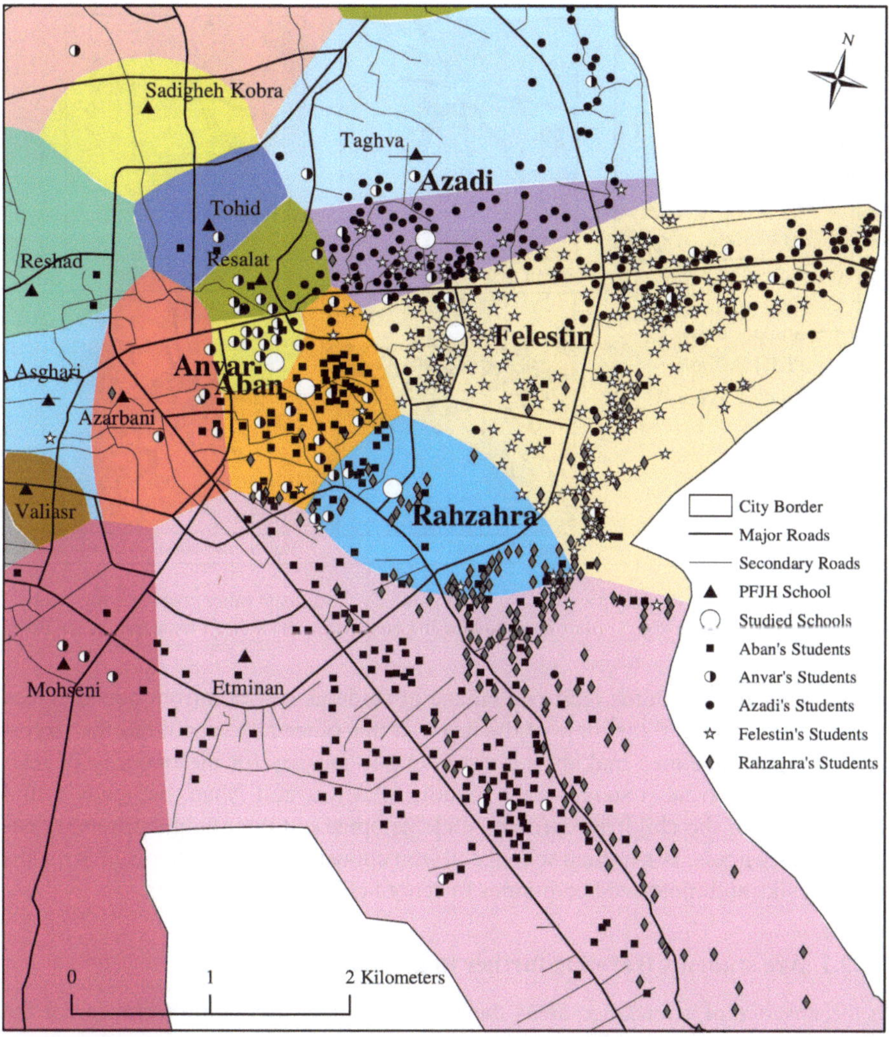

Fig. 8.4 An overlay of proposed school attendance areas over the current residential distribution of students attending the five studied PFJH schools in Rasht

to draw students from a high number of adjacent or further away attendance areas. Following the construction of the proposed attendances areas, analysis concluded that students from between four to six different attendance areas would travel away from their local designated schools to attend the aforementioned ones.

One question that follows is whether or not a high percentage of female students from these outside areas mix with the local resident students at the sampled schools. The answer is a resounding "yes". The results indicate that an average of 65% of students enrolled at the studied schools are in fact from other attendance

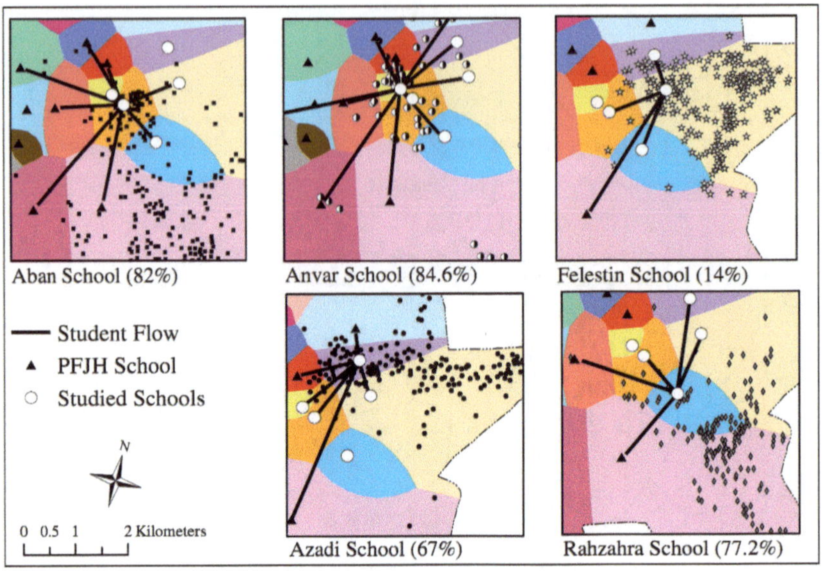

Fig. 8.5 Spatial flows of students attending each studied school from other area showing the proposed school attendance areas. *Note*: The numbers in parentheses show the percentages of the flows

areas. The school with the least percentage of students enrolled from outside areas was Felestin, a natural outcome of the fact that this outer-city school has the largest defined attendance area making it a considerable journey for the majority of residents. At the two most sought after schools of Anvar and Aban, the study found that 82–84% of the children represent students opted out their neighborhood attendance area schools. For Iranian school children choosing further schools, this means longer walks and motorized commutes to school (Fig. 8.5).

8.3.2.1 Are students traveling further to get to better schools?

In the absence of attendance areas there is widespread freedom of choice for the PFJH students in Rasht. This phenomenon arises within the context of a school allocation system based on residence that is not enforced. Though the Rasht Board of Education administrators encourage enrolment in schools closest to homes, there is no mechanism to put the policy into effect. As mentioned before, the study shows that a high number of female students opted out of their local neighborhood schools to get to their preferred ones. Two interrelated questions arise: (1) are parents choosing better public schools that might offset commuting inconveniences?, and (2) are their chosen schools better than the closer ones? Studies (Teelken, Driessen, & Smit, 2005; Srivastava, 2007) found that in order to have official and informed parental school choice there must be three main elements:

1. Schools that are "different" and distinct from one another so as to allow parents to decide which one best meets their educational needs;

2. Formal policies and funding mechanisms that support the provision of viable school choices; and
3. Information on each choice so that parents make decisions based on an assessment of academic performance and school features.

In the study area, over 80% of pupils enrolled in two of the public schools are from outside the official attendance areas. Hence a more fundamental question is: When students opt out of their neighborhood schools to attend the more popular Aban and Anvar public schools, are the elements at work? Although discussion of school choice in developing countries occurs, the primary focus often rests upon the availability of private schools, with scant literature available on the impact of public school choice. The next section will present a comparative review of the distinct choices, policy support and information available on public school choice within Iran and the US, a country with perhaps the longest experience in school choice.

In the US, government grants or funding mechanisms create public schools which are both educationally distinct and of potentially higher quality. Under the No Child Left Behind (NCLB) scheme in the US, for instance, $100 million in grants are allocated to enable school districts to open magnet schools (US Department of Education, 2007). Technically the extra funding translates into considerable funding advantages with these schools spending an average of $200 more per student than regular public schools (Yu & Taylor, 1997). The extra funding allows these public schools to offer students from other school zones a thematic curriculum and unique methods of instruction. Direct federal grants are not the only route to increased resources and making schools more diverse. Variety in educational choices also arises indirectly as a function of school funding systems that exacerbate difference and allow some districts to accrue more funding than others. In the US, federal money accounts for a mere 6% of school educational funds. The remainder is a mix of State (50%) and local funding (44%). Studies of the Detroit (Strate & Wilson, 1993) and Chicago (Cullen, Jacob, & Levitt, 2005) public schools showed that local funding results in the reality of "wealthy districts" capable of academically outperforming poorer ones and that district to district variations in tax bases can make for a clear-cut decision to opt out of neighborhood schools. Like with magnet schools, the additional travel for their children is justified by the promise of a better quality education.

Surplus funding to outspend conventional public schools is not the only route to achieving educational distinction in the public sector. Charter schools can also offer a unique alternative to regular public schools. Freed from much of the "red tape" and regulatory restraints of regular public schools, charters have increased opportunities to innovate and be educationally diverse in their approaches. As documented in previous studies, benefits of charters include lower average student to teacher ratios, unconventional teachers, exposure to specialized curriculum, and educational outputs such as higher scores on standardized testing (Finn, Manno, & Bierlein, 1996; Boyd, Hare, & Nathan, 2002). Though no doubt useful in providing choices for parents, the option of magnets, charters, or affluent districts are still not yet available in the Iranian education system. Like other economically developing countries all of Iran's public junior high schools fall under the single category of "regular" public schools. Since the government is the central source for education budgets, local (provincial) funding does not play a role in giving rise to wealthy attendance areas as in the US.

A case in point is that the two schools with the highest number of students from outside attendance areas are located in an area of the city with lower income groups (Azimi, 2004). There is also no special government funding equivalent to the NCLB set aside to create alternative public schools like magnets. Although all public schools are potential "choice" schools for Iranian parents, there is no national voucher scheme where funding to schools follows student choice. Instead, government-run junior high schools receive funding from the national budget which often falls short of covering full school costs. Though league tables, brochures, and even achievement data report cards are available to gauge the academic performance of public schools in the US (US Department of Education, 2007) and UK (Gorard, Fitz, & Taylor, 2001), no published data on academic performance was available for public schools in the study area. Why then do parents choose further public schools?

Studies show that in addition to published data on academic competence, parents base their choices between public schools on non-academic considerations. In open enrolment systems these include school environments, its facilities, lower class sizes, and curriculum. In an informal open enrolment system without unique magnets or charters, perhaps such characteristics are also central to the choice of public schools. To answer this question, first-hand observations and interviews with school staff of Aban and Anvar, the two schools with the highest outside area enrolments were conducted to see if their environments and educational facilities, class sizes and curriculum were in any way better than the other three (Rhazara, Felistin, Azadi).

Despite attracting the most students from beyond the local area, observations revealed that Aban and Anvar did not deliver any points of distinction in any of the four areas described above. Class sizes are a pertinent illustration. Rather than lower student teacher ratios, Aban and Anvar had ratios close to the studied school average (33 students per teacher). In fact, Aban actually had the highest with an average of 38 students per teacher, meaning that even less attention could be devoted to individual students there. It is also worth mentioning that in line with the Ministry of Education policy on public schools, there were also no deviations from the standard curriculum or in teaching methods. First hand observations of facilities at Aban and Anvar also revealed similar quality between the studied schools. Like most public schools in Iran, the facilities displayed the same shortcomings such as poorly maintained buildings, cramped school yards and unusable amenities (for example, restrooms).

Observations of the school environment data yielded more of the same. It has been observed that neighborhoods with higher socio-economic characteristics can make for lower levels of violence and overall safer environments (Cullen et al., 2005). To gauge the school environment then, first hand observations focused on the socio-economic level of students at the schools. Without published census data on the income level of families in Rasht city, determining the socio-economic level of students was a difficult task. Consequently, observations of the general appearance of the school children formed the basis for the categorization of income groups. Pupils were observed with respect to factors such as their appearance, cleanliness and quality of their school clothes. We found that students attending Aban and Anvar were of a largely homogenous socio-economic make up and between middle to lower middle class. Since observations of the students at Felistin, Razahara and Azadi yielded identical results; it was not possible to conclude that high income groups made up noteworthy portions of the student population.

Our analysis revealed that in addition to the lack of published information such as league tables on school performance to help differentiate between them, there were also no visible non-academic distinctions in terms of school characteristics. Perhaps the greatest puzzle raised by our analysis then is why so many students choose to opt out of their nearest schools, despite the fact that the research was able to identify little tangible benefit of doing so. Aban and Anvar, after all, maintained their status as the two most reputable and sought after schools in the area. Another is how information on perceived differences among the public schools (in this case PFJHs) was disseminated. One possible explanation is that instead of using the more formal knowledge produced by schools themselves or published as examination results or league tables, parents relied on "grapevine knowledge" "Grapevine knowledge" can be characterized in terms of rumor, rather than gossip (Ball & Vincent, 1998). In the absence of other more reliable sources of information, rumor was a means for filling in missing information or explaining the inexplicable.

Responses to our questionnaires lend support to this idea. When asked how they arrived at their decision to enroll in either Aban or Anvar, parents often indicated that decisions were based in part on word of mouth recommendations from relatives, friends and elementary school teachers. The grapevine knowledge driven school choice for Aban may have been based on the fact that it is also the oldest and longest established PFJH in the study area. In addition, parents noted in the questionnaire that the experience of the teachers is one reason why they consider Aban to be of better quality and therefore more reputable. So the "hot knowledge" on better teachers, while understandable, also yields problems since it is difficult to measure in the study area. One sensible measure of teacher quality is teacher training and teacher education (Boyd et al., 2002). Despite the perception of special teachers, interviews with the Rasht Board of Education revealed that teachers at these schools did not have any more experience or training than any of the other studied PFJHs. Even if teachers at a particular school are educationally diverse and innovative in their techniques, a yearly teacher rotation policy in Rasht still forces them to "rotate out" to different schools. Within the context of an ongoing perception about better teachers and name recognition at Aban and Anvar, it may be rational for parents seeking to associate themselves with quality education to attend a school that is an established name, even though in reality there was no hard evidence that this was true. Without quantifiable statistics on teacher training, how students perform on standardized tests or data on teacher qualifications, parents may not be in a position to accurately gauge educational quality.

8.3.2.2 High Traffic Accident Rates in Iran

Though parents in the study area exercised a degree of choice in an informal open enrolment system, opting out of neighborhood schools did not mean better quality of schooling. Another concern was that school-aged children in Iran face far greater risks of traffic dangers. Unlike the developed world where travel on buses or using taxis in the public and private sectors is relatively safe increased commutes raises serious concerns in economically developing countries. Deaths from road traffic injuries and, in particular, motor traffic accidents are often referred to as a "hidden epidemic," accounting for one in every 50 deaths worldwide. In 2000, this ratio translated into 1.26 million deaths worldwide (PAHO, 2004). Without safer transportation options, developing countries in fact bear the major brunt of the "epidemic" with a disproportionate share of these fatalities (Murray & Lopez, 1996). In 1998, developing countries accounted for more than 85% of all deaths due to road traffic crashes globally and more significantly 96% of all children killed (Krug, 1999).

The accident statistics for Iran make reduction of commutes for children an equally pressing concern. Iran currently has one of the highest traffic accident rates in the world, representing the second leading cause of death in the country. One Iranian dies in a traffic related accident every 30 min. Risk of death is not confined to motorized commutes as one third of all traffic related deaths are pedestrian fatalities (Zare & Nuri, 2001). In addition, the mortality rates in Iran, a country with a relatively young population, are highest among young people with traffic injuries as the leading cause of injury related deaths. If broken down by the age group that includes the population that is the concern of this research, people between the ages of 10 and 20 made up 31% of all of these fatalities (Montazeri, 2004). Thus, school aged children included in this group face considerable risk. Longer commutes by foot or in motorized commutes to schools put the school children in even greater peril. As these risks were not offset by tangible educational benefits, the research determined that enforcing the formal policies in place by the Iranian government was the best course of action to ensure safe commutes. The next section will describe how the implementation of attendance areas can reduce commute time and distances to reduce the traffic risks.

8.3.3 Minimizing Student Travel Time and Distance Through Proposed Attendance Areas

To provide a comprehensive picture of current walking commutes to PFJH schools in Rasht, we extracted the current average home-school walking distance and time from the questionnaire forms. It shows that the daily journey on foot from home can be as low as 1.4 km for students travelling to Aban PFJH School and as high as 2 km for those walking to the schools of Azadi or Felestin. Students currently travel an average distance of 1.7 km on foot from their homes. Like the distance measurement of commutes on foot, times also vary ranging anywhere from an average of 13 min

Table 8.3 Average current travel times and distances and estimated maximum projected travel times and distances for the studied schools

Travel time (min)/distance (km)		Studied schools					
		Aban	Anvar	Azadi	Felestin	Rahzahra	Average
Current average travel:							
On foot	Time	13.1	13.2	13.4	14.8	15.7	14.0
	Distance	1.4	1.7	2.0	1.9	1.5	1.7
By vehicle	Time	21.3	26.6	21.5	20.5	16.5	21.2
	Distance	4.7	2.7	2.7	2.8	3.7	3.3
Estimated maximum travel within the proposed attendance areas:							
On foot	Time[a]	11.2	4.6	11.4	27.2	17.7	14.4
	Distance	1.2	0.6	1.7	3.5	1.7	1.7

[a]Estimated maximum travel time on foot to school calculated by the equation of MTt = (MTd.Tt)/Td where MTd = estimated maximum travel distance to schools, Tt = current average travel time on foot, and Td = current average travel distance on foot

for the Aban School to almost 16 min for Rahzahra, the highest commuting time (Table 8.3).

Although the term "vehicle commute" might typically prompt images of more distant travel from the outskirts of town, most vehicle journeys to the five studied schools were from locations that were relatively close. The average travel distance for motorized vehicles, the majority of which are commutes on buses and taxis, was no more than a 3.3 km journey for the studied schools. When compared to the length of commutes on foot, current journeys by vehicles were only twice the average distance that most students walked to the studied schools. Why then, were students living relatively close to the current average walking distance not going on foot? The responses from the questionnaires showed that despite the fact that walking to school might still have been possible, parental fears about the dangers inherent in longer walks precluded most from allowing their children to go such distances on foot – choosing commutes by safe rented vehicles instead. This makes sense within the context of the numerous environmental dangers that children on foot have to contend with. Walking further distances, after all, would entail increased exposure to the possibility of being hit by cars during heavy traffic or falling victim to the sexual and verbal harassment prevalent in Rasht streets. As a consequence, preferring not to take any chances with the safety of their children most parents living beyond an average distance of 1.7 km opt for rented vehicles for commutes instead. The next phase of the discussion is a comparison of current commuting in a parental-choice system with projected commutes within the proposed attendance areas to see if inconveniences were reduced by the proposed school attendance areas.

As seen in Table 8.3, the reduction of the walking commute is evidenced by the fact that within the proposed attendance areas for the studied schools the absolute maximum distances that most students would have to travel on foot on average would not exceed what is now the average walking commute for students. It is also worth pointing out that since a majority of students will be able to travel within a 1.7 km average journey on foot without buses or taxis, estimated times and distances for vehicle travel are also unnecessary and thus omitted from the table. Given the goal of reducing the current reliance on vehicle commutes, the length of the projected maximum commute within the constructed areas is a promising step towards increased convenience for students. Recall here that due to concern for their children, parents turned to vehicle commutes out of necessity whenever walking commutes would exceed a walk of 1.7 km beyond the relative protection of the local neighborhood. With commutes limited to a maximum of 1.7 km within the constructed areas it is possible to allay the "stranger danger" fears about safety that drive the current choice of highly inconvenient and potentially life-threatening vehicle commutes.

A comparative analysis of current commutes and of commutes within the delimited attendance areas lends clear support to the conclusion that an overwhelming majority of students will be able to make more convenient commutes on foot over shorter distances. The estimated commuting distances within the attendance areas showed that students travelling to the vast majority of schools in the study area will be able to travel under the current walking commute distance. As shown in Fig. 8.6 with the exception of Felestin, the percentage of students who would have travel more than 1.7 km within the delimited boundaries to the studied PFJH schools drops to zero. This result is highly relevant to achieving the outcome of increased student travel safety.

As related above, some students in the study area residing in the more remote sections in the city would still have to commute by vehicle within the proposed attendance areas. Given that the goal of this research is dealing with any travel on inconvenient vehicles, recommendations for a safer, more serviceable alternative need to be provided to replace continued travel in the hazardous taxis and buses of Rasht. The solution to the problem of finding convenient low risk transport is best answered by looking to the popular standardized commuting option already in use in developed countries – free school bus services. This mode of convenient, reliable

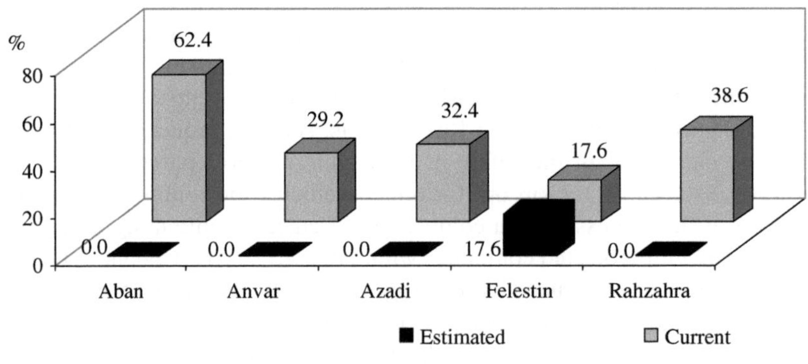

Fig. 8.6 Current percentages of students travelling more than 1.7 km versus estimated percentage who will travel the same distance to the studied schools within the proposed attendance areas

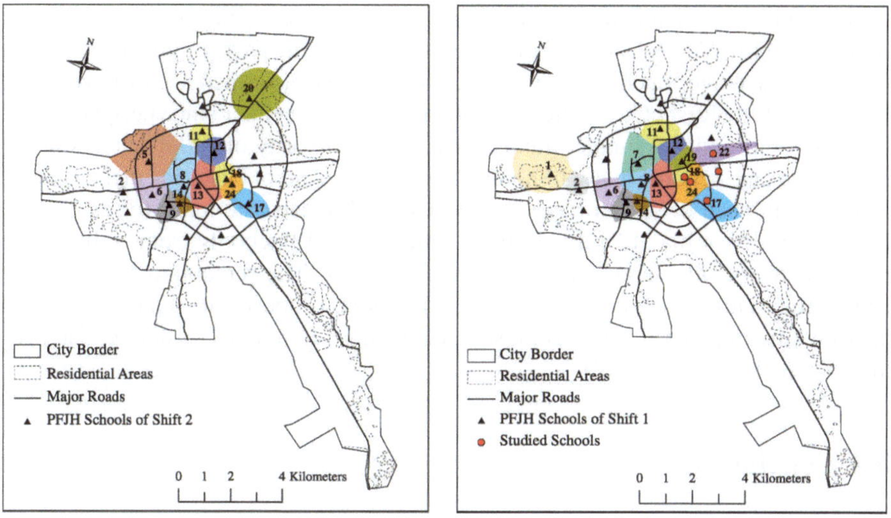

Fig. 8.7 Shift 1 and Shift 2 Rasht PFJH schools that will not need school buses and their proposed attendance area showing studied school location and city residential areas – 1: Aemmeh, 2: Balalzadeh, 5: Narjes, 6: Enghelab, 7: Reshad, 8: Asghari, 9: Lotfi, 11: Sadigheh Kobra, 12: Tohid, 13: Azarbani, 14: Valiasr, 17: Rahzahra, 18: Anvar, 19: Resala, 20: Esmat, 22: Azadi, 24: Aban

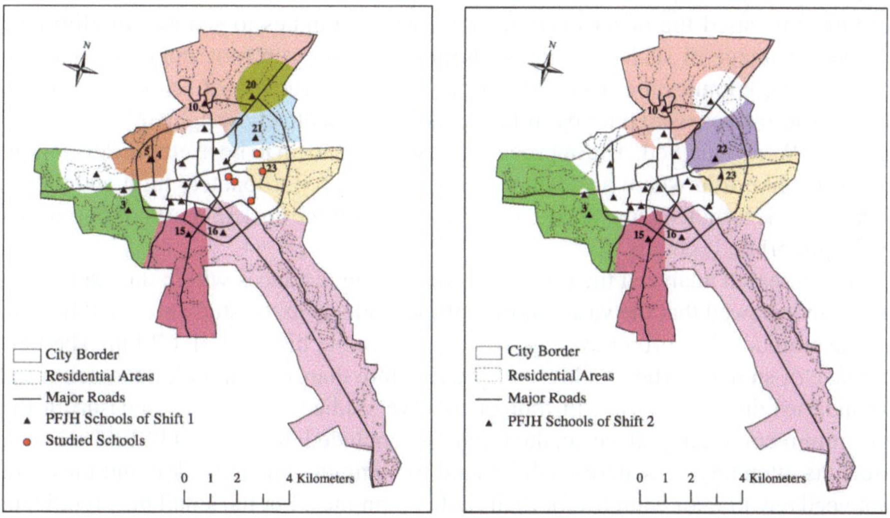

Fig. 8.8 Shift 1 and Shift 2 Rasht PFJH schools that will need school buses for commuters more than 1.7 km and their proposed attendance areas showing studied school locations and city residential areas – 3: Amupur, 4: Taleghani, 5: Narjes, 10: Ghods, 15: Mohseni, 16: Etminan, 20: Esmat, 21: Taghva, 22: Azadi, 23: Felestin

travel would be well suited to the small but significant percentage of students in the study area (17.6%) that attend Felestin, where they still have to travel more than the 1.7 km average. Such services will not be excessively expensive or beyond the resources of the Rasht City Education Office in this case either since peripheral city schools compromise a smaller number and thus a minority of PFJH schools in the city. In the end, schools that will not need to provide school buses for their students and those in need of school buses were illustrated in separate maps for each shift. Figure 8.7 shows that 15 schools from Shift 1 and 13 schools from Shift 2 will not need to provide school buses for their students. The remaining nine schools from Shift 1 and six schools from Shift 2 with larger proposed attendance areas can provide a free bus service to ensure safe, convenient commutes to shuttle distances that exceed 1.7 km (Fig. 8.8).

8.4 Conclusions

The school choice in the absence of defined attendance areas led to the outcome of Rasht city PFJH students taking longer commutes than necessary. An investigation into the specific length of commutes found that the commute for 43.6% of students currently enrolled at the five studied PFJH schools is more than 2 km. Unlike the relatively safe travel along sidewalks, a trip to school of 2 km or even longer is a lengthy journey for female students of Rasht. Inordinately long commutes, in turn, increased the occurrence of motorized commutes to school. To eliminate the disadvantage of such commutes, shorter commutes on foot and elimination of motorized commutes are necessary through the construction of school attendance areas. These areas were built by utilizing the multiplicatively weighted Voronoi diagram (MWVD) method. The approach utilized was shown to be useful in developing countries like Iran where accident fatalities are disproportionately high and public schools lack sufficient educational diversity to offset the societal costs of opting out of neighborhood schools.

The research analyzed the estimated commuting distances within the attendance areas and showed that the vast majority of the students in the study area will be able to travel under the current average walking commute distance of 1.7 km. The percentage of students who would have a travel more than 1.7 km within the delimited boundaries dropped to zero in four of the five studied schools. By measuring the maximum commuting distance, the research concluded that of the 43 PFJH schools, students attending 28 schools will be able to commute under 1.7 km and thus will not need any kind of vehicles for their daily commutes. For the small but significant percentage of students in the study area that attend the remaining 15 schools where they would still have to travel more than the 1.7 km average, the research suggested the introduction of a free school bus service. Finally, to ensure that all schools are equipped with the same resources so that parents willingly enroll their children to their local allocated schools within the constructed attendance areas, educational policies which ensure that the most qualified teachers and staff circulate to all PFJH schools within the city are also necessary.

Acknowledgment The authors would like to thank Dr. Mu Lan (Department of Geography, University of Georgia) for providing us her WVD18 software, a GIS-based MW-Voronoi package.

References

Ahmadi Nejad Masouleh, F. (2006). *A geographical study of school attendance areas using the multiplicatively weighted Voronoi method: A case of Rasht City, Iran.* Ph.D. Dissertation, University of Tsukuba, Japan.

Azimi, N. (2004). *Restructuring urban morphology in the context of globalizing economy: A case study of Rasht, Iran* (Research Report). Rasht, Iran: University of Guilan.

Ball, J. S., & Vincent, C. (1998). I heard it on the Grapevine: Hot knowledge and school choice. *British Journal of Sociology of Education, 19*, 377–400.

Boyd, W. L., Hare, D., & Nathan, J. (2002). *What really happened? Minnesota's experience with statewide public school choice programs* (Center for School Change Research Report). Minneapolis, MN: University of Minnesota, Hubert H. Humphrey Institute of Public Affairs.

Clarke, S., & Surkis, J. (1968). An operations research approach to the desegregation of school systems. *Socio-Economic Planning Science, 1*, 259–272.

Cullen, J., Jacob, B., & Levitt, S. (2005). The impact of school choice on student outcomes: An analysis of the Chicago public schools. *Journal of Public Economics, 89*, 729–760.

Finn, C. E., Manno, B. V., & Bierlein, L. A. (1996). *Charter schools in action: What have we learned?* (Educational Excellence Network Research Report). Washington, DC: Hudson Institute.

Ghosh, A., & Rushton, G. (1987). *Spatial analysis and location-allocation models.* New York: Van Nostrand Reinhold.

Gorard, S., Fitz, J., & Taylor, C. (2001). School choice impacts: What do we know?. *Educational Researcher, 30*, 18–23.

Hubbard, R. (1970). *The Spatial Pattern and Market Areas of Urban Settlement in Jamaica.* Kingston: Department of Geography, University of the West Indies, Research Note No. 1.

Huff, D. L., & Lutz, J. M. (1979). Ireland's urban hierarchy. *Economic Geography, 55*, 196–212.

Iran Management and Planning Organization (1996). *Selected statistics of Gilan, Iran.* Tehran: Iran Management and Planning Organization.

Iran Ministry of Education (2004). *Educational statistics.* Tehran: Iran Ministry of Education.

Jennergren, P. L., & Obel, B. (1980). A study in the use of linear programming for school planning in Odense. *Journal of the Operational Research Society, 31*, 791–799.

Krug, E. (1999). *Injury: A leading cause of the global burden of disease.* Geneva: World Health Organization. Department for Disability, Injury Prevention and Rehabilitation, Social Change and Mental Health Cluster.

Maxfield, D. W. (1972). Spatial planning of school districts. *Annals of the Association of American Geographers, 62*, 580–590.

McDaniel, R. D. (1975). Case study of the use of the transportation algorithm for school districting under federal integration guidelines. *Socio-Economic Planning Science, 9*, 271–272.

Montazari, A. (2004). Road-traffic-related mortality in Iran: A descriptive study. *Public Health, 118*, 110–113.

Mu, L. (2004). Polygon characterization with the multiplicatively weighted Voronoi diagram. *Professional Geographer, 56*, 223–239.

Murray, C., & Lopez, A. (1996). *The global burden of disease.* Cambridge, MA: Harvard University Press.

Okabe, A., Boots, B., Sugihara, K., & Chiu, S. N. (2000). *Spatial tessellations: Concepts and applications of Voronoi diagrams.* Chichester, UK: Wiley.

PAHO (Pan American Health Organization) (2004). Deaths from motor vehicle traffic accidents in selected countries of the Americas, 1985–2001. *Epidemiology Bulletin, 25*, 2–5.

SCI. (2005). *Estimated population of counties based on census 1996*. Tehran: Statistical Centre of Iran.

Srivastava, P. (2007). *Neither voice nor loyalty: School choice and the low-fee private sector in India*. New York, Columbia University, National Center for the Study of Privatization in Education, Research Publications Series, Occasional Paper No 134.

Strate, J. M., & Wilson, C. A. (1993). Public opinion on school choice: The Detroit metropolitan area. *Urban Review, 25*, 123–137.

Teelken, C., Driessen, G., & Smit, F. (2005). Frictions between formal education policy and actual school choice: Case studies in an international comparative perspective. *International Review of Education, 51*, 35–58.

US Department of Education (2007). *No child left behind act: Helping families by supporting and expanding school choice*. Retrieved December 1, 2008, from http://www.ed.gov/nclb/choice/schools/choicefacts.pdf

Von Hohenbalken, B., & West, D. S. (1984). Predation among supermarkets: An algorithmic locational analysis. *Journal of Urban Economics, 15*, 244–257.

Wood, L. J. (1974). Spatial interaction and partitions of rural market space. *Tijdschrift voor Economische en Social Geografie, 65*, 23–34.

Woodall, M., Cromley, R. G., Semple, R. K., & Green, M. B. (1980). The elimination of racially identifiable schools. *Professional Geographer, 32*, 412–420.

Yu, C. M., & Taylor, W. L. (1997). *Difficult choices: Do magnet schools serve children in need?* Washington, DC: Citizens' Commission on Civil Rights.

Zare, M., & Nuri, H. (2001). *Analysis of road traffic related fatalities in Iran*. Tehran: Center of Disease Management, Work Report (in Persian).

Part III
Land Use and Land Cover Change

Chapter 9
Accuracy of Land Use and Land Cover Mapping Methods

Rajesh B. Thapa and Yuji Murayama

9.1 Introduction

Knowledge of land use and land cover is important for many planning and management activities and is considered an essential element for modelling and understanding the earth as a system. Land cover maps are presently being developed from local to national to global scales. The use of panchromatic, medium-scale aerial photographs to map land use has been practiced since the 1940s (Lillesand, Kiefer, & Chipman, 2008). Remote sensing has long been used to map urban growth and urban morphology, and implies the mapping of the form, land uses, and density of urban areas, each having an associated shape, configuration, structure, pattern, and organization of land use. Sometimes simply mapping an urban or non-urban dichotomy is important; while sometimes detailed morphologic mapping is needed, where the positions of buildings and roads or the extraction of the three-dimensional topographical aspects of urban areas are needed. Satellite imagery has the unique ability to provide synoptic views of large areas at a given time that is not possible using conventional survey methods.

The term *land cover* refers to the type of feature present on the surface of the earth. Forest, lakes, concrete highways, ice sheet are all examples of land cover types. While the *land use* refers to the human activity associated with a specific piece of land. For example, a tract of land on the fringe of an urban area may be used for single-family housing. Depending on the level of mapping detail, its land use could be described as urban use, residential use, or single-family residential use. The same tract of land would have a land cover consisting of roofs, pavement, grass, and trees. Thus for a planner, a knowledge of both land use and land cover (LULC) is necessary for planning and land management activities.

R.B. Thapa (✉)
Division of Spatial Information Science, Graduate School of Life and Environmental Sciences, University of Tsukuba, Tsukuba, Ibaraki, Japan
e-mail: thaparb@yahoo.com; thaparb@gmail.com

Y. Murayama, R.B. Thapa (eds.), *Spatial Analysis and Modeling in Geographical Transformation Process*, The GeoJournal Library 100, DOI 10.1007/978-94-007-0671-2_9,

Continues advancement of remote sensing technologies and the increasing availability of high resolution earth observation satellite data provide great potential for acquiring detailed spatial information to identify LULC of urban regions. Remote sensing (RS) data coupled with fieldwork information and geographic information systems (GIS) have been recognized as effective tools in quantitatively measuring spatial patterns of LULC over relatively large geographic scales. Transitions in architecture and building density, vegetation and intensive socioeconomic activities at the block level in cities often transform the urban landscape towards heterogeneity. Therefore, the urban environment represents one of the most challenging areas for remote sensing analysis due to the high spatial and spectral diversity of surface materials (Thapa & Murayama, 2009a). In recent years, a series of earth observation satellites have provided abundant data at high resolutions (0.6~2.5 m; QuickBird, IKONOS, OrbitView, SPOT and ALOS) to moderate resolutions (15~30 m; ASTER, IRS and LANDSAT) for urban area mapping. Remote sensing data from these satellites have specific potential for detailed and accurate mapping of urban areas at different spatiotemporal scales. The high resolution imagery provides data for monitoring urban infrastructures, whereas moderate resolution imagery can provide synoptic measures of urban growth, surface temperature and more. A wide range of urban remote sensing applications from both sensors is available to date (Carlson & Arthur, 2000; Miller & Small, 2003; Maktav, Erbek, & Jurgens, 2005; Gatrell & Jensen, 2008; Thapa & Murayama, 2009b). These include quantifying urban growth and land use dynamics, population estimation, life quality improvement, urban infrastructure characterization, monitoring land surface temperature, air quality and vegetation, and topographic mapping. Having the potential to monitor human activities at the earth surface, however, the information acquired from remote sensing data could be an additional resource in developed economies, while it might be the only alternative in the developing countries.

Despite advances in satellite imaging technology, computer-assisted image classification is still unable to produce LULC maps and statistics with high enough accuracy. Image analysis techniques are evolving rapidly, but many operational and applied remote sensing analyses still require extracting discrete thematic land surface information from satellite imagery using classification-based techniques. Several image classification techniques, from automated to manual digitization, can be found in the literature. However, these have spanned a broad range of land-surface types and sensors. Very few studies (Carvalho, 2006; Lee & Warner, 2006; Lo & Choi, 2004; Nangendo, 2007; Ozkan & Sunar-Erbek, 2005; Prenzel & Treitz, 2005) have compared different image classification methods with different satellite sensors to determine how the organization of information inherent to the classification scheme influences classification accuracy.

Automated classification procedures of satellite imagery have been based mainly on multi-spectral classification techniques (per-pixel classifiers). These procedures assign a pixel to a class by considering its statistical similarities, in terms of reflectance, with respect to a set of classes. The unsupervised classification approach provides an automated platform for image analysis, mainly based on surface reflectance and generally ignoring basic land cover characteristics (i.e. shape and size) of landforms. The supervised classification approach can preserve the basic land cover characteristics through statistical classification techniques using a number of well-distributed training pixels. However, the maximum likelihood classifier often used in supervised classification has been proven ineffective at identifying land uses at urban fringe areas due to the heterogeneity of urban land cover (Johnsson, 1994). Suburban residential areas form a complex mosaic of trees, lawns, roofs, concrete, and asphalt roadways. Such a complex urban environment develops mixed pixel problems, often causing misclassification of remote sensing images. In this case, the fuzzy supervised classification approach helps to reduce mixed pixel problems in the heterogeneous earth surface by using a membership function (Zhang & Foody, 2001). However, classification techniques that combine more than one classification procedure improve remote sensing-based mapping accuracies (Thapa & Murayama, 2009b).

Considering the complexity of the urban landscape and the importance of spectral and radiometric resolution to LULC classification accuracies, we discuss the benefits of four approaches: unsupervised, supervised, fuzzy supervised, and GIS post-processing. These approaches can address a wide range of mapping problems in urban frontiers and provide alternatives to improve mapping accuracies for urban planners. The objectives of this study are to derive LULC maps using four different mapping approaches and to compare the accuracies of the approaches in mapping urban area using Advanced Land Observation Satellite data. Tsukuba city was selected to test the mapping approaches. This city is an interesting place to study remote sensing applications as it includes both heterogeneous and homogeneous anthropogenic landscape patterns.

9.2 Methods

9.2.1 Study Area: Tsukuba City, Urban Frontier of Tokyo

Geographically, Tsukuba city is situated within the geographic coordinates 35°59′42″ to 36°14′2″ North latitudes and 140°0′2″ to 140°10′39″ East longitudes, northeast of the Tokyo metropolitan fringe (Fig. 9.1). We considered a rectangular shape strategy covering the city and its adjacent hinterlands to remove administrative biases in mapping spatial patterns. The study site covers 55,075 ha of land. The coverage has homogeneous (i.e. paddy field, water, etc.) and heterogeneous (residential, parks, etc.) landscapes. It includes both dense and sparse types of landscape development.

The agricultural landscape of Tsukuba in the 1960s has been transformed into a modern city; in Japan the city is known as Science city. The city is well-planned and developed with a special purpose: to promote science by establishing educational institutes and national-level research institutes. Therefore, it carries a unique perspective of development by absorbing a significant number of educated populations rather than the industrial population. A high-speed train system (Tsukuba Express) was established in 2005. This transportation system makes it easy to commute and

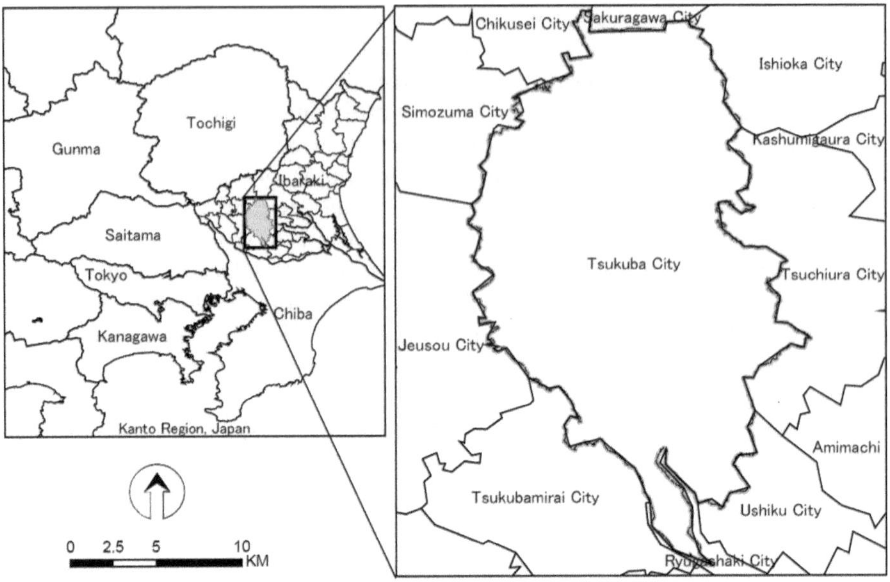

Fig. 9.1 Study area, Tsukuba city, Japan

reduces the travel time to the Tokyo centre. Due to the establishment of state of the art facilities, improved life quality and reduction in travel time to Tokyo, Tsukuba is becoming the centre of attraction for the residents, even for those who are working in different parts of Tokyo. The population in the business core of Tsukuba and its vicinity is growing, with a density of 730 persons per square kilometre as of 2008; this is 25 heads higher than in 2005 (Statistics Bureau, 2008). New residential and commercial zones are being built. Rapid changes in landscape can be observed, even at a monthly or bimonthly basis.

Physically, the study area is part of the flat Tsukuba-Inashiki Plateau, 20–30 m above sea level, covered with the Kanto Loam Layer. Mt. Tsukuba (elevation of 877 m), one of the major mountains in the Kanto region, is located to the north of the study site (Thapa & Murayama, 2007). Four major rivers (Kokai, Sakura, Higashi Yata, and Nishi Yata) irrigate the area from north to south. Forests and agriculture fields in suburban areas provide natural green spaces to city dwellers. The average annual temperature was 14.2°C with annual precipitation of 1612 mm in the year 2006 (Japan Meteorological Agency, 2006). The city gets fairly cold in the winter; snow falls about twice a year.

9.2.2 Data Sources

9.2.2.1 Remote Sensing Image Data

We used an ALOS (Advanced Land Observing Satellite) multi-spectral Advanced Visible Near Infra Red 2 (AVNIR2) sensor image acquired on 4th August 2006

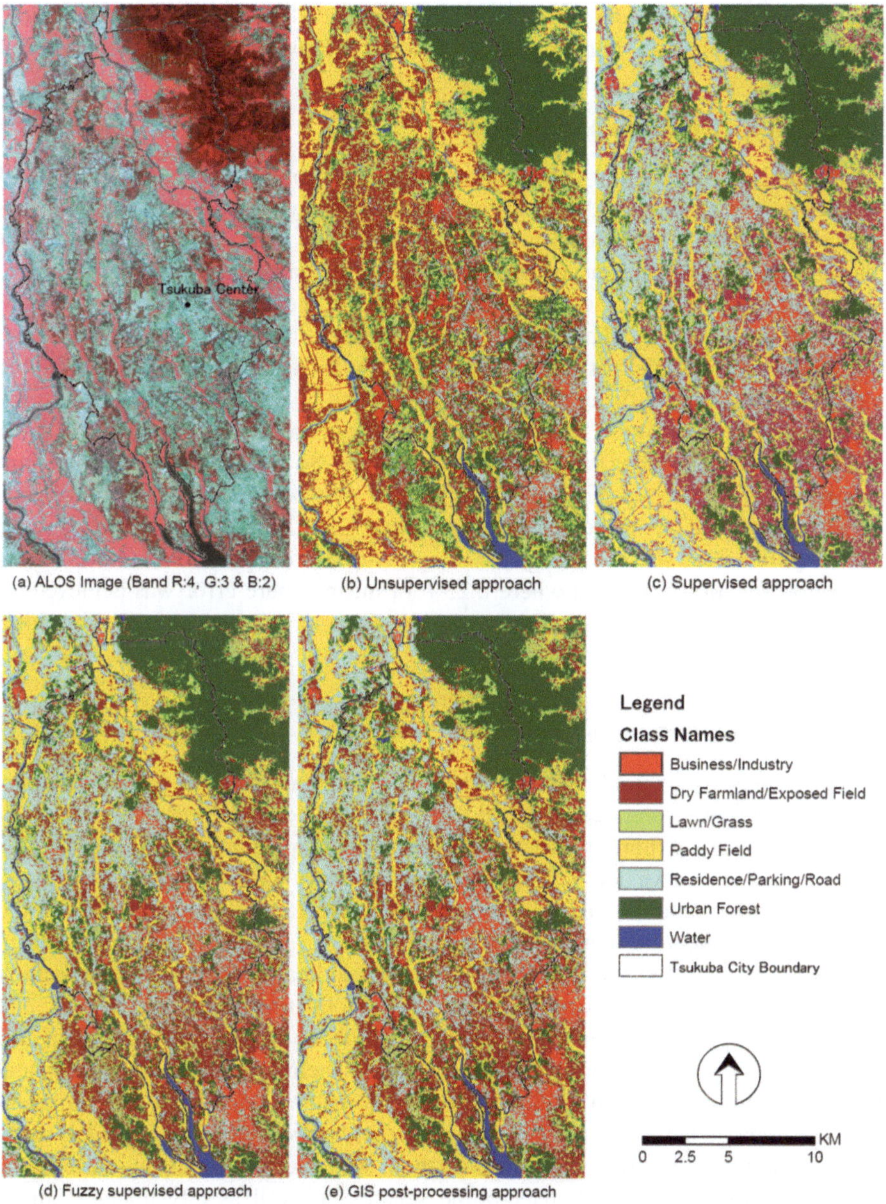

Fig. 9.2 Image data and land use/cover maps produced by the classification approaches

(Fig. 9.2a). The ALOS (locally known as "Daichi") is a new satellite, launched in 2006 (JAXA, 2006). The ground coverage (swath width) of the sensor at nadir is 70 km. This image consists of visible and near-infrared bands (wavebands: Band 1 [blue, 0.42–0.50 μm], Band 2 [green, 0.52–0.60 μm], Band 3 [red, 0.61–0.69 μm], Band 4 [near infrared, 0.76–0.89 μm]. The spatial resolution of the image is 10 m.

This image was selected for this study as it provided suitable cloud-free spatial coverage with relatively high spatial and spectral resolutions.

9.2.2.2 Image Geometry Rectification

Raw digital images usually contain geometric distortions of which cannot be used directly as a map base without subsequent processing. The sources of these distortions range from variations in the altitude, attitude, and velocity of the sensor platform to factors such as panoramic distortions, earth curvature, atmospheric refraction, relief displacement, and nonlinearities in the sweep of a sensor's IFOV (Lillesand et al., 2008). Therefore, accurate registration of multi-spectral remote sensing data is essential for analysing LULC conditions of a particular geographic location. There can be systematic distortions caused by earth rotation during imaging and random distortion due to unknown systematic distortions. In this study, the systematic distortions wers corrected at source. So we need to correct random distortions which was carried out using a road network map with the local projection system (i.e., Transverse Mercator, Tokyo GRS 1980 datum). Thirty ground control points were used to rectify the image. A first-order polynomial linear transformation function was used where 0.23 root mean square error was achieved. A nearest neighbourhood re-sampling algorithm was applied, since this does not alter the radiometric values of individual pixels.

9.2.2.3 Ground Reference Data

In image analysis, ground reference data play important roles to determine information classes, interpret decisions, and assess accuracies of the results. Substantial reference data and a thorough knowledge of the geographic area are required at this stage. In this study, we adopted both methods (primary and secondary) for collecting ground truth data. Intensive fieldwork was conducted in December 2006 as the primary data collection method. A Personal Digital Assistant (PDA) with built-in Global Positioning System (GPS) equipped with a data entry form and navigation map and a handheld digital camera were used for collecting the geographic data and recording perspective views of the locations for laboratory analysis. A total of 100 geographic locations in points and polygons and their corresponding biophysical attributes were collected in the field. The locations of the collected data represent both the homogeneous and the heterogeneous landscape environments of the study area. In the secondary data collection method, we used higher resolution imagery acquired from airborne and space-borne sensors, as well as city planning maps and other documents. A QuickBird satellite image with 0.6 m resolution acquired in October 2006 and colour aerial photographs with 0.5 m resolution acquired in November 2005 of selected areas were used. Using all of these data, detailed ground reference data of the study area were prepared to support the land use class scheming, image classification and subsequent accuracy assessments. Furthermore, the ground reference data used for the image classification were invalid for mapping accuracy assessment purposes.

9.2.3 Scheming LULC Classification

Classification schemes provide frameworks for organizing and categorizing information that can be extracted from image data (Thapa & Murayama, 2007). A proper classification scheme includes classes that are both important to the study and discernible from the data on hand. The USGS devised a LULC classification system for use with remote sensor data in the mid-1970s (Anderson, Hardy, Roach, & Witmer, 1976). The basic concept and structure of this system are still valid today. A number of recent LULC mapping efforts follow these basic concepts and although their mapping units varied. With the consideration of this basic concept, image enhancement, contrast stretching and false colour composites were worked out to improve the visual interpretability of the image by increasing the apparent distinction between the features. Knowledge-based visual interpretation, texture and association analysis were done at the preliminary stage. Furthermore, field survey data, aerial photographs and the QuickBird image, city planning maps and documents were carefully analyzed while preparing the classification scheme.

A false colour composite (Bands: 4, 3 and 2 as Red, Green and Blue, respectively) of the AVNIR2 image of the study site used as input data is shown in the Fig. 9.2a. The false colour image clearly shows the water bodies in black, paddy fields in pink, vegetation in dark red and urban surface materials as light bluish. It is difficult to distinguish dry farm land and exposed field in the false colour image, but this is distinguishable in true colour (Bands: 3, 2 and 1 as Red, Green and Blue, respectively). Separation of asphalt surface from the image is easier, but the association of the surface in the study area makes it difficult to consider it an entity of the urban land use pattern. Most of the roads and parking areas are built out of asphalt, and are associated with residential, facility and industrial areas. Therefore, after analysing the fieldwork information, we decided to combine the roads, parking lots and residential land uses as one class labeled residence/parking/road. The land use categories recommended by the Geographical Survey Institute, sole authority of land use mapping in Japan (GSI, 2008), are also reviewed. After analyzing these information sources, we decided to extract seven types of land use and land cover classes as thematic information from the image (Table 9.1).

Table 9.1 Land use/cover schemes

No.	Classes	Definition
1.	Urban forest (UF)	Natural vegetation and planted trees
2.	Lawn/grass (LG)	Lawn, grass and bush
3.	Paddy field (PF)	Paddy field
4.	Dry farmland/exposed field (DF)	None irrigated land, vegetables and fruits area
5.	Facility/industry (FI)	Large space house
6.	Residence/parking/road (RP)	Small houses, back/front yards, parking area, road
7.	Water (WA)	Lake, river, wetland

9.2.4 Image Classification

Land cover classes are typically mapped from digital remotely sensed data through digital image classification and interpretation. Image classification process replaces visual analysis of the image data with quantitative techniques for automating the identification of features in a scene. This normally involves the analysis of multispectral image data and the application of statistically based decision rules for determining the land cover identity of each pixel in an image. The overall objective of the image classification procedure is to automatically categorize all pixels in an image into land cover classes or themes (Lillesand et al., 2008). In this study, four approaches (unsupervised, supervised, fuzzy supervised and GIS post-processing) were used for image classification and mapping of the urban area. However, the land use prediction methods are constrained by the spatial resolution of satellite imagery, the mapping approach, and expert knowledge of the study area (Thapa & Murayama, 2008).

9.2.4.1 Unsupervised Classification Approach

The unsupervised classification approach is an automated classification method that creates a thematic raster layer from a remotely sensed image by letting the software identify statistical patterns in the data without using any ground truth data. This approach involves algorithms that examine the unknown pixels in an image and aggregated them into a number of classes based on the natural groupings or clusters present in the image values. The basic principle is that values within a given land cover type should be close together in the measurement space, whereas data in different classes should be comparatively well separated. Clusters are defined with a clustering algorithm that uses all pixels in the input image for analysis. The land cover classes that result from unsupervised classification are spectral classes, the identity of those classes will not be initially known. After the classification is complete, the analyst then employs a posteriori knowledge to label the spectral classes into information classes. Initially, thirty spectral clusters were formed to separate the image information into a more readable form. The Iterative Self-Organizing Data Analysis Technique (ISODATA) was used to cluster the image pixels into groups. Many more clusters than actual classes (i.e., schemed class in Table 9.1) were chosen because the exact number of spectral classes in the dataset was unknown. These thirty clusters were carefully judged using expert knowledge and ground reference data. Spectrally similar classes of identical land cover types were merged. These merged clusters were evaluated to whether they belonged to the land use information classes listed in Table 9.1. Finally, a labelling process was carried out to generate a thematic urban land use and land cover map.

9.2.4.2 Supervised Classification Approach

In this approach, the spatial patterns in the image dataset are evaluated by the computer using predefined decision rules to determine the identity of each pixel.

Supervised classification requires input from an analyst in order to automate the classification algorithm to associate pixel values with the correct land cover category (Jansen, 2005; Lillesand et al., 2008). Training data, a set of statistics that describe the spectral response pattern for each land cover type, is needed for this classification. Training data must be both representative and complete. This means that the image analyst must develop training statistics for all spectral classes constituting each information class to be discriminated by the classifier. For example, in a final classification output, one might wish to delineate an information class called water. If the image under analysis contains only one water body and if it has uniform spectral response characteristics over its entire area, then only one training area would be needed to represent the water class. If, however, the same water body contained distinct areas of very clear water and very turbid water, a minimum of two spectral classes would be required to adequately train on the feature. If multiple water bodies occurred in the image, training statistics would be required for each of the other spectral classes that might be present in the water-covered areas. Accordingly, the single information class water might be represented by four or five spectral classes. In turn, the four or five spectral classes would eventually be used to classify all the water bodies occurring in the image. We identified homogeneous sample pixels as training pixels in the image that can be used as representative samples for each land use category to train the algorithm to locate similar pixels in the image. For each land use and land cover type (Table 9.1), five to ten areas of interest were prepared as the signatures of training samples. The training areas were created in order to discriminate the individual classes. The ground reference data were used to prepare the training signatures. After obtaining satisfactory discrimination between the classes during spectral signature evaluation, supervised classification with the Maximum Likelihood Classifier (MLC) was run using all four bands of the image. This classifier quantitatively evaluates both the variance and the covariance of the category spectral response patterns when classifying an unknown pixel (Shalaby & Tateishi, 2007).

9.2.4.3 Fuzzy Supervised classification Approach

The fuzzy supervised classification approach works using a membership function, where a pixel's value is determined by whether it is closer to one class than another (Jensen, 2005). This approach considers that each pixel might belong to several different classes without definite boundaries. Therefore, it can deal with the mixed pixel problem or the more heterogeneous features representation problem. This is also similar to application of maximum likelihood classification; the difference being that fuzzy mean vectors and covariance matrices are developed from statistically weighted training data. Like supervised classification, fuzzy classification requires training but instead of delineating training areas that are purely homogeneous, a combination of pure and mixed training sites may be used. In this approach, we prepared five to ten training areas for each LULC class (Table 9.1). Instead of delineating training areas that are purely homogeneous, a combination of pure and mixed training sites was used. Mixtures of various feature types defined the fuzzy training class weights. A classified pixel was then assigned a membership grade with respect to its membership in each information class. Two maps (multilayer class map and distance map) were generated. Fuzzy convolution was then performed to create a single classification layer by calculating the total weighted inverse distance of all the classes in a 3 × 3 window of pixels. This operation assigns the centre pixel in the class with the largest total inverse distance summed over the entire set of fuzzy classification layers (Leica Geosystems, 2005). Classes with very small distance values remain unchanged, while classes with higher distance values may change to a neighbouring value if there are a sufficient number of neighbouring pixels with class values and small corresponding distance values. The convolution method has a built-in function that creates context-based classification to reduce speckle or salt-and-pepper noise in the classification map.

9.2.4.4 GIS Post-processing Approach

A combination of more approaches in mapping provides better results than just using a single approach. In this study, we propose a GIS post-processing approach that combines the advantages of all three approaches (unsupervised, supervised, and fuzzy supervised) to produce an improved LULC map. Here, the maps derived from the unsupervised and supervised approaches were combined using the GIS overlay function. Then, we extracted common land use pixels from the map considering the land use clusters that were identified by both approaches as the best results. The resulting map was carefully evaluated and revealed that the most likely homogeneous features were represented by common pixels, but the more heterogeneous features were left empty. Many studies suggest (Jensen, 2005; Wang, 1990; Zhang & Foody, 2001) that heterogeneous landscape can be better identified by the fuzzy approach. Therefore, the remaining empty pixels were filled by the LULC pixels derived from the fuzzy supervised approach taking into consideration fuzzy strength and rigidness to deal with the heterogeneous landscape. This GIS post-processing approach can represent both homogeneous and heterogeneous areas of the city.

The classified data often manifest a salt and pepper noise due to the inherent spectral variability encountered by a classifier when applied on a pixel by pixel basis. In this case, it is often desirable to smooth the classified output to show only the dominant classification. Post-classification smoothing was applied by a 3 × 3 majority filter in the maps generated from unsupervised, supervised, and GIS post-processing approaches before the accuracy assessment.

9.3 Accuracy Assessment

An image classification is not complete until its accuracy is assessed (Lillesand et al., 2008). Accuracy assessment is a general term for comparing predicted (i.e., classification) results to geographical reference data that are assumed to be true. The accuracy of thematic maps derived by image classification analyses is often compared in remote sensing studies. This comparison is typically achieved by a basic subjective assessment of the observed difference in accuracy but should be undertaken in a statistically rigorous fashion. Most common means of expressing classification accuracy is the preparation of a classification error matrix, also called a confusion matrix, or a contingency table in some literatures. Error matrices compare, on a category by category basis, the relationship between known reference data, i.e., ground truth, and the corresponding results of an automated classification. Therefore, a set of reference pixels representing geographic points on the classified image is required for the accuracy assessment. Randomly selected reference pixels lessen or eliminate the possibility of bias (Congalton, 1991). A random stratified sampling method was used to prepare the ground reference data. This sampling method allocates the sample size for each land use based on its spatial extent (Shalaby & Tateishi, 2007). A total of 300 reference pixels were prepared for each map as ground truth, using the source data as discussed earlier in the ground reference data section. The minimum representation threshold for each land use land cover class was set to 30.

An error matrix was prepared for each resulting thematic map. The matrix provided the correspondence between the predicted and the actual classes of membership for an independent testing dataset. It made possible to derive a range of quantitative measures of classification accuracy. Four measures (producer's, user's and overall accuracies, and Kappa statistic) of accuracy assessment were computed to evaluate the accuracy of the thematic maps. The producer's accuracy represents the measure of omission (exclusion) errors that corresponds to those pixels belonging to the class of interest that the classifier has failed to recognize. The user's accuracy, on the other hand, refers to the measure of commission (inclusion) errors that correspond to those pixels from other classes that the classifier has labelled as belonging to the class of interest. The overall accuracy is the percentage of correctly classified samples. The Kappa coefficient expresses the proportionate reduction in error generated by a classification process. Kappa accounts for all elements of the confusion matrix and excludes agreement that occurs by chance. Consequently, it provides a more rigorous assessment of classification accuracy (Congalton, 1991).

9.4 Results and Discussion

Each approach produced one thematic land use and land cover map (Fig. 9.2b–e). The selection of classification approaches often has an impact in quantitative spatial extent of land uses (Table 9.2). The land use area statistics derived from the supervised, fuzzy supervised and GIS post processing approaches showed small differences in spatial extent as compared to the unsupervised approach. Automated clustering technique (unsupervised) often failed or overestimated the heterogeneous landscapes, mainly in residential and suburban area. High contrast is observed between the spatial statistics of unsupervised and other approaches, especially in the residence/parking/road class. This may be due to the complexity of the urban environment, which forces the classifier to overestimate land use and land cover area. In this case, the supervised and fuzzy supervised approaches have shown good results

Table 9.2 Impact of classification approaches in spatial extent of land use/cover

	Approaches (area in %)			
Land use/cover	Unsupervised	Supervised	Fuzzy supervised	GIS post-processing
Urban forest	20.97	19.17	17.82	18.74
Lawn/grass	18.66	13.10	13.71	13.24
Paddy field	19.75	19.58	18.41	19.27
Dry farmland/exposed field	28.44	21.22	23.25	22.85
Facility/industry	3.92	5.95	6.01	5.59
Residence/parking/road	7.17	19.62	19.51	18.98
Water	1.09	1.36	1.28	1.33
Total	100.00	100.00	100.00	100.00

because these approaches use signatures of particular surface materials to train the algorithm. However, all classifiers showed very few differences in the spatial extent (within ±2%) of the classes, i.e. urban forest, water and paddy field classes. Natural land covers are separable in all classification processes, so there is no significant change for the corresponding spatial statistical representations. All the classification approaches were able to present dry farmland/exposed field as a major land use in the study area.

Tables 9.3, 9.4, 9.5, and 9.6 were prepared to analyze the accuracy of the mapping results. The reference sample size of land uses in each accuracy assessment table differed depending on its spatial representation at the surface. The same sample size of reference data of the urban forest class derived from the unsupervised approach

Table 9.3 Error matrix of the unsupervised approach for land use/cover classification

	Reference data								
Classified data	UF	LG	PF	DF	FI	RP	WA	Total	U. Acc %
UF	39	1	1	3	1	4	0	49	80
LG	2	33	7	3	0	1	1	47	70
PF	1	2	44	0	0	1	0	48	92
DF	2	3	1	32	2	17	0	57	56
FI	0	0	0	1	24	8	0	33	73
RP	1	0	0	5	5	24	1	36	67
WA	0	0	0	0	0	0	30	30	100
Total	45	39	53	44	32	55	32	300	
P. Acc %	87	85	83	73	75	44	94		

Overall classification accuracy = 75.33%

Overall kappa statistics = 0.71

U. Acc., user's accuracy; P. Acc., producer's accuracy, see Table 9.1 for other abbreviations (number unit is in pixel)

Table 9.4 Error matrix of the supervised approach for land use/cover classification

	Reference data								
Classified data	UF	LG	PF	DF	FI	RP	WA	Total	U. Acc %
UF	44	2	0	1	0	1	0	48	92
LG	0	33	7	1	0	1	0	42	79
PF	2	3	41	2	0	0	0	48	85
DF	1	3	1	40	1	3	1	50	80
FI	0	0	0	2	29	3	0	34	85
RP	0	4	0	8	2	34	0	48	71
WA	0	0	0	0	0	0	30	30	100
Total	47	45	49	54	32	42	31	300	
P. Acc %	94	73	84	74	91	81	97		

Overall classification accuracy = 83.67%

Overall kappa statistics = 0.80

U. Acc., user's accuracy; P. Acc., producer's accuracy, see Table 9.1 for other abbreviations (number unit is in pixel)

Table 9.5 Error matrix of the fuzzy supervised approach for land use/cover classification

	Reference data								
Classified data	UF	LG	PF	DF	FI	RP	WA	Total	U. Acc %
UF	43	2	1	0	0	0	0	46	93
LG	3	30	5	3	0	1	0	42	71
PF	0	1	45	0	0	1	0	47	96
DF	0	1	1	43	0	6	1	52	83
FI	0	0	0	1	33	1	0	35	94
RP	2	0	0	1	3	40	2	48	83
WA	0	0	0	0	1	0	29	30	97
Total	48	34	52	48	37	49	32	300	
P. Acc %	90	88	87	90	89	82	91		

Overall classification accuracy = 87.67%

Overall kappa statistics = 0.85

U. Acc., user's accuracy; P. Acc., producer's accuracy, see Table 9.1 for other abbreviations (number unit is in pixel)

was not necessarily the same as the sample size of the corresponding class derived from the supervised approach.

In this study, the GIS post-processing approach appeared to be the best approach. This approach showed an overall accuracy of 89.33% (Table 9.6), which is close to the overall accuracy (87.67%) of the fuzzy supervised approach (Table 9.5). The fuzzy approach dealt with mixed pixel problems and the heterogeneous representation of land surface features in residential and park areas in the city. The supervised and unsupervised approaches produced lower accuracies (83.67% (Table 9.4) and

Table 9.6 Error matrix of the GIS post-processing approach for land use/cover classification

	Reference data								
Classified data	UF	LG	PF	DF	FI	RP	WA	Total	U. Acc %
UF	47	0	0	0	0	0	0	47	100
LG	0	27	10	4	0	1	0	42	64
PF	0	1	46	0	0	1	0	48	96
DF	0	3	1	44	2	1	0	51	86
FI	0	0	0	0	33	0	1	34	97
RP	1	1	0	2	2	41	1	48	85
WA	0	0	0	0	0	0	30	30	100
Total	48	32	57	50	37	44	32	300	
P. Acc %	98	84	81	88	89	93	94		

Overall classification accuracy = 89.33%

Overall kappa statistics = 0.87

U. Acc., user's accuracy; P. Acc., producer's accuracy, see Table 9.1 for other abbreviations (number unit is in pixel)

75.33% (Table 9.3), respectively). Due to the various surface materials in the complex urban system, the unsupervised approach formed several class clusters in images, creating difficulties in interpretation. However, the unsupervised approach provided better insight to identify large space objects in the image (e.g., commercial complexes and industrial plants).

The kappa indices presented a somewhat clearer picture. The kappa coefficient shows that the GIS post-processing approach can reduce most of the errors during the classification process. The kappa for the approach (Table 9.6) was 0.87 (87% reduction of error), which is a bit better than the kappa for the fuzzy supervised approach of 0.85 (see, Table 9.5), with a difference of 1.66% in overall classification accuracy between them. Looking at the kappa, we observed a bigger difference between the supervised and unsupervised approaches, with kappa indices of 0.80 (Table 9.4) and 0.71 (Table 9.3), respectively, an increase of 8.34% in accuracy. This signifies that the supervised approach performed better than the unsupervised in mapping of urban area. The overall accuracy and kappa only represent an average result. It is still difficult to determine which approach projected refined mapping results for which surface materials at the class level.

GIS post-processing and fuzzy supervised approaches exhibited high (over 80%) producer's accuracies or low omission errors (Tables 9.5 and 9.6) in all classes; similar patterns were observed in user's accuracies (low commission error), except in the Lawn/Grass class. Urban forest, paddy field, business/industry and water classes have very high user's accuracies (over 89%) in both approaches. With the exception of paddy fields, the user's accuracies of these classes were also similar. Somewhat higher errors of commission occurred in paddy fields. Both the GIS post-processing and the fuzzy supervised approaches did best in extracting urban and natural land covers as both of their producer's and user's accuracies were high.

Because of the mixture of surface materials (i.e., roof tiles, concrete, asphalt and vegetation) in residential areas, the unsupervised (Table 9.3) and supervised (Table 9.4) approaches scored the user's and producer's accuracies poorly in the residential/parking/road class. Due to the capacity of dealing with heterogeneous surface features, mapping of such features by the fuzzy supervised approach improved greatly, showing over 82% producer's and user's accuracies in residential land use classifications. The GIS post-processing approach, with a user's accuracy of 85%, appeared to be slightly better than the fuzzy supervised approach (user's accuracy of 83%), an increase of 2% for extracting residential land use. In other words, for the GIS post-processing approach, the producer's accuracy for residential land use was 93%, an improvement of 11% over the fuzzy supervised approach. In both cases, the GIS post-processing approach seems superior for addressing the mapping issues of residential land uses. The user's accuracies show that the supervised approach estimates the lawn/grass class better than the other approaches. A high contrast is observed between the producer's and user's accuracies of the lawn/grass class in all approaches. In the more homogeneous land covers (i.e., paddy field, natural vegetation and water bodies), both the unsupervised and supervised approaches exhibited good producer's and user's accuracies. The unsupervised approach performed better than the supervised approach in clustering the paddy field, with 92% of user's and 83% of producer's accuracy; the supervised approach had 7% lower in user's and 1% higher in producer's accuracies.

In the supervised approach, despite a producer's accuracy of 91% for the business/industry class, there was actually 85% user's accuracy, which means at least 6% of the business/industry land use was classified wrong (Table 9.4). For the unsupervised approach, the producer's accuracy for this class was 75%, while 73% user's accuracy was actually business/industry land use, making a difference of 2% of wrong classification (Table 9.3). Here, the unsupervised classifier seems slightly better than the supervised classifier in dealing with big homogeneous parcels characterized by business/industry land use. The supervised approach exhibited a lower difference between the user's and producer's accuracies for urban forest, paddy fields and water, which compared well even with the fuzzy and GIS post-processing approaches.

9.5 Conclusions

The urban landscape of Tsukuba city is diverse and complex, comprising both homogeneous and heterogeneous surface features, causing problems of spectral variability in the satellite image data. Assimilating spectral and radiometric properties of image data is more important than spatial resolution in improving computer-assisted land use and land cover classification accuracy. In order to improve mapping accuracies from remotely sensed data, relying on only one approach is not enough. In this study, we examined four approaches (unsupervised, supervised, fuzzy supervised and GIS post-processing) and their accuracies to extract land use and land cover information using an AVNIR2 sensor image of the ALOS satellite. The combination of fieldwork, satellite image data and analysis techniques really improved mapping accuracies. We found that the spatial statistics of land use/cover derived from remotely sensed images mostly depend on the adaptation of mapping approaches. The accuracy assessment report showed that the GIS post-processing approach can predict land use and land cover of the complex urban environment more accurately. The urban woodland, water, business/industry and paddy field mapped by the GIS post-processing were observed more accurately compared to the other approaches. The fuzzy supervised approach presented slightly more accurate results than the traditional supervised approach. This method also shows great potential for dealing with heterogeneous surface features in urban residential areas showing very low difference in the errors of omission and commission. The supervised approach exhibited a lower difference between the user's and producer's accuracies for the vegetation, paddy field and water classes compared to other approaches. In fact, the unsupervised approach greatly helped us understand the land cover structure and identify homogeneous clusters in the imagery. Each classification approach has its special characteristics and benefits to analyse remotely sensed images and mapping of the earth surface. This study explored the strengths of the four approaches for mapping urban area from the AVNIR2 sensor of ALOS, which may significantly help urban planners understand and interpret complex urban characteristics more precisely, as they often cite problems of mapping techniques and spatial resolution (Carlson, 2003).

References

Anderson, J. R., Hardy, E. E., Roach, J. T., & Witmer, R. E. (1976). *A land use and land cover classification system for use with remote sensor data*. Washington, DC: US Geological Survey Professional Paper No. 964.

Carlson, T. N. (2003). Applications of remote sensing to urban problems. *Remote Sensing of Environment*, *86*, 273–274.

Carlson, T. N., & Arthur, T.S. (2000). The impact of land use – land cover changes due to urbanization on surface microclimate and hydrology: A satellite perspective. *Global and Planetary Change*, *25*, 49–65.

Carvalho, J., Soares, A., & Bio, A. (2006). Improving satellite images classification using remote and ground data integration by means of stochastic simulation. *International Journal of Remote Sensing*, *27*, 3375–3386.

Congalton, R. G. (1991). A review of assessing the accuracy of classifications of remotely sensed data. *Remote Sensing of Environment*, *37*, 35–46.

Gatrell, J. D., & Jensen, R. R. (2008). Sociospatial applications of remote sensing in urban environments. *Geography Compass*, *2*, 728–743.

GSI (2008). *Detailed digital information: 10 meter grid land use*. Geographical Survey Institute. Retrieved July 1, 2008, from http://www.gsi.go.jp/ENGLISH/index.html

JAXA (2006). *About ALOS*. Earth Observation Research Center. Retrieved December 1, 2006, from http://www.eorc.jaxa.jp/ALOS/about/about_index.htm

Jensen, J. R. (2005). *Introductory digital image processing: A remote sensing perspective*. Upper Saddle River, NJ: Prentice Hall.

Johnsson, K. (1994). Segment-based land-use classification from SPOT satellite data. *Photogrammetric Engineering and Remote Sensing*, *60*, 47–53.

Lee, J. Y., & Warner, T. A. (2006). Segment based image classification. *International Journal of Remote Sensing*, *27*, 3403–3412.

Leica Geosystems (2005). *ERDAS Field Guide*. Norcross, Georgia: Leica Geosystems Geospatial Imaging, LLC.

Lillesand, T. M., Kiefer, R. W., & Chipman, J. W., (2008). *Remote sensing and image interpretation*. New York: Wiley.

Lo, C. P., & Choi, J. (2004). A hybrid approach to urban land use/cover mapping using Landsat 7 Enhanced Thematic Mapper Plus (ETM+) images. *International Journal of Remote Sensing*, *25*, 1687–2700.

Maktav, D., Erbek, F. S., & Jurgens, C. (2005). Remote sensing of urban areas. *International Journal of Remote Sensing*, *26*, 655–659.

Miller, R. B., & Small, C. (2003). Cities from space: Potential applications of remote sensing in urban environmental research and policy. *Environmental Science & Policy*, *6*, 129–137.

Nangendo, G., Skidmore, A. K., & Oosten, H. (2007). Mapping East African tropical forests and woodlands – A comparison of classifiers. *ISPRS Journal of Photogrammetry and Remote Sensing*, *61*, 393–404.

Ozkan, C., & Erbek, F. S. (2005). Comparing feature extraction techniques for urban land-use classification. *International Journal of Remote Sensing*, *26*, 747–757.

Prenzel, B., & Treitz, P. (2005). Comparison of function- and structure-based schemes for classification of remotely sensed data. *International Journal of Remote Sensing*, *26*, 543–561.

Shalaby, A., & Tateishi, R. (2007). Remote sensing and GIS for mapping and monitoring land cover and land-use changes in the Northwestern coastal zone of Egypt. *Applied Geography*, *27*, 28–41.

Statistics Bureau (2008). Population, population change, area and population density. *Director-General for Policy Planning and Statistical Research and Training Institute*.

Thapa, R. B., & Murayama, Y. (2008). Land evaluation for peri-urban agriculture using analytical hierarchical process and geographic information system techniques: A case study of Hanoi. *Land Use Policy*, *25*, 225–239.

Thapa, R. B., & Murayama, Y. (2009a). Urban mapping, accuracy, & image classification: A comparison of multiple approaches in Tsukuba City, Japan. *Applied Geography*, *29*, 135–144.

Thapa, R. B., & Murayama, Y. (2009b). Examining spatiotemporal urbanization patterns in Kathmandu valley, Nepal: Remote sensing and spatial metrics approaches. *Remote Sensing*, *1*, 534–556.

Wang, F. (1990). Improving remote sensing image analysis through fuzzy information representation. *Photogrammetric Engineering and Remote Sensing*, *56*, 1163–1169.

Zhang, J., & Foody, G. M. (2001). Fully-fuzzy supervised classification of sub-urban land cover from remotely sensed imagery: Statistical and artificial neural network approaches. *International Journal of Remote Sensing*, *22*, 615–628.

Chapter 10
Urban Dynamics Analysis Using Spatial Metrics Geosimulation

Yaolong Zhao and Yuji Murayama

10.1 Introduction

The past century was such a period of rapid urbanization all over the world, in which most people quickly congregated in the urban areas. The urban population in the world was estimated at 2.4 billion in 1995 and is expected to double by 2025 (Antrop, 2000). While the urban areas taken by the huge population account for only 2% of the Earth's land surface (Grimm, Grove, Pickett, & Redman, 2000), land-use and land-cover changes caused by the rapid urbanization have greatly impacted the local (McKinney, 2006; Paul & Meyer, 2001; Lin & Ho, 2003) and global environmental changes (Grimm et al., 2000; Lambin et al., 2001). Therefore, to effectively understand the spatial processes of urbanization and explore the extent of future urban land-use changes, has attracted many scientists' attention coming from different disciplines (Alberti & Waddell, 2000; Batty, 1989; 1994).

Cities are among the most complex structures created by the human societies. Their complex system is characterized by the complex patterns of land-use. However, the phenomena are not easily experimented with on the ground. Realistic but synthetic computer simulations by modeling the complex land-use dynamics based on GIS can be built as a laboratory for exploring ideas and plans that we would not otherwise be able to effect on the ground (Clarke, Hoppen, & Gaydos, 1997; White, Engelen, & Uljee, 1997). This method represents the implementation of links between land-use pattern and land-use process. Spatial metrics, which come from landscape field and are used to characterize the spatial pattern and composition of landscapes, have been argued as one impactful tool to link urban land-use pattern and dynamic process when coupled with remote sensing (Parker,

---

Y. Zhao (✉)
School of Geography, South China Normal University, Guangzhou, Guangdong, PR China
e-mail: yaolongzhao@gmail.com

This chapter is improved from "Yaolong Zhao and Yuji Murayama (2006), Urban dynamics analysis using spatial metrics: A case study of Yokohama city, Tsukuba Geoenvironmental Sciences, 2, 9–18", with permission from University of Tsukuba, Geoenvironmental Science Program.

Y. Murayama, R.B. Thapa (eds.), *Spatial Analysis and Modeling in Geographical Transformation Process*, The GeoJournal Library 100, DOI 10.1007/978-94-007-0671-2_10,

Evans, & Meretsky, 2001; Herold, Goldstein, & Clarke, 2003; 2005). Especially, Herold, Couclelis, and Clarke (2005) have discussed the role of spatial metrics in the analysis and modeling urban growth. However, most of the literature, which emphasizes the importance of spatial metrics in linking between pattern and process, consider the urban area as one object, and generally classify the study area into binary categories of land-use – built-up area and non-built-up area. As urban area is the dynamic composition of a variety of land-use categories, such as industrial, residential, commercial and so on, it should be essential to analyze the dynamic patterns of urban land-use at high-resolution scale and multi-classification system. Only a few scientists have discussed this kind of issue. This research focuses on this topic and tries to interpret the differences in analysis results at different spatial scales.

In the next section, we briefly review the history of spatial metrics and the current status in urban dynamics analysis. Section 10.3 presents the study area and the data set. The limitation of remote sensing in high-resolution urban analysis is discussed. We emphasize the significance of the data set "Detailed Digital Information (10 m Grid Land Use) of Metropolitan Area" (DDIMA10m) of Tokyo in empirical study for high- resolution urban analysis. The analysis and results are described in Section 10.4. Section 10.5 presents some concluding remarks.

10.2 Spatial Metrics

Spatial metrics come from the concept of landscape metrics which were developed in the late 1980s and incorporated measures from both information theory and fractal geometry (Mendelbrot, 1983; Shannon & Weaver, 1964) based on a categorical, patch-based representation of the landscape. Patches are defined as homogenous regions for a specific landscape property of interest, such as "industrial land", "park" or "high-density residential area". Landscape metrics are used to quantify the spatial heterogeneity of individual patches, all patches belonging to a common class, and the landscape as a collection of patches. When applied to multi-scale or multi-temporal datasets, the metrics can be used to analyze and describe the changes in the degree of spatial heterogeneity (Dunn, Sharpe, Guntensbergen, Stearns, & Yang, 1991; Wu, Jelinski, Luck, & Tueller, 2000). In the application to urban area field, Herold et al. (2003) pointed out the approaches and assumptions might be more generally described as "spatial metrics".

The interest in using spatial metric concepts for the analysis of urban environments is starting to grow. In 1997, Geoghegan et al. firstly explored spatial metrics in modeling land and housing values (Geoghegan, Wainger, & Bockstael, 1997). Alberti and Waddell (2000) substantiate the importance of spatial metrics in urban modeling. They propose specific spatial metrics to model the effects of the complex spatial pattern of urban land-use and land-cover on social and ecological processes. Parker et al. (2001) summarize the usefulness of spatial metrics with respect to a variety of urban models and argue for the contribution of spatial metrics in helping link economic processes and patterns of land-use. Herold et al. (2003) proposed the integration approach of remote sensing and spatial metrics in spatiotemporal analysis and modeling of urban growth. In 2005, Herold et al. systematically analyzed the role of spatial metrics in the analysis and modeling urban growth and argued that spatial metrics definitely deserve a place in the urban dynamics research agenda.

While many literatures have discussed the usefulness of spatial metrics in urban research, most of them have focused on just two categories of landscape heterogeneity: built-up area and non-built-up area. We assume that it should also be very useful for high resolution analysis of urban land-use changes. Considering the rapid urban growth and the availability of high resolution land-use data set, Yokohama city is a data-rich area to study the dynamics of spatial and temporal urban land-use change.

FRAGSTATS, a public domain spatial metrics program, was developed in the mid-1990s and has been continuously improved (McGarigal, Cushman, Neel, & Ene, 2002). FRAGSTATS provides a large variety of metrics at class, patch and landscape levels. Table 10.1 describes the subset of available metrics used in this research. A more detailed description including the specific mathematical equations of all of the metrics can be found in McGarigal et al. (2002).

Class area (CA) is the measure of the area of all the categories of urban land-use. Change of CA across time can present the dynamic changes of urban land-use structure. The number of patches (NP) metric quantifies the number of individual areas for all the categories of urban land-use. The dynamics of NP coupled with CA can describe the fragmentation of one category of urban land-use. The edge density (ED) is a measure of the total length of the edge of land-use patches. In certain extent, it can present the spatial complexity of the urban land-use pattern. The largest patch index (LPI) describes the percentage of total landscape area comprised by the largest patch. As such, it is a simple measure of dominance and presents the extent of the aggregation of one category of urban land-use. The mean nearest neighbor distance (ENNMN) represents the average minimum distance between the individual urban land-use category blocks. Hence, it is a measure of the extent of disperses.

The fractal dimension describes the complexity and the fragmentation of a patch by a perimeter-area proportion. The value of the fractal dimension falls into the interval between 1 and 2. Low values are derived when a patch has a compact rectangular form with a relatively small perimeter relative to the area. If the patches are more complex and fragmented, the perimeter increases and yields a higher fractal dimension. The area weighted fractal dimension improves the measure of class patch fragmentation because the structure of smaller patches is often determined more by image pixel size than by characteristics of natural or manmade features found in the landscape (Milne, 1991).

The splitting index (SPLIT) is based on the cumulative patch area distribution and is interpreted as the effective mesh number or number of patches with a constant patch size when the corresponding patch type is subdivided into S patches, where S is the value of the splitting index. SPLIT increase as the focal patch type is increasingly reduced in area and subdivided into smaller patches. All the metrics were calculated for each land-use category using the software of FRAGSTATS.

Table 10.1 Spatial metrics used in this study, adopted from McGarigal et al. (2002)

Metric	Description	Units	Range
CA – class area	CA equals the sum of the areas (m^2) of all patches, divided by 10,000 (to convert to hectares).	Hectares	CA>0, no limit
NP – number of patches	NP equals the number of patches in the landscape.	None	NP>=1, no limit
ED – edge density	ED equals the sum of the lengths (m) of all edge segments involving the patch type, divided by the total landscape area (m^2), multiplied by 10,000 (to convert to hectares).	Meters per hectare	ED>=0, no limit
LPI – largest patch index	LPI equals the area (m^2) of the largest patch of the corresponding patch type divided by total area (m^2), multiplied by 100 (to convert to a percentage).	Percent	0<LIP<=100
ENNMN – Euclidian mean nearest neighbor distance	ENNMN equals mean value of the distance (m) over all patches to the nearest neighboring urban patch, based on shortest edge-to-edge distance from cell center to cell center.	Meters	ENNMN>0, No limit
FRACAM – area weighted mean patch fractal dimension	Area weighted mean value of the fractal dimension values of all patches, the fractal dimension of a patch equals two times the logarithm of patch perimeter (m) divided by the logarithm of patch area (m^2); the perimeter is adjusted to correct for the raster bias in perimeter.	None	1<=FRACAM<=2
SPLIT – splitting index	SPLIT equals the total landscape area (m^2) squared divided by the sum of patch area (m^2) squared, summed across all patches of the corresponding patch type.	None	1<=SPLIT<=number of cells in the landscape area squared
CONTAG – contagion	CONTAG measures the overall probability that a cell of a patch type is adjacent to cells of the same type.	Percent	0<=CONTAG<=100
SHDI – Shannon's diversity index	SHDI equals minus the sum, across all patch types, of the proportional abundance of each patch type multiplied by that proportion.	Information	0<=SHDI

10.3 Study Area and Data Set

The study area includes all the 18 administration wards of Yokohama city with about 434 km^2 (Fig. 10.1). We used the data set of DDIMA10m of Tokyo, which was released in 1998 by the Geographical Survey Institute.

Not as land cover defined by Barnsley et al. (2001) as "the physical materials on the surface of a given parcel of land (e.g. grass, concrete, tarmac, water)", land-use refers to "the human activity that takes place on, or makes use of that land (e.g. residential, commercial, industrial)". Remote sensing techniques have already showed their value in mapping urban areas, and as data sources for the analysis and modeling of urban growth and land-use change (Batty & Howes, 2001; Clarke, Parks, & Crane, 2002; Treitz & Rogan, 2004). Remote sensing provides spatially consistent data sets that cover large areas with both high spatial detail and high temporal frequency. Batty and Howes (2001) have emphasized the importance of remote sensing as a "unique view" of the spatial and temporal dynamics of the processes of urban growth and land-use change. However, land-use is an abstract concept, constituting a mix of social, cultural, economic and policy factors, which has little physical importance with respect to reflectance properties, and hence has a limited relationship to remote sensing (Treitz & Rogan, 2004). That may be one of the reasons why the study area was always divided in many literatures into two types – built-up area and non-built-up area – only using remote sensing technique, as such detailed land-use information cannot be detected easily. The data set DDIMA10m of Tokyo provides abundant and detailed urban land-use classifications including a variety of socio-economic information.

The study area was originally classified into 17 land-use categories. We did not alter the classification system for urban area, just grouped the non-built-up area into one category in a multi-category system. To compare the results of the analysis

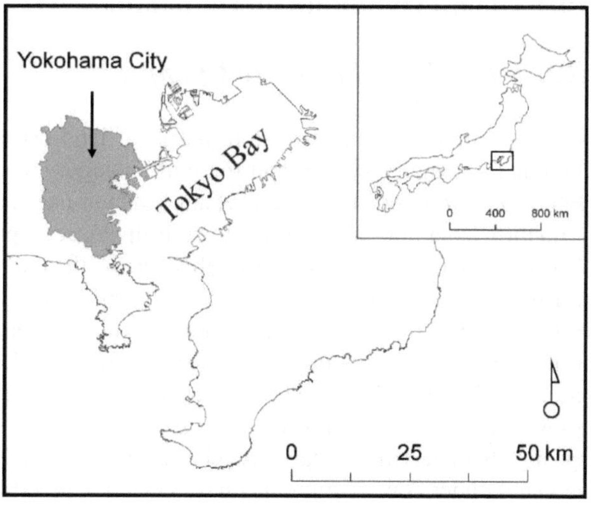

Fig. 10.1 Study area

Table 10.2 Land-use classification systems

Categories in the original data set	Categories in the multi-category system	Categories in the binary-category system
A. Woods B. Paddy field C. Dry fields	1. Non built-up	I. Non-built-up
D. Under construction	2. Under construction	II. Built-up
E. Vacant	3. Vacant	
F. Industrial	4. Industrial	
G. Low storey residential	5. Low storey residential	
H. Densely developed low storey residential	6. Densely developed low storey residential	
I. Medium and high storey residential	7. Medium and high storey residential	
J. Commercial and service industrial	8. Commercial	
K. Road	9. Road	
L. Park	10. Park	
M. Public	11. Public	
N. Special	12. Special	
O. River, lake and pond P. Sea	13. Water	III. Water

between macro-scale and micro-scale, we arranged a binary-category system for this area. All the classification systems are shown in Table 10.2. Figure 10.2 presents the land-use pattern of Yokohama city in 1974 and 1994, and the land-use change patterns from 1974 to 1994 at 5 years increment in binary-category system.

10.4 Urban Dynamics of Yokohama City

10.4.1 Land Use Structure Changes

We calculated the transformation matrix of land-use from 1974 to 1994 for both the binary-category system and the multi-category system (Tables 10.3 and 10.4). Table 10.3 shows that the general change of land-use in this period exhibits the transformation from non-built-up area to built-up area comprising of 5204.77 ha or about 74.6% of the total changes. During this period, the land-use mainly took the characteristic of urban growth. More importantly, the change of land-use not only took the transformation from non-built-up area to built-up area, but also from built-up area to non-built-up area which constitutes 825.07 ha or 12.8% of the total changes. It indicates the self-adjustment or somewhat decay of the city at certain places. Urban growth also occurred at former water areas, especially the sea, presenting one of the special characteristics of land-use change in this area. Moreover, some vacant areas and land under construction may have been reclaimed for agricultural use.

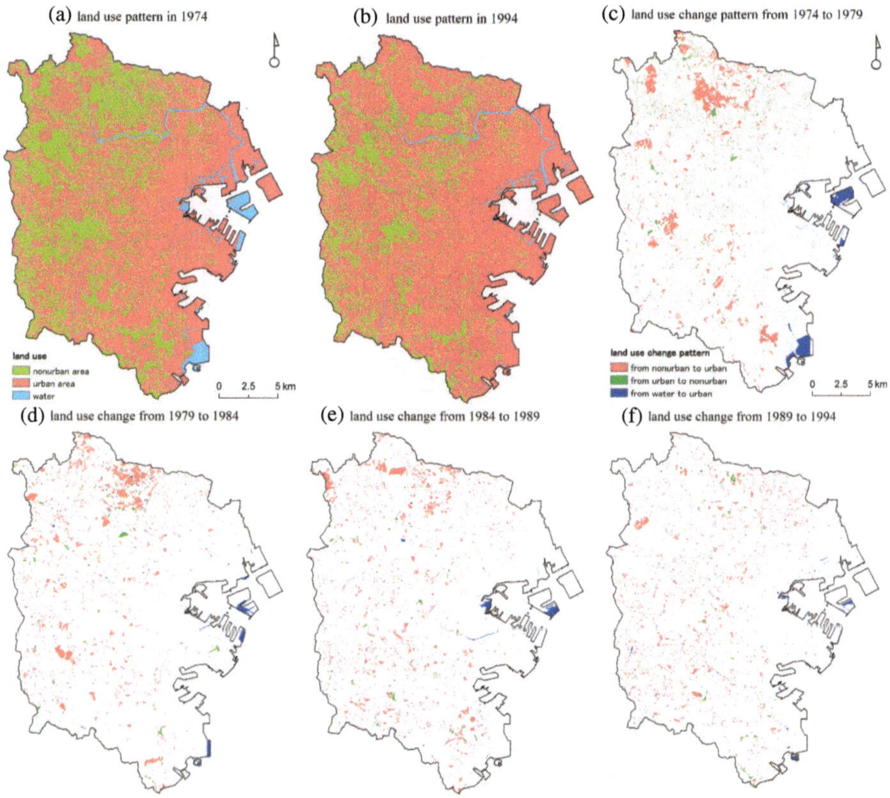

Fig. 10.2 Land-use of Yokohama city in 1974 (**a**) and 1994 (**b**), and land-use change pattern from 1974 to 1979 (**c**), 1979 to 1984 (**d**), 1984 to 1989 (**e**), and 1989 to 1994 (**f**) in the binary-category system

Table 10.3 Transformation matrix of land-use in the binary-category system from 1974 to 1994

	Land-use in 1994 (ha)		
Land-use in 1974 (ha)	Non built-up	Built-up	Water
Non built-up	8928.74	5204.77	30.35
Built-up	825.07	26921.18	20.17
Water	5.77	894.48	733.09

The growth of built-up area came from the interaction and competition of sub-categories of urban area. The analysis above could not show the significance. Table 10.4 illustrates the dynamics among the sub-categories as well as the dynamics between the non-built-up area and sub-categories of the urban area. We can find that most of transformation of land-use from non-built-up area to built-up area took

Table 10.4 Transformation matrix of land-use in the multi-category system from 1974 to 1994

Land-use in 1974	Land-use in 1994: Non built-up	Under construction	Vacant	Industrial	Low-resi[a]	Densely-resi[b]	Medium-resi[c]	Commercial	Road	Park	Public	Special	Water
	(ha.)	(ha.)	(ha.)	(ha.)	(ha.)	(ha.)	(ha.)	(ha.)	(ha.)	(ha.)	(ha.)	(ha.)	(ha.)
Non built-up	8928.74	426.82	1411.3	85.87	1060.37	19.59	367.92	276.25	811.77	254.36	489.71	0.81	30.35
Under construction	129.52	64.43	240.01	275.9	330.12	2.79	152.27	119.37	334.93	114.23	248.18	0.02	1.02
Vacant	212.23	36.64	1351.38	90.65	895.17	23.62	140.92	216.29	243.49	88.03	218.23	0.08	5.55
Industrial	11.36	34.16	58.25	1665.45	15.77	0.37	54.44	86.93	32.27	20.4	25.29	0.01	2.19
Low-resi[a]	125.92	38.56	227.54	17.78	7107.19	4.38	96.67	70.49	131.97	16.77	25.38	0.09	2.75
Densely-resi[b]	8.89	1.34	9.14	0.88	24.14	407.11	3.68	4.61	5.26	0.66	1.25	0.03	0.04
Medium-resi[c]	10.43	6.18	11.03	0.46	10.4	0.87	831.75	4.22	7.77	2.22	1.88	0.00	0.02
Commercial	43.75	16.1	71.95	16.03	86.14	7.81	36.73	1673.25	54.3	7.9	29.54	0.09	1.44
Road	183.29	29.29	144.58	28.77	499.96	21.75	71.86	164.34	3621.92	20.29	38.67	0.31	5.33
Park	27.46	2.79	11.38	2.2	12.56	0.73	9.22	9.21	46.33	1069.45	20.19	0.02	0.7
Public	63.37	17.38	88.91	11.46	43	1.95	20.75	56.79	82.47	17.44	2146.55	0.11	1.13
Special	8.85	3.07	30.52	0.01	6.43	0.05	8.2	8.98	11.48	5.58	7.99	274.61	0.00
Water	5.77	105.11	51.41	121.72	17.65	0.47	10.03	158.77	154.08	89.23	189.73	0.28	733.09

[a]Low-resi stands for low storey residential
[b]Densely-resi stands for densely low storey residential
[c]Medium-resi stands for medium and high storey residential

place within vacant and low storey residential and very little in the industrial. This shows the demand for open space in the urban area to create a living charm in Yokohama as the suburb of the Tokyo metropolitan area.

10.4.2 Analysis of Spatial and Temporal Urban Land-Use Pattern

We selected land-use categories of industrial, residential (low storey residential, densely developed low storey residential, and medium and high storey residential), and commercial in order to catch the main characteristics of the urban area. Figures 10.3, 10.4, 10.5 and 10.6 present the urban dynamics in terms of the selected spatial metrics.

Figure 10.3 shows the histogram of the dynamics of CA for each land-use category. It reflects the change in the urban land-use structure. Built-up area in the binary-category system kept increasing over the whole period, as discussed in the previous subsection. For the multi-category system, although the area of all land-use categories increased in this period, the increase rates were different.

The spatial diagrams of metrics of NP, LPI, ED, FRACAM, and ENNMN, for different land-use categories are shown in Fig. 10.4. Figure 10.4a shows that although the area of all the land-use categories increased, the trends of the NP value were different. In the multi-category system, the value of NP for low storey residential and commercial categories decreased rapidly from 1974 to 1979, and then increased gradually till 1994. It means that these land-use categories grew dispersively from 1974 to 1979 and got compact gradually after 1979. The value of NP for other land-

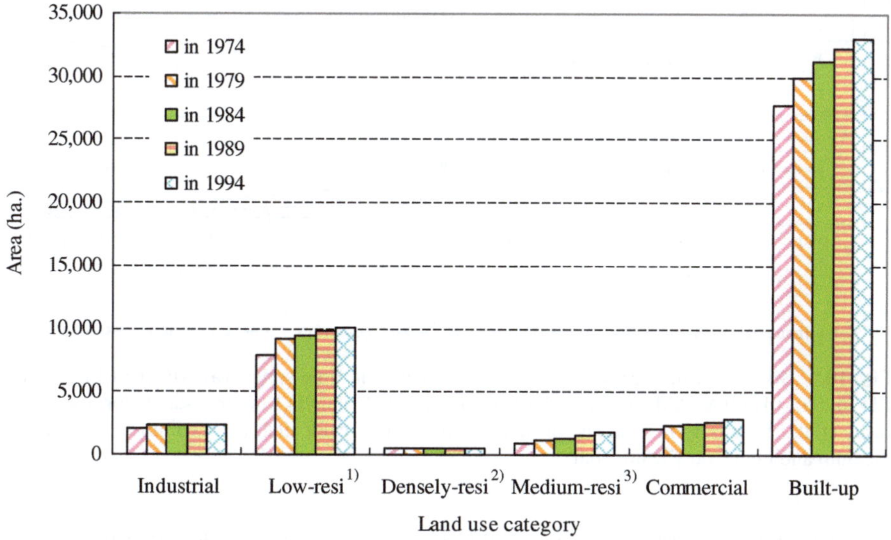

Fig. 10.3 CA by land-use category (1974–1994). Note: (1) Low-resi stands for low storey residential, (2) densely-resi stands for densely low storey residential, (3) medium-resi stands for medium and high storey residential

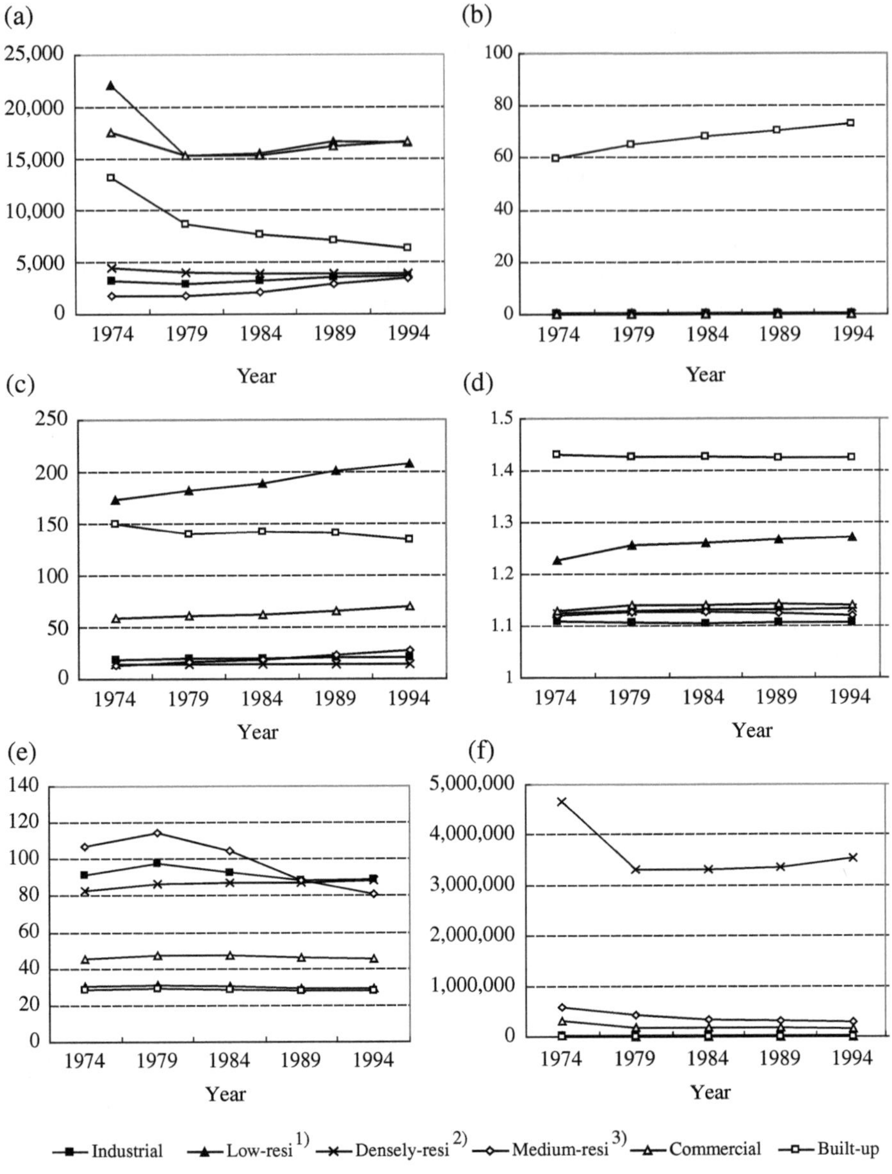

Fig. 10.4 Six spatial metrics of each land-use category in time series. (**a**) NP; (**b**) LPI (%); (**c**) ED (m/ha); (**d**) FRACAM; (**e**) ENNMN (m); (**f**) SPLIT; Note: (1) Low-resi stands for low storey residential, (2) densely-resi stands for densely low storey residential, (3) medium-resi stands for medium and high storey residential

use categories did not obviously change in this period as the increases in the area were not so much. The value of NP for built-up area in binary-category system has decreased all along, indicating that urban growth mainly took place at the fringe of urban area and got close to the urbanized area.

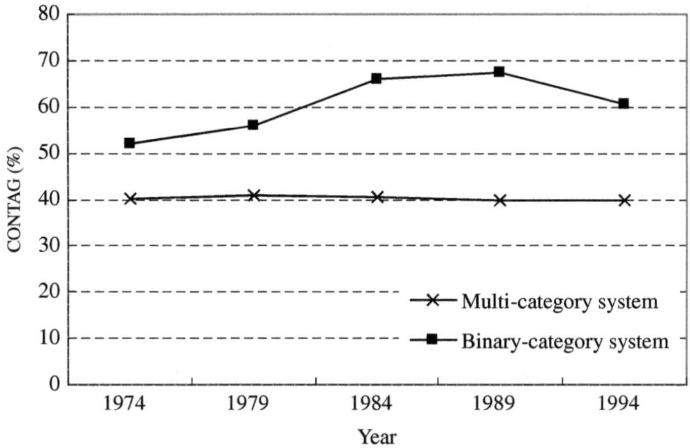

Fig. 10.5 CONTAG of two land-use classification systems in time series

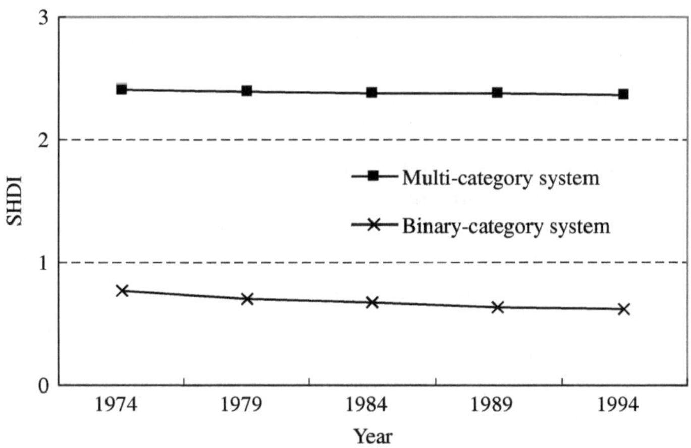

Fig. 10.6 SHDI of two land-use classification systems in time series

The LPIs of all of the land-use categories in multi-category system have not changed in this period (Fig. 10.4b). This implies that the growth of these land-use categories did not show the characteristics of dominance. The LPI of built-up area in binary-category system increased gradually however, indicating the centralization of urban growth. The land-use categories in multi-category system did not show this kind of characteristics.

The dynamics of ED and FRACAM describe the complexity of urban landscapes (Fig. 10.4c). Low storey residential increased gradually in both the value of ED and FRACAM, indicating the increasing complexity. Values of ED and FRACAM metrics of other land-use categories did not change significantly. Figure 10.4d illustrates that the value of ENNMN for medium and high storey residential had increased a little from 1974 to 1979 then declined gradually while the value of other land-use categories only changed slightly. It indicates that from 1974 to 1979, medium and high storey residential was settled away from the existing place of these categories due to city planning or developers' decision. From 1979, new developed places of these categories turned close to the existing ones. This phenomenon shows the characteristics of aggregation of urban land-use changes in this period. The value of SPLIT for densely low storey residential, which was smaller than the other land-use categories in area (Fig. 10.3), was greatly higher than that of other land-use categories. Coupled with Fig. 10.4a, b, we can see that the densely low storey residential grew mostly at neighborhood area from 1974 to 1979. This has caused the rapid decline of SPLIT metrics. The values of SPLIT for built-up area in binary-category system were nearly 2; this was quite small. It substantiates the characteristic of aggregation in the process of urban growth as shown using NP and LPI metrics.

CONTAG and SHDI can be used to investigate the general characteristics of the whole landscape. The values of CONTAG for binary-category system were bigger than that of multi-category system in time series as shown in Fig. 10.5. This validates the function of CONTAG in representing the heterogeneity of landscape. In binary-category system, as the numbers of categories were not many and built-up area tended to aggregate, the area of most patches was large. In multi-category system, urban area was divided into more categories and became more heterogeneous. For binary-category system, from 1974 to 1989, the value of CONTAG had increased gradually, indicating that with urban growth the built-up area connected into bigger blocks and the landscape became homogeneous. From 1989 to 1994, the landscape of the city became more heterogeneous as the most of urban growth did not connect with existing built-up area. For multi-category system, the value of CONTAG has declined a little across the time series. It indicates that the growth of the urban land-use categories has dispersively occurred in non-built-up area and the landscape gradually became more heterogeneous.

The SHDI metric is a popular measure of diversity of landscape. From Fig. 10.6 we can find that the values of SHDI for multi-category system across the time series were larger than that of the binary-category system as the information in multi-category system was abundant. Nevertheless, the values of the SHDI metric in both systems seldom changed across the time series. It means that the SHDI metric is not sensitive in representing urban dynamics in the short period.

10.5 Conclusions

This study has presented a detailed analysis using spatial metrics to interpret urban dynamics at two scales of land-use classification in a case of Yokohama city. The results validated the effectiveness of spatial metrics in linking land-use pattern with land-use process in the detailed land-use classifications and the urban area as one category, as investigated in literatures (Parker et al., 2001; Herold et al., 2003). However, even for the same place, the characteristics of urban dynamics differ with land-use classification system reflecting the effect of spatial scale. This indicates that differences in understanding the process of urban dynamics exist at different scales and analyzing urban dynamics at multi-scale using spatial metrics would contribute to the comprehensive interpretation of urban dynamics, and improve the construction of spatial model for urban dynamics. More empirical case studies will be needed on this phenomenon.

Although spatial metrics have been applied in some cases of urban growth or sprawl analysis (Herold et al., 2003, 2005; Torrens, 2006), there are some fundamental problems which should be further discussed, such as selection of metrics described by Herold et al. (2005). This study analyzed urban dynamics at multi-categories system using multiple spatial metrics. The authors found that some metrics show similar characteristics in representing the process of urban land-use categories. For instance, both of ED and FRACAM, which are defined differently, can be applied to present the dynamics of the complexity of urban landscapes. Moreover, we found that the value of SHDI is not sensitive in representing urban dynamics for a short period. These findings can offer some useful information for discussion of selecting metrics.

It is generally recognized that in the field of landscape ecology, spatial pattern and spatial scale are inseparable in theory and reality. Spatial pattern occurs on different spatial scales, and spatial scale affects spatial pattern to be observed (Qi & Wu, 1996; Turner, O'Neill, Gardner, & Milne, 1989). Spatial scale also affects the interpretation of urban land-use pattern as well as land-use process. Zhao and Murayama (2005, 2006) have systematically investigated the effect of spatial scale on the result of urban land-use pattern analysis using spatial autocorrelation indices. In this study, the results came from the original 10 m $\times$ 10 m spatial resolution of land-use cells. How spatial scale affects the results of urban dynamic analysis in terms of spatial metrics would be a valuable extension to the current study.

Acknowledgements National Natural Science Foundation of China, No.40901090, 70863014; Foundation of Japan Society for the Promotion of Science (JSPS), No.19.07003; Talents Introduced into Universities Foundation of Guangdong Province of China, No. 2009–26.

References

Alberti, M., & Waddell, P. (2000). An integrated urban development and ecological simulation model. *Integrated Assessment, 1*, 215–227.

Antrop, M. (2000). Changing patterns in the urbanized countryside of Western Europe. *Landscape Ecology, 15*, 257–270.

Barnsley, M. J., Moller-Jensen, L., & Barr, S.L. (2001). Inferring urban land use by spatial and structural pattern recognition. In J. P. Donnay, M. J. Barnsley, & P. A. Longley (Eds.), *Remote sensing and urban analysis* (pp. 115–144). London: Taylor and Francis.

Batty, M. (1989). Urban modeling and planning: Reflections, retrodictions and prescriptions. In B. Macmillan (Ed.), *Remodeling geography* (pp. 147–169). Oxford: Basil Blackwell.

Batty, M. (1994). A chronicle of scientific planning: The Anglo-American modeling experience. *Journal of the American Planning Association, 60*, 7–12.

Batty, M., & Howes, D. (2001). Predicting temporal patterns in urban development from remote imagery. In J. P. Donnay, M. J. Barnsley, & P. A. Longley (Eds.), *Remote sensing and urban analysis* (pp. 185–204). London: Taylor and Francis.

Clarke, K. C., Hoppen, S., & Gaydos, L. (1997). A self-modifying cellular automaton model of historical urbanization in the San Francisco Bay area. *Environment and Planning B*, *24*, 247–261.

Clarke, K. C., Parks, B. O., & Crane, M. P. (2002). *Geographic information systems and environmental modeling*. Upper Saddle River, NJ: Prentice Hall.

Dunn, C. P., Sharpe, D. M., Guntensbergen, G. R., Stearns, F., & Yang, Z. (1991). Methods for analyzing temporal changes in landscape pattern. In M. G. Turner & R. H. Gardner (Eds.), *Quantitative methods in landscape ecology: The analysis and interpretation of landscape heterogeneity* (pp. 173–198). New York: Springer.

Geoghegan, J., Wainger, L. A., & Bockstael, N. E. (1997). Spatial landscape indices in a hedonic framework: An ecological economics analysis using GIS. *Ecological Economics*, *23*, 251–264.

Grimm, N. B., Grove, J. M., Pickett, S. T. A., & Redman, C. L. (2000). Integrated approaches to long-term studies of urban ecological systems. *Bioscience*, *50*, 571–584.

Herold, M., Couclelis, H., & Clarke, K. C. (2005). The role of spatial metrics in the analysis and modeling of urban land use change. *Computers, Environment and Urban Systems*, *29*, 369–399.

Herold, M., Goldstein, N. C., & Clarke, K. C. (2003). The spatiotemporal form of urban growth: Measurement, analysis and modeling. *Remote Sensing of Environment*, *86*, 286–302.

Lambin, E. F., Turner, B. L., Geist, H. J., Agbola, S. B., Angelsen, A., Bruce, J. W., et al. (2001). The causes of land-use and land-cover change: Moving beyond the myths. *Global Environmental Change*, *11*, 261–269.

Lin, G. C. S., & Ho, S. P. S. (2003). China's land resources and land-use change: Insights from the 1996 land survey. *Land Use Policy*, *20*, 87–107.

McGarigal, K., Cushman, S. A., Neel, M. C., & Ene, E. (2002). *Spatial pattern analysis program for categorical maps*. Amherst, MA: University of Massachusetts

McKinney, M. L. (2006). Urbanization as a major cause of biotic homogenization. *Biological Conservation*, *127*, 247–260.

Mendelbrot, B. B. (1983). *The fractal geometry of nature*. New York: W.H. Freeman and Company.

Milne, B. T. (1991). Lessons from applying fractal models to landscape patterns. In M. G. Turner & R. H. Gardner (Eds.), *Quantitative methods in landscape ecology: The analysis and interpretation of landscape heterogeneity* (pp. 199–235). New York: Springer.

Parker, D. C., Evans, T. P., & Meretsky, V. (2001). *Measuring emergent properties of agent-based landuse/landcover models using spatial metrics*. Proceedings of the 7th annual conference of the international society for computational economics, Yale University.

Paul, M. J., & Meyer, J. L. (2001). Streams in the urban landscape. *Annual Review of Ecology Systematics*, *32*, 333–365.

Qi, Y., & Wu, J. G. (1996). Effects of changing spatial resolution on the results of landscape pattern analysis using spatial autocorrelation indices. *Landscape Ecology*, *11*, 39–49.

Shannon, C., & Weaver, W. (1964). *The mathematical theory of communication*. Urbana: University of Illinois Press.

Torrens, P. M. (2006). Simulating sprawl. *Annals of the Association of American Geographers*, *96*, 248–275.

Treitz, P., & Rogan, J. (2004). Remote sensing for mapping and monitoring land-cover and land-use change-an introduction. *Progress in Planning*, *61*, 269–279.

Turner, M. G., O'Neill, R. V., Gardner, R. H., & Milne, B. T. (1989). Effects of changing spatial scale on the analysis of landscape pattern. *Landscape Ecology*, *3*, 153–162.

White, R., Engelen, G., & Uljee, I. (1997). The use of constrained cellular automata for high-resolution modeling of urban land-use dynamics. *Environment and Planning B*, *24*, 323–343.

Wu, J., Jelinski, E. J., Luck, M., & Tueller, P. T. (2000). Multiscale analysis of landscape heterogeneity: Scale variance and pattern metrics. *Geographic Information Sciences*, *6*, 6–19.

Zhao, Y., & Murayama, Y. (2005). Effect characteristics of spatial resolution on the analysis of urban land-use pattern: a case study of CBD in Tokyo using spatial autocorrelation index. In Y. Murayama, & G. Du (Eds.), *Cities in global perspective: Diversity and transition* (pp. 585–594). Tokyo: IGU Urban Commission.

Zhao, Y., & Murayama, Y. (2006). Effect of spatial scale on urban land-use pattern analysis in different classification systems: An empirical study in the CBD of Tokyo. *Theory and Applications of GIS, 14*, 29–42.

Chapter 11
Modeling Deforestation Using a Neural Network-Markov Model

Duong Dang Khoi and Yuji Murayama

11.1 Introduction

Three well-known global changes are increasing carbon dioxide in the atmosphere, alterations in the biochemistry of the global nitrogen cycle and continuing land-use/land-cover change (Vitousek, 1994). Land-use/cover change (LUCC) generates many environmental consequences globally and locally, such as the release of green-house gases, the loss of biodiversity and the sedimentation of lakes and streams (Walker, 2004). In particular, it is recognized as the major driver of the loss of biodiversity and ecosystem services (Sala et al., 2000). The effect of LUCC on bio-diversity may be greater than climate change, nitrogen deposition, biotic exchange, and elevated carbon dioxide concentration at the global scale (Sala et al., 2000). Deforestation is known as one of the most important elements in LUCC. Large-scale deforestation is occurring in the tropical forests, which contain most of the species in the world (Myers, Mittermeier, Mittermeier, da Fonseca, & Kent, 2000). Globally, deforestation has been occurring at an alarming rate of 13 million hectares per year (FAO, 2006). In Vietnam, two-thirds of the territory was primary forest until the mid-20th century (Poffenberger & Nguyen, 1998). Though forest cover in the country as a whole was 40.7% in 1943, it declined to 27.7% by 1993 (Do Dinh Sam, 1994). Primary forest was deforested to its lowest levels in the late 1980s and early 1990s (Meyfroidt & Lambin, 2008). Vietnam's deforestation rate was the highest among low-income countries over the period from 1965 to 1989 (World Bank, 1992). This trend still continued for the period from 1990 to 2005. The primary forest area per total forest area for the entire country declined from 4.1% in 1990 to 0.7% in 2005 (FAO, 2006). However, from the mid-1990s until now, there has been an increase in reforestation across the country (Meyfroidt & Lambin, 2008).

D.D. Khoi (✉)
Division of Spatial Information Science, Graduate School of Life and Environmental Sciences, University of Tsukuba, Tsukuba, Ibaraki, Japan
e-mail: khoi_tn@yahoo.com

This chapter is improved from "Duong Dang Khoi and Yuji Murayama (2010), Forecasting areas vulnerable to forest conversion in the Tam Dao National Park Region, Vietnam, Remote Sensing, 2, 1249–1272".

Y. Murayama, R.B. Thapa (eds.), *Spatial Analysis and Modeling in Geographical Transformation Process*, The GeoJournal Library 100, DOI 10.1007/978-94-007-0671-2_11,

Deforestation not only reduces forest area but also alters landscape configuration. Therefore, protected areas should be established to maintain the large, contiguous areas of forests for the protection of threatened species. Globally, 11.2% of the total forest area had been designated for the conservation of biological diversity in 2005 (FAO, 2006). In Vietnam, protected areas were established in most of the representative ecological zones for the period from 1995 to 2005. The country's protected areas now account for 14.7% of the total forest area (FAO, 2006). Many protected areas in the country are experiencing forest changes (ICEM, 2003). The management of the remaining forests within protected areas is very difficult to achieve because the livelihoods of local residents in the surrounding areas often heavily depend on agriculture and the extraction of forest products (TDMP, 2005). From a protected area management perspective, there is a need to identify the areas vulnerable to forest conversion in order to prioritize conservation efforts. One way to achieve this identification is to use remote sensing data and spatial models to map forest change patterns. Satellite remote sensing plays a key role in mapping and predicting forest changes (Linkie, Smith, & Leader-Williams, 2004; Giriraj, Irfan-Ullah, Murthy, & Beierkuhnlein, 2008). Satellite imagery provides an accurate measure of forest cover and deforestation (Turner et al., 2003). Changes in land use, derived from remotely sensed data, can be related to landscape and location attributes. A model can be established to describe the relationship between the dependent variable (forest cover change) and independent location variables (Lambin, 1997). Then, the model can be used for predicting the spatial patterns of forest cover changes.

Tam Dao National Park (TDNP) is a protected area in Vietnam. It contains the last remaining primary forest. It is endowed with rich biodiversity and is known to host a number of rare and endemic animal species. Yet, the park has been experiencing considerable forest changes due to population pressure in the surrounding areas. As a result, several species are in danger. For example, 45 rare animal species are known to be threatened by habitat destruction (Khang et al., 2007). Much of the primary forest has been cleared for cropland. These forest changes are exerting an increasing pressure on biodiversity conservation efforts. Different protection measures have been introduced to control forest logging, but illegal logging is still a significant threat to the remaining forest areas (TDMP, 2005). Modeling forest conversion can be an important instrument for understanding forest cover dynamics in the TDNP region. Forest change models can provide a better understanding of the factors that drive forest changes, they can generate future forest cover scenarios, and they can support the design of policy responses to forest changes (Lambin, 1994). Forest change is associated with multiple factors. The relationships between change and its driving factors can be very complex and are often non-linear (Mas, Puig, Palacio, & Lopez, 2004), requiring an appropriate modeling approach that accounts for such complex non-linear relationships.

Forest change can be predicted using empirical and simulation models (Lambin, 1997). The multi-layer perceptron neural network (MLPNN) with an integrated Markov model (M) (hereby referred to MLPNN-M) is a recently developed approach for spatiotemporal dynamic modeling of forest change (Eastman, 2009). The MLPNN allows the integration of the driving factors of forest change, whereas the Markov model controls the temporal dynamics of forest change.

A multiregression approach often performs poorly when the relationships between variables are non-linear and some variables must be transformed. Conversely, the MLPNN models are good at dealing non-linear relationships and do not require the transformation of variables (Lek et al., 1996). It is generally recognized that the MLPNN models can perform better in land change modeling (Mas et al., 2004; Lek et al., 1996; Zhou & Civco, 1996). In a recent study, the MLPNN was found to be better than logistic regression and other empirical modeling tools, such as empirical probabilities and empirical likelihoods, in land change modeling (Eastman, 2009). This chapter aims to predict areas vulnerable to forest conversion in the TDNP region using remote sensing data and the MLPNN-M model. Multi-temporal satellite images were used to detect changes in forest cover, and then the MLPNN-M model was applied to predict changes in forest cover in the near future. Predicting forest change patterns provides important information for identifying priority areas for conservation and forest management in the TDNP region. These predictions may improve the efficiency of efforts to protect the remaining primary forest in the study area.

11.2 Methods

11.2.1 Study Area

The study area covers a region of 141,238 ha that includes the TDNP (35,000 ha) and the buffer zone. The area is located in three provinces, namely Vinh Phuc (Binh Xuyen and Tam Dao district), Tuyen Quang (Son Duong district) and Thai Nguyen (Dai Tu and Pho Yen district) in the northern part of Vietnam (Fig. 11.1). The TDNP is considered to be one of the best and largest examples of rainforest habitat in Vietnam. It is known to host a variety of insects, butterflies, birds, medical plants and rare animal species (Ghazoul, 1994). Furthermore, the TDNP supports some of the highest levels of recorded insect diversity in Vietnam. A recent biological survey identified 1,436 plant species and 1,141 animal species in the park (Khang et al., 2007). The region is characterized by a tropical monsoon climate with a mean annual rainfall of around 2,600 mm, and most of the rainfall occurs from April to October. The elevation of the TDNP varies from 100 to 1,580 m above the mean sea level. The TDNP is the last remaining primary forest close to Hanoi, the capital of Vietnam. In the early 1960s, the population density in the area was relatively low; therefore, primary forest was dominant. In the 1970s, primary forest was gradually deforested by slash and burn farming. Due to the biological values of Tam Dao region, the area was recognized as a nature reserve in 1977. Before 1985, forest logging took place at low level, but in the early 1990s, the intensity of logging increased in response to an increased demand for timber (Khang et al., 2007). In 1996, Tam Dao nature reserve was declared to be a national park. The decision to establish the park has halted commercial forest logging; however, illegal forest logging still exists. Intensive population pressure and weakly enforced management have seriously degraded the park's natural resources and resulted in the destruction of most low-lying forest areas. The park is still threatened by deforestation (Fig. 11.2) due

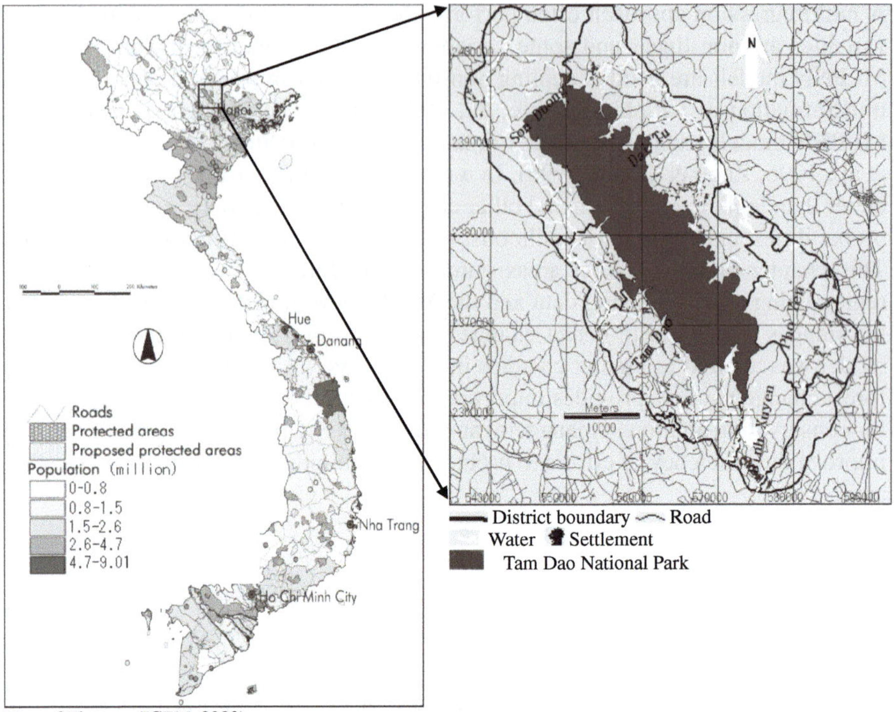

Map of Vietnam (ICEM, 2003)

Fig. 11.1 Tam Dao national park region

Fig. 11.2 Forest clearing for agricultural use in the TDNP region (Photo by author, 2009)

to the high level of firewood extraction and agricultural encroachment. Most of the 200,000 inhabitants in the surrounding areas of the park generate their incomes from small-scale farming and timber extraction. The high incidence of poverty and the poor awareness of conservation are major challenges to forest management and biodiversity conservation efforts (TDMP, 2005).

11.2.2 Multi-layer Perceptron Neural Network and Markov Model

We implemented the model MLPNN-M within the Land Change Modeler (LCM) available in IDRISI Taiga GIS and Image Processing software (Eastman, 2009) to predict primary and secondary forest conversions. In general, two forest cover maps derived from satellite imagery from two different dates were used to predict a forest cover map for a third date. The prediction process can be characterized by the estimation of forest conversion potentials followed by the forest conversion prediction stage (Fig. 11.3). Firstly, observed forest changes were used as the dependent variables and spatial variables were used as the independent variables (Table 11.2) to train the MLPNN and then estimate the primary and secondary forest conversion potential maps. Secondly, forest conversions were predicted using a competitive land allocation algorithm similar to the multi-objective land allocation (MOLA) algorithm. The MOLA looks through all conversions to list the host classes that lose some amount of land and the claimant classes that acquire some amount of land from each host. The quantities of conversions were determined by the Markovian

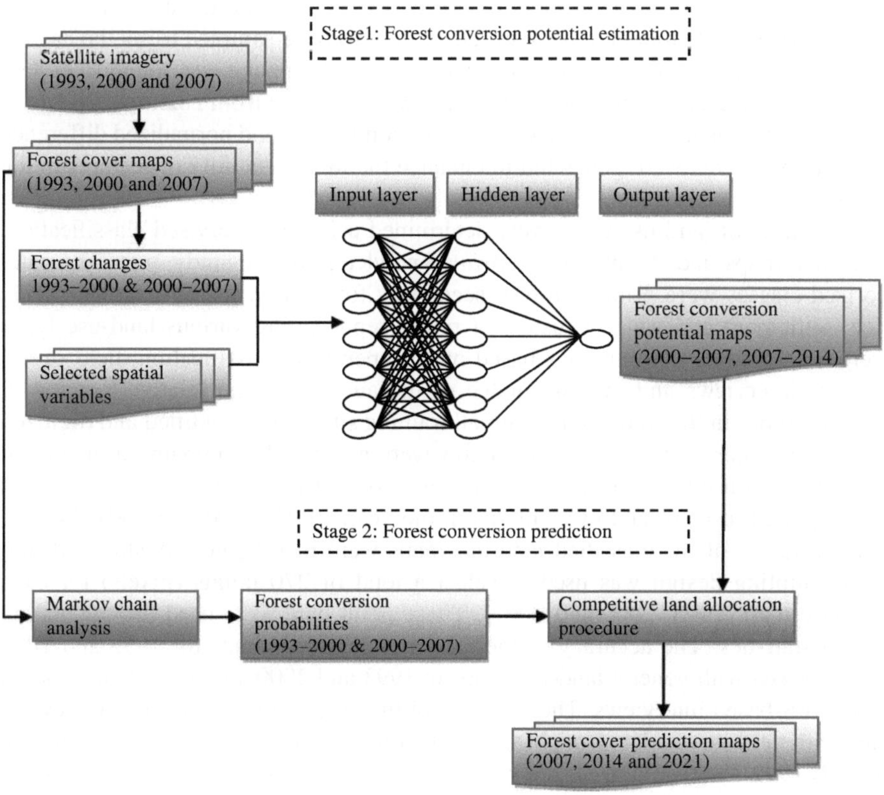

Fig. 11.3 Flowchart of the MLPNN-M model for predicting forest conversion

conversion probabilities. After this, a multi-objective allocation was run to allocate land for all claimants of a host class. The results of the reallocation of each host class were then overlaid to produce a final prediction map (Eastman, 2009). Detailed descriptions of the multi-objective land allocation algorithm can be found in Eastman, Jin, Kyem, and Toledano (1995).

11.2.2.1 Observed Changes in Forest Cover

Changes in forest cover are often related to the vicinity of land uses. Therefore, we interpreted land-use patterns to capture such variables for forecasting the location of changes in forest cover. Multi-temporal Landsat satellite images (path 127 and row 045) from December 27, 1993, October 04, 2000, and November 11, 2007, were obtained from the Global Land Cover Facility (http://www.landcover.org), University of Maryland. The digital maps of general land use (1993 and 2000), topography, and the road network were gathered from the TDNP office as reference data for land-use classification. In addition, field observations and interviews were conducted in March 2009 to supplement the reference data. Each LANDSAT satellite image was rectified to a common UTM/WGS84 coordinate system based on a 1:50,000 scale topographic map. These data were re-sampled using the nearest neighbor algorithm. The root mean square error of the image was found to be less than one pixel. The resolutions of all images were adjusted from 28.5 m × 28.5 m to 30 m × 30 m. Then, contrast stretching, color composites and normalized difference vegetation indexes were generated to enhance the interpretability of features in the images.

A number of land-use types were determined using unsupervised classification, reference maps and the authors' a *priori* knowledge of the study area. The categorized classes were primary forest, secondary forest, rain-fed agriculture, paddy rice, settlement and water. Clusters of pixels representing various land-use types were identified as training sites based on unsupervised classification, field observations, interviews and the knowledge of the authors on the relative location of land-use types in the study area. After all training sites were identified and digitized by the on-screen method, class signatures were generated. A maximum likelihood classifier was used to classify these images into land-use maps. These procedures were applied to map land-use patterns in the years 1993, 2000 and 2007. Finally, the accuracies of the classified land-use maps were investigated. A stratified random sampling design was used to select a total of 270 points (pixels) for each land-use map (1993, 2000 and 2007). These point data were used for calculating Kappa statistics. The accuracy of the classified land-use maps for 1993 and 2000 was assessed with general land-use maps in 1993 and 2000 and point data gathered from GPS-based interviews. The accuracy of the land-use map for 2007 was evaluated with field survey data from March 2009. Finally, changes in forest cover for the periods 1993–2000 and 2000–2007 were detected using cross-tabulation technique. Cross-tabulation analysis was used to compare forest-cover maps. This analysis was implemented using the LCM change analysis procedure.

11.2.2.2 Selection of Spatial Variables

Spatial variables were selected based on the availability of reliable data and the ability to express the data as a spatially explicit variable. The spatial variables expected to compose a considerable share of the factors driving past and future forest cover changes in the area. The statistical summary and spatial distribution of the variables are presented in Table 11.1 and Fig. 11.4. These variables are often highlighted in deforestation studies, such as in Pijanowski, Brown, Shellito, and Manik (2002), and land-use change models, such as in Li and Yeh (2002) and Dendoncker, Roundsevell, and Bogaert (2007).

The conversion of forest often relates to physical accessibility variables. Accessibility to a road is a significant factor of deforestation. For example, the role of road access was highlighted in predicting the location of deforestation in many areas, such as the Basho Valley, Northern Pakistan (Ali, Benjaminsen, Hammad, & Dick, 2005), Northern Thailand (Cropper, Puri, & Griffiths, 2001) and the Congo Basin (Wilkie, Shaw, Rotberg, Morelli, & Auzel, 2000). The location of water affects to the location of cultivation; therefore, the proximity to water is closely related to deforestation. Permanent cultivation in the area seemed to be concentrated close to water. In addition to road and water access, forest conversion also depends on the type of land-use in the neighborhood. For instance, Ludeke, Maggio, and Reid (1990) found a strong relationship between deforestation and proximity to forest edge in a given period in Honduras. In this study, several of these variables were included: proximity to primary forest, proximity to secondary forest and proximity to settlement. The proximity to primary forest, secondary forest and settlement was measured as the shortest distance from each location to the nearest primary forest, secondary forest and settlement, respectively. Furthermore, we included the proximity to cropland, which was measured as the shortest distance from each location to the nearest cropland. Some studies found a strong relationship between deforestation and the expansion of cultivation in the mountains of northern Vietnam (Meyfroidt & Lambin, 2008; De Koninck, 1999). Topography often influ-

Table 11.1 Statistical summary of spatial variables

Spatial variable	Mean	S.D.	Min.	Max.
Elevation (m)	86	194	0	1,581
Slope (°)	6.2	11.5	0	58
Proximity to road (m)	403	802	0	5,237
Proximity to water (m)	616	1,050	0	6,191
Proximity to primary forest in 2000 (m)	477	1,038	0	8,101
Proximity to primary forest in 2007 (m)	770	1,673	0	10,040
Proximity to secondary forest in 2000 (m)	146	406	0	4,248
Proximity to secondary forest in 2007 (m)	246	696	0	6,598
Proximity to settlement in 2000 (m)	1,169	1,818	0	8,517
Proximity to settlement in 2007 (m)	1,113	1,747	0	8,489
Proximity to cropland in 2000 (m)	122	392	0	3,506
Proximity to cropland in 2007 (m)	130	288	0	3,586

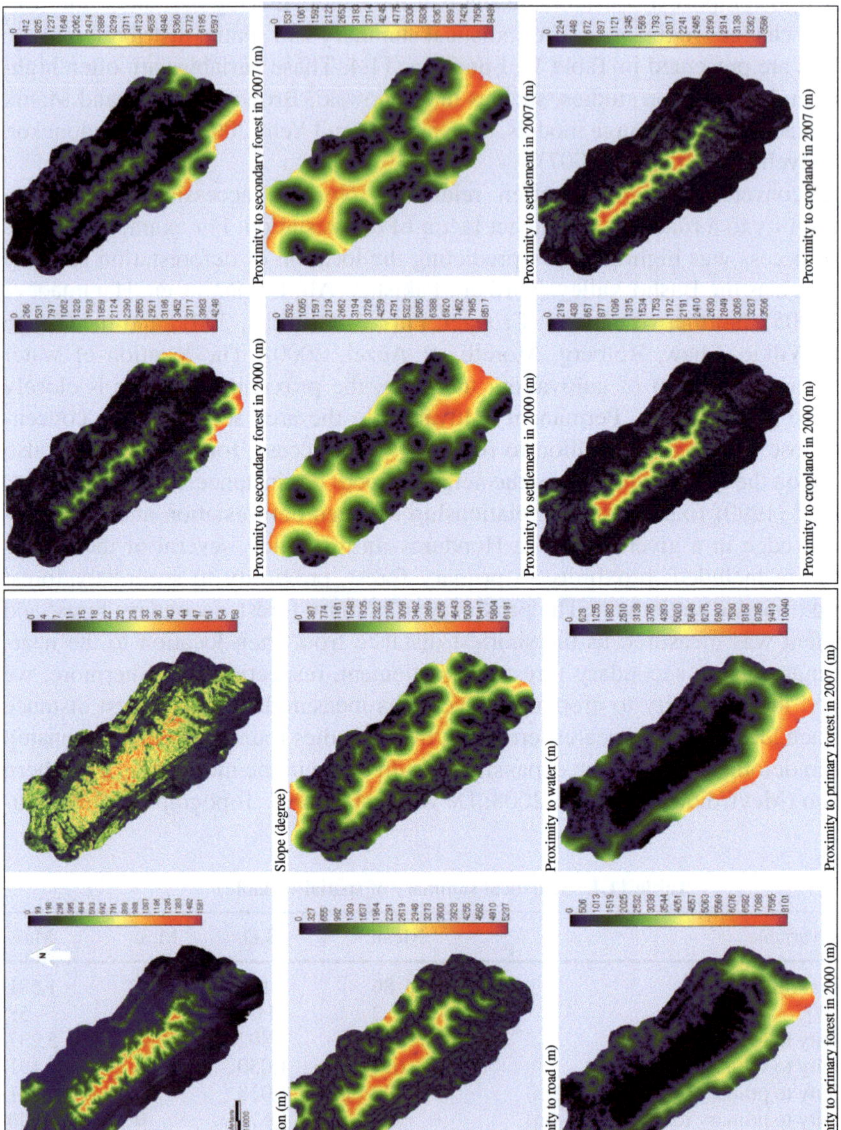

Fig. 11.4 Spatial variables

ences the spread and extent of forest conversion. For example, a case study in Costa Rica (Sader et al., 1998) found that as the slope gradient increased, deforestation decreased. In this study, topographic variables, including elevation and slope, were created from a contour map with a scale of 1:50,000 and contour interval of 20 m. This map was collected from the TDNP office.

The issue of correlated variables and data redundancy is minor because the neural network is good at solving these problems (Li & Yeh, 2002). In this study, we examined the nature of the association between observed forest changes and spatial variables using Cramer's V coefficient (Eastman, 2009). The quantitative variables were binned into 256 categories to conduct the test (Eastman, 2009) (Table 11.2). A Cramer's value close to 1 indicates a higher potential explanatory value of the variable; however, it does not guarantee a strong performance because it cannot account for the mathematical requirements and the complexity of the relationship. However, a variable can be discarded if the Cramer's V coefficient is less than 0.15 (Eastman, 2009).

Table 11.2 The relationship between observed forest changes and spatial variables

1993–2000	Cramer's V	2000–2007	Cramer's V
Conversion from primary forest to secondary forest			
Proximity to settlement in 2000	0.3459	Proximity to settlement in 2007	0.3302
Proximity to water	0.5903	Proximity to water	0.6431
Slope	0.7053	Slope	0.6843
Elevation	0.7161	Elevation	0.7680
Proximity to road	0.8082	Proximity to road	0.8582
Proximity to primary forest in 2000	0.9132	Proximity to primary forest in 2007	0.9347
Conversion from primary forest to cropland			
Proximity to settlement in 2000	0.2911	Proximity to settlement in 2007	0.2525
Proximity to road	0.3204	Proximity to road	0.2750
Elevation	0.4289	Elevation	0.4471
Slope	0.5014	Slope	0.4938
Proximity to water	0.5700	Proximity to water	0.5701
Proximity to cropland in 2000	0.6552	Proximity to cropland in 2007	0.6986
Proximity to primary forest in 2000	0.8139	Proximity to primary forest in 2007	0.8087
Conversion from secondary forest to cropland			
Proximity to settlement in 2000	0.3811	Proximity to settlement in 2007	0.3536
Proximity to road	0.4101	Proximity to road	0.3652
Elevation	0.5089	Elevation	0.5473
Slope	0.6012	Slope	0.5835
Proximity to water	0.6803	Proximity to water	0.6903
Proximity to cropland in 2000	0.7554	Proximity to cropland in 2007	0.7961
Proximity to secondary forest in 2000	0.8935	Proximity to secondary forest in 2007	0.8743

Notes: The test was conducted in the LCM of IDRISI Taiga. The Cramer's V coefficients were tested with the p value of less than 0.05.

11.2.2.3 Forest Conversion Potential Estimation

We trained the MLPNN as a network with the three layers: an input layer with the number of nodes equal to the number of spatial variables; a hidden layer with the same number of nodes; and an output layer with one node representing a conversion potential map (Fig. 11.3). The neural network is trained to derive the appropriate connection weights between the input layer and hidden layer and between the hidden layer and the output layer for classifying unknown pixels. The training process starts by iteratively presenting input data to the network. The connection weights are adjusted during network training to minimize the difference (error) between the network output and the desired output (Kanellopoulos & Wilkinson, 1997).

In the study area, selective logging directly converts primary forest into secondary forest. Another pathway is the conversion of primary forest into cropland by shifting cultivation. The third way is the conversion of secondary forest into cropland. Therefore, we estimated three forest conversion potential maps for the prediction of forest cover. Each of the three conversions was trained individually. Then, we estimated the 2007 conversion potential maps for the prediction of forest cover in 2007 (model validation) and the conversion potential maps for the prediction of forest cover in 2014 and 2021. For the 2007 forest conversion potential maps, elevation, slope, proximity to road, proximity to water and the dynamic variables (proximity to primary forest, proximity to secondary forest, proximity to settlement and proximity to cropland) for the year 2000 were presented to the MLPNN for training as independent variables while the 1993–2000 forest changes were presented as the dependent variables (Table 11.2). With the same procedure, spatial variables and the 2000–2007 forest changes were presented to the network for training to estimate the 2014 and 2021 forest conversion potential maps. The dynamic variables were recalculated for the years 2007 and 2014. We followed the MLPNN automatic dynamic training mode where all training parameters were automatically changed to better model the data. A detailed of the MPL training procedure can be found in (Eastman, 2009).

In general, the training results indicated a quick decline in the root mean square (RMS) error after 1,000 iterations, and the RMS error was mostly stable from 3,000 to 5,000 iterations. The RMS error flattened with little decline after 5,000 iterations; therefore, we stopped the training of the network after 5,000 iterations with a minimum loss of accuracy. According to (Eastman, 2009), the accuracy rate of training should be achieved in the vicinity of 80%. Therefore, we terminated network training when the accuracy rate exceeded the minimum level. Once the network was trained, new data could be run through it.

11.2.2.4 Prediction of Forest Conversion for Identifying Vulnerable Areas

The prediction procedure used by the IDRISI's LCM is based on a competitive land allocation procedure similar to the MOLA (Eastman, 2009). The MOLA combines the predictions of the location and the quantity of land cover change. For the prediction of the 2007 forest cover, the MOLA looks through the three forest conversion potential maps from 2000 to 2007 and the quantity of area for each conversion. These forest conversion potential maps were produced by MLPNN. The quantities of area were estimated using Markov chain analysis. The purpose of using the Markov chain is to determine the amount of change that may occur to some point in the future. A Markovian process is one in which the state of a land-cover is identified by knowing its previous state and the probability of conversion from each state to another (Eastman, 2009). During the MOLA process, IDRISI's Markov module was employed to produce the 1993–2000 forest conversion probability matrix (Table 11.4) based on the forest-cover maps of 1993 and 2000. In the matrix, the diagonal represents the self-replacement probabilities, whereas the off-diagonal values show the probability of a change occurring from one land cover to another. The MOLA allocated land for each category. For example, in order to allocate the primary forest to cropland, the MOLA used both the conversion potential map from the primary forest to cropland and the quantity of the conversion. Using this conversion potential map, the MOLA allocated the pixels with the highest potential to cropland according to the amount. Other forest conversions were done in the same way. Finally, the predicted forest cover map of 2007 was generated by overlaying all results of the MOLA procedure. By using the same prediction procedure, the forest cover maps of 2014 and 2021 were predicted. The forest cover map of 2014 was predicted using the forest conversion potential maps from 2007 to 2014 and the 2000–2007 forest conversion probabilities (Table 11.6). The forest cover map of 2021 was predicted using the forest conversion potential maps from 2014 to 2021 and the 2000–2007 forest conversion probabilities.

In order to apply the MLPNN-M model for the prediction of forest cover in the study area, the model needs to be validated. The purpose of model validation is to assess the predictive ability of the model for predicting changes in forest cover in the study area. The calibration data were separated from the validation data. The 1993–2000 forest cover maps and the 2000 spatial variables were used to calibrate the model. The 2007 actual forest cover map was only used for model validation. After the model was validated, forest cover scenarios were then predicted for the years 2014 and 2021 based on the assumption of forest conversions following the 2000–2007 Markovian dynamics. These prediction maps of future forest cover were used to identify areas vulnerable to forest conversions.

11.3 Results

11.3.1 Observed Changes in Forest Cover

The LANDSAT images for the years 1993, 2000 and 2007 were classified into primary forest, secondary forest, rain-fed agriculture, paddy rice, settlement and water (Fig. 11.5). The overall accuracy of the classified maps for the years 1993, 2000 and 2007 ranged from 86.67 to 90.01%, and Kappa indices varied from 0.83 and 0.87. In this study, we focused on forecasting forest conversions; therefore, these classified

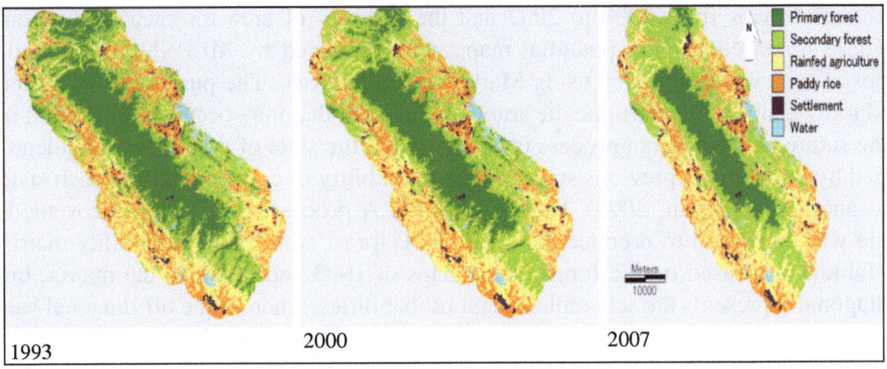

Fig. 11.5 Land-use maps derived from Landsat in 1993, 2000 and 2007

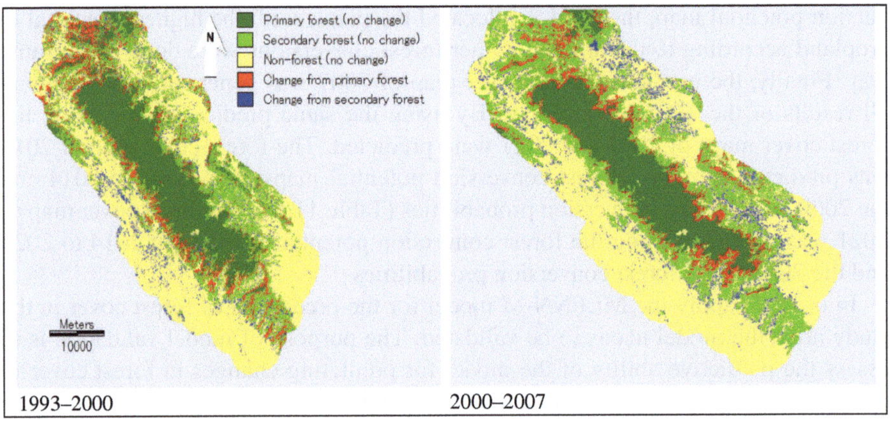

Fig. 11.6 Forest persistence and change for the periods of 1993–2000 and 2000–2007

maps were aggregated into primary forest, secondary forest and non-forest areas (Figs. 11.6 and 11.7).

Figure 11.6 shows the spatial patterns of forest changes for the periods of 1993–2000 and 2000–2007. The total primary forest loss for the period of 1993–2000 was 7,870 ha, equivalent to 20.59% of the primary forest area in 1993. For the period of 2000–2007, the primary forest loss was 4,893 ha, equivalent to 16.12% of the primary forest area in 2000. Some primary forest was converted into secondary forest. As a result, secondary forest for the first period increased by 3,970 ha, equivalent to 9.51% of the secondary forest area in 1993. In the second period, secondary forest increased by 385 ha, equivalent to 0.88% of the secondary forest area in 2000 because the conversion of primary forest into secondary forest was reduced substantially. The low conversion may be attributable to better management of the primary forest. As a result of the conversion of primary and secondary forest into non-forest areas, non-forest areas increased over the periods. The increase in non-forest areas for 1993–2000 and 2000–2007 were 4,080 and 4,508 ha, respectively.

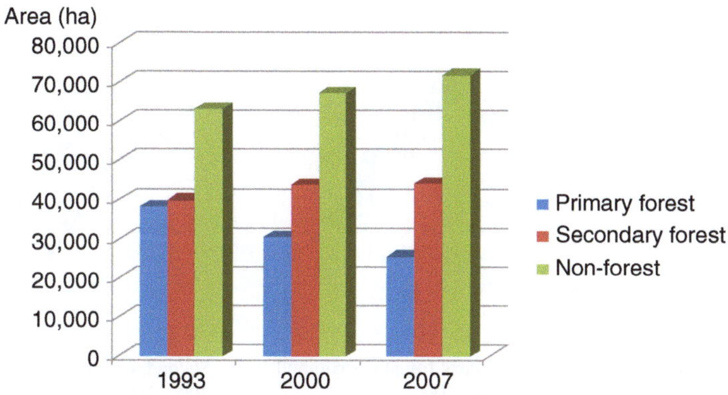

Fig. 11.7 Areas of primary forest, secondary forest and non-forest

11.3.2 Model Validation

Figure 11.8 shows the results of the training and the projection of the conversion potential maps for the year 2007. A forest conversion potential map consists of pixels with continuous scores varying from 0 to 1. A higher score pixel indicates a higher potential for forest change for that pixel. The higher potential areas for primary forest conversion are visible across the primary forest edges within the park and the surrounding areas. Similarly, the higher potential areas for secondary forest conversion are visible on the edges of the existing secondary forest segments. The forest conversion probability matrix was then estimated using the maps of forest cover for 1993–2000 (Table 11.3). Figure 11.9 presents the predicted map of forest cover in 2007 using the 2007 forest conversion potential maps and the 1993–2000 Markovian conversion probabilities. This output was used for model validation.

A validation technique that compares a predicted model with a null model was applied for model validation (Pontius, Huffaker, & Denman, 2004). Basically, this technique considers the agreement between two pairs of maps according to percent correct criterion. In this study, the first comparison was between the 2000 actual forest cover map and the 2007 actual forest cover map (the null model). The second comparison was between the 2007 predicted forest cover map and the 2007 actual forest cover map (the predicted model). Finally, the predicted model was compared with the null model. The components of agreement and disagreement of the two models were calculated using the Validate Module within the IDRISI software (Table 11.4).

For the agreement components, both the two models have some similar characteristics. The largest component of agreement was due to location, followed by due to chance and due to quantity. Overall, the percent correct of the predicted model (96%) was greater than the percent correct of the null model (92%). Therefore, the prediction model performed better than the null model at the 30-m resolution. According to (Pontius et al., 2004), a prediction model should be used in an area where the model predicts as well as or better than the null model. Therefore, the

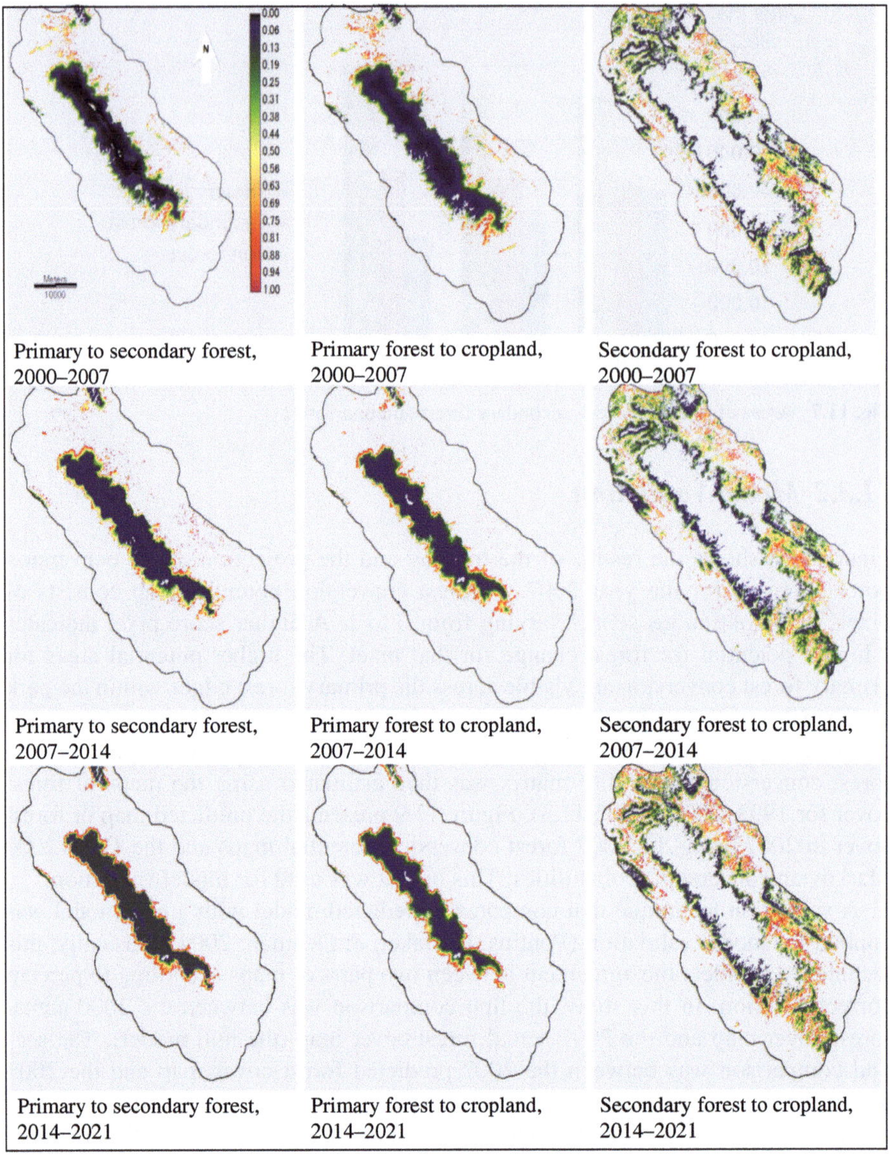

Primary to secondary forest, 2000–2007 | Primary forest to cropland, 2000–2007 | Secondary forest to cropland, 2000–2007

Primary to secondary forest, 2007–2014 | Primary forest to cropland, 2007–2014 | Secondary forest to cropland, 2007–2014

Primary to secondary forest, 2014–2021 | Primary forest to cropland, 2014–2021 | Secondary forest to cropland, 2014–2021

Fig. 11.8 Forest conversion potential maps consisting of pixels with continuous scores varying from 0 to 1 (the legend is the same in all conversion potential maps)

model can be used for predicting forest cover in the region. By individual class, the non-forest class had the best agreement, followed by the primary forest and the secondary forest. The model appeared to predict contiguous patterns better than fragmented patterns. Both the non-forest and primary forest were characterized by contiguous patterns, but the secondary forest showed fragmented patterns. These

Table 11.3 Forest conversion probability matrix for 1993–2000

Category	Primary forest	Secondary forest	Non-forest
Primary forest	0.7941	0.1959	0.0100
Secondary forest	–	0.9072	0.0928
Non-forest	–	–	1.0000

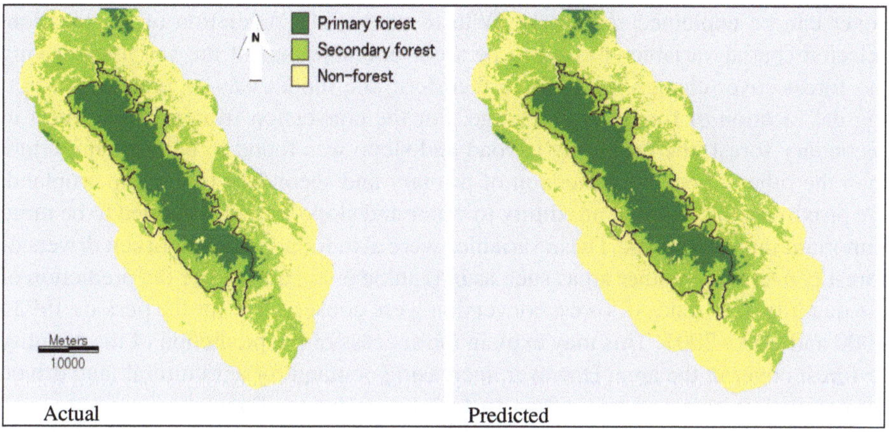

Fig. 11.9 Actual *versus* predicted forest cover in 2007

Table 11.4 Agreement and disagreement of the null and predicted models at 30-m resolution (% landscape)

Components of agreement and disagreement	The null model	The predicted model
Agreement due to chance	33	33
Agreement due to quantity	6	7
Agreement due to location	53	56
Disagreement due to location	3	3
Disagreement due to quantity	5	1

characteristics may explain why the accuracy of the predicted secondary forest is less than the others.

The disagreement due to location and quantity is important in evaluating the accuracy in quantity and location of the predicted forest cover. In particular, these components help to improve the prediction. For the null model, the disagreement due to quantity was greater than the disagreement due to location. On the other hand, for the prediction model, the disagreement due to quantity was less than the disagreement due to location. This result showed that the MLPNN-M model was more accurate at predicting the quantity than the location of forest cover in the region. The disagreement due to location can be improved by enhancing the forest conversion potential maps because the forest conversion potential maps alone determine the location of forest conversion. This can be undertaken by considering additional explanatory variables. For example, soil variables such as soil organic carbon and soil fertility may improve the forest conversion potential maps because forest is mainly converted into agricultural land in the region. However, soil data are not available in the area.

The success of the model in predicting the location and the quantity of forest cover can be explained separately. With respect to the prediction of the location, selected spatial variables proved to be a considerable part of the variables driving the forest cover change in the area; therefore, the model was accurate at predicting the location of forest cover change. For the conversion from primary forest to secondary forest, the proximity to road and slope was found to be more important than the others. For the conversion of primary and secondary forest into cropland, the proximity to cropland, proximity to water and slope were determined to be more important than the others. These variables were also found to be important drivers of forest conversion in other areas such as in (Linkie et al., 2004). For the prediction of the quantity, the trends of forest conversion were conservative for the periods 1993–2000 and 2000–2007. This may explain the success of the prediction of the quantity of forest cover in the area. However, increasing demand for agricultural land driven by population pressure may affect to forest conversion. Population pressure may accelerate in the future; therefore, it is hard to infer whether the model predicts the correct quantity in the future.

11.3.3 Areas Vulnerable to Future Forest Conversions

The forest conversion potential maps for the years 2014 and 2021 are presented in Fig. 11.8. Table 11.5 indicates the estimations of the forest conversion probabilities for 2000–2007. These inputs were combined within the model to simulate the forest cover patterns up to 2014 and 2021 (Fig. 11.10). Within the study area, the MLPNN-M model predicts that the remaining primary forest will decrease from 18.03% in 2007 to 15.10% in 2014 and 12.66% in 2021. The secondary forest areas will decline only slightly from 31.17% in 2007 to 30.88% in 2014 and 30.18% in 2021 because a large portion of primary forest is converted into secondary forest. The non-forest areas increase from 50.81% in 2007 to 54.01% in 2014 and 57.16% in 2021 as

Table 11.5 Forest conversion probability matrix 2000–2007

Category	Primary forest	Secondary forest	Non-forest
Primary forest	0.8379	0.1527	0.0094
Secondary forest	–	0.9026	0.0974
Non-forest	–	–	1.0000

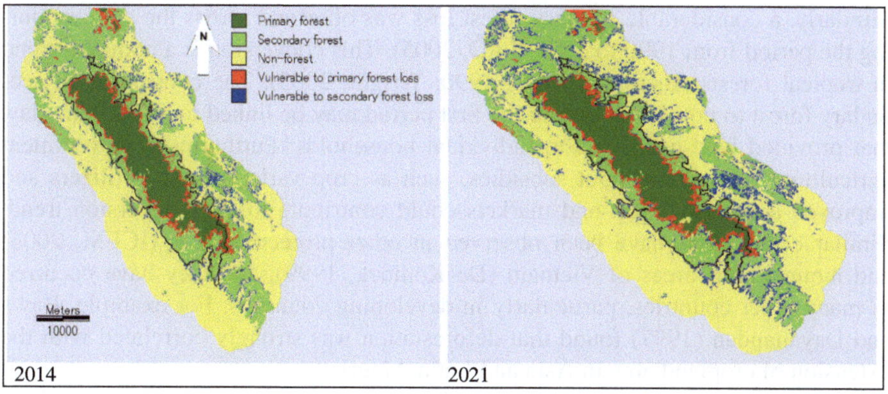

Fig. 11.10 Predicted forest cover and areas vulnerable to forest changes in 2014 and 2021

Table 11.6 Forest cover for 1993–2007 and predicted forest cover in 2014 and 2021 (percentages)

Category	1993	2000	2007 (Real)	2007 (Predicted)	2014	2021
Primary forest	27.06	21.49	18.03	16.82	15.10	12.66
Secondary forest	28.21	30.89	31.17	32.60	30.88	30.18
Non-forest	44.73	47.62	50.81	50.58	54.01	57.16

a result of the conversion of both primary and secondary forest into these areas (Table 11.6).

In order to visualize the spatial patterns of forest changes, overlay analysis was conducted to highlight such areas. The forest change patterns are shown in Fig. 11.10. The decline in primary forest is projected to be 4,127 ha in 2014 and 7,585 ha in 2021, equivalent to 16.21 and 29.79% of the 2007 remaining primary forest, respectively. The conversion of secondary forest into non-forest areas (cropland) is predicted to be 4,287 ha in 2,014 and 8,535 ha in 2021, equivalent to 9.74 and 19.39% of the secondary forest in 2007, respectively. Many areas within the park appear to be vulnerable to conversion. This area may require intensified protection measures if the remaining primary forest is to be maintained in the future. The areas susceptible to secondary forest conversion often overlay with the areas near the edges of secondary forest in the buffer zone.

11.4 Discussion

The conversion of primary and secondary forest for the periods from 1993–2000 and 2000–2007 was observed within the strictly protected primary forest and the buffer zone. The primary forest loss for 1993–2000 was bigger than that for 2000–2007. Thus, primary forest loss still continues by illegal forest logging (TDMP, 2005).

Similarly, a considerable primary forest loss was observed across the country during the period from 1990 to 2005 (FAO, 2005). This conversion is a common trend in tropical forests (Brown & Lugo, 1990; Wright, 2005). The conversion of secondary forest to non-forest during the first period may be linked to a 1993 land law that provided land-use rights to individual households. Furthermore, government agricultural production input subsidies, such as crop varieties and fertilizers and improved access to credit and markets could contribute to this conversion trend. Similar conversions have been observed in other protected areas (ICEM, 2003) and mountainous areas of Vietnam (De Koninck, 1999), and they have occurred in many other countries, particularly in developing countries. For example, Bawa and Dayanandan (1997) found that deforestation was strongly correlated with the extension of cropland area in Asia and Latin America.

There may be many driving factors of forest conversion, and they may vary from place to place. In our case study, selected spatial variables composed a considerable share of the factors driving forest changes. In particular, the accessibility variables seemed to be more important than the topographical ones. Many of these factors have been found to be important in other areas. For example, Merten and Lambin (1997) identified proximity to road, town and forest/non-forest edge as important drivers of forest change in southern Cameroon. Elevation and proximity to road were highlighted as important factors of forest change in the lowlands of Sumatra, Indonesia (Linkie et al., 2004). Elevation, slope, proximity to road, settlement and proximity to forest/non-forest edge were the key factors of forest change in southeast Mexico (Mas et al., 2004). Aside from these biophysical factors, socio-economic factors are often recognized as underlying driving forces of forest cover changes (Bawa & Dayanandan, 1997; Geist & Lambin, 2002) and also play an important role in changing landscape. Although these underlying factors are the main pressures on forest conversions, their effect frequently comes from outside the forested areas. For example, the population in the secondary forest edge may have less influence on the conversion of secondary forest to tea plantations than populations outside these areas. This conversion may be caused by tea demand that originates from places further away. Therefore, our empirical analysis was based solely on the site factors of forest conversions.

The predicted forest changes are based on the assumption that forest changes will follow the 2000–2007 Markovian dynamics. In this area, the trends used in our predictions were largely driven by population pressure, and may be conservative. Increasing population, the high incidence of poverty and the poor awareness of conservation among local residents have contributed to the loss of the primary forest and the conversion of secondary forest into cropland (TDMP, 2005). Forest conversion tended to occur in land suitable for agriculture. Shifting cultivation and commercial tea plantations were causes of cropland expansion into primary and secondary forest areas in the past. In particular, commercial tea plantations exist within the boundary of the park. Tea plantations may continue to extend into the primary forest area in the future because the area is highly suitable for tea plantations.

The identification of the areas vulnerable to forest changes is fundamental in the TDNP and has important implications for biodiversity conservation in the region.

One of the most important applications would be to relate the spatial patterns of forest changes to the spatial distribution of species. This is particularly important for large protected areas. Surveys on the distribution of plant species in the area showed that the remaining primary forest within the park is the most structurally complex and richest in plant species composition, particularly in the areas from 350 to 800 m above mean sea level (Kuznetsov, 2005). According to our predictions, forest loss is likely to occur within this range of elevation. The loss of the remaining primary forest will threatens the survival of many species in the region. In particular, cultivation within primary forest areas drastically altered the composition and abundance of plant species (Ghazoul, 1994). In addition, the conversion of secondary forest into cropland indicates increasing pressure on the steep land areas in the surrounding areas, and may cause severe land degradation in the future due to soil erosion. Continuing soil degradation may pose a threat to the natural resource-based local economy. From a protected area management perspective, the prediction maps of forest change patterns can help protected area managers identify where conservation and forest management efforts should be focused. This approach is particularly significant in Vietnam because limited finance resources for protected areas require focused efforts for conservation. Most of the government funding for protected areas is spent on salaries of forest rangers and not on development activities for affected populations. At a larger scale, the prediction of forest change patterns can aid long-term sustainable forest management.

11.5 Conclusions

This study investigated the conversion of primary and secondary forest using remote sensing and the MLPNN-M model in Vietnam's TDNP region. We parameterized the MLPNN-M model to simulate the conversion of primary forest and secondary forest in the near future. The rates and driving factors of forest changes were identified using remote sensing data. Then, these data were used to calibrate the model for projecting forest change patterns. The results of model validation showed that selected spatial variable proved to be a considerable part of the variables driving forest conversion in the area; therefore, the model was accurate at predicting the location of forest cover in 2007. However, the more successful component of the model was its prediction of the quantity of forest cover in 2007. Based on the 2007 model validation scenario, the forest cover in 2014 and 2021 was simulated to identify areas that are vulnerable to conversion of primary and secondary forest. According to our model predictions, a considerable portion of primary forest within the park is threatened by forest clearance, which indicates that intensified protection measures are required to prevent further loss of primary forest. The secondary forest in the steep areas in the buffer zone of the park is likely to be converted into agricultural land, and these areas are susceptible to soil erosion. The methodology and results produced in this study can be a vital tool for monitoring the remaining primary and secondary forest in the TDNP region. The monitoring process can be implemented by regularly updating Landsat-derived maps of forest cover and predicting forest change patterns. Moreover, the prediction maps can be used to focus biological conservation efforts. The methodology can also be used more widely for conservation planning and management in other protected areas that are experiencing forest changes in Vietnam. The model predicted quite reasonably the spatial patterns of forest changes in the study area; however, reforestation was not taken into account. Recently, deforestation and reforestation has been occurring simultaneously in the TDNP region. Therefore, future studies should attempt to include the reforestation process in order to further understand the dynamics as well as the patterns of forest changes in this area.

References

Ali, J., Benjaminsen, A. T., Hammad, A. A., & Dick, B. O. (2005). The road to deforestation: An assessment of forest loss and its causes in Basho Valley, Northern Pakistan. *Global Environmental Change*, *15*, 370–380.

Bawa, K. S., & Dayanandan, S. (1997). Socioeconomic factors and tropical deforestation. *Nature*, *386*, 562–563.

Brown, S., & Lugo, A. E. (1990). Tropical secondary forests. *Journal of Tropical Ecology*, *6*, 1–32.

Cropper, M., Puri, J., & Griffiths, C. (2001). Predicting the location of deforestation: The role of roads and protected areas in North Thailand. *Land Economics*, *77*, 172–186.

De Koninck, R. (1999). *Deforestation in Vietnam*. Ottawa, ON: International Development Research Center. Retrieved September 15, 2008, from http://www.idrc.ca/en/ev-9318-201-1-DO_TOPIC.html.

Dendoncker, N., Roundsevell, M., & Bogaert, P. (2007). Spatial analysis and modeling of land use distributions in Belgium. *Computers, Environment and Urban Systems*, *31*, 188–205.

Eastman, J. R. (2009). *IDRISI taiga, guide to GIS and remote processing* (pp. 234–256). Worcester: Clark University.

Eastman, J. R., Jin, W., Kyem, P. A. K., & Toledano, R. (1995). Raster procedures for multi-criteria/multi-objective decisions. *Photogrammetric Engineering and Remote Sensing*, *61*, 539–547.

FAO, 2006. *Global forest resources assessment 2005: Progress toward sustainable forest management*. FAO. Retrieved September 25, 2008, from http://www.fao.org/DOCREP/008/a0400e/a0400e/a0400e00.htm

Geist, H. J., & Lambin, E. F. (2002). Proximate causes and underlying driving forces of tropical deforestation. *BioScience*, *52*, 143–150.

Ghazoul, J. (1994). *Tam Dao nature reserve: Results of a biological survey*. Hanoi: Society for Environmental Exploration, UK and Xuan Mai Forestry College.

Giriraj, A., Irfan-Ullah, M., Murthy, M. S. A., & Beierkuhnlein, A. (2008). Modelling spatial and temporal forest cover change patterns (1973–2020): A case study from South Western Ghats (India). *Sensors*, *8*, 6132–6153.

Haines-Young, R. (2009). Land use and biodiversity relationships. *Land Use Policy*, *26*, 178–186.

ICEM (International Centre for Environmental Management). (2003). *Vietnam national report on protected areas and development* (pp. 19–47). Indooroopilly, QLD: ICEM.

Kanellopoulos, I., & Wilkinson, G. G. (1997). Strategies and best practice for neural network image classification. *International Journal of Remote Sensing*, *18*, 711–725.

Khang, N. D., Hoe, H., Duc, H. D., Thin, N. N., Tien, D. D., Lanh, V. L., et al. (2007). *Tam Dao national park* (pp. 9–56, in Vietnamese). Hanoi, Vietnam: Agricultural Publishing House.

Kuznetsov, A. N. (2005). *Rapid botanical assessment of Tam Dao national park: Detailed botanical survey final report*. Tam Dao: Tam Dao National Park.

Lambin, E. F. (1994). *Modelling deforestation processes: A review (TREES Series B: Research Report No. 1, EUR 15744 EN)*. Luxembourg: European Commission.

Lambin, E. F. (1997). Modelling and monitoring land-cover change processes in tropical regions. *Progress in Physical Geography, 21*, 375–393.

Lek, S., Delacoste, M., Baran, P., Dimopoulos, I., Lauga, J., & Aulanier, S. (1996). Application of neural networks to modelling non-linear relationships in ecology. *Ecological Modelling, 90*, 39–52.

Li, X., & Yeh, A. G. O. (2002). Neural-network-based cellular automata for simulating multiple land use changes using GIS. *International Journal of Geographical Information Science, 16*, 323–343.

Linkie, M., Smith, R. J., & Leader-Williams, N. (2004). Mapping and predicting deforestation patterns in the lowlands of Sumatra. *Biodiversity and Conservation, 13*, 1809–1818.

Ludeke, A. K., Maggio, R. C., & Reid, L. M. (1990). An analysis of anthropogenic deforestation using logistic regression and GIS. *Journal of Environmental Management, 1*, 247–259.

Mas, J. F., Puig, H., Palacio, J. L., & Lopez, A. S. (2004). Modelling deforestation using GIS and artificial neural networks. *Environmental Modeling and Software, 19*, 461–471.

Merten, B., & Lambin, E. F. (1997). Spatial modeling of tropical deforestation in southern Cameroon: Spatial disaggregation of diverse deforestation processes. *Applied Geography, 17*, 143–162.

Meyfroidt, P., & Lambin, F. E. (2008). The causes of the reforestation in Vietnam. *Land Use Policy, 25*, 182–197.

Myers, N., Mittermeier, R. A., Mittermeier, C. G., da Fonseca, G. A. B., & Kent, J. (2000). Biodiversity hotspots for conservation priorities. *Nature, 403*, 853–858.

Pijanowski, B. C., Brown, D. G., Shellito, B. A., & Manik, G. A. (2002). Using neural networks and GIS to forescast land use changes: A land transformation model. *Computers, Environment and Urban Systems, 26*, 553–575.

Poffenberger, M., & Nguyen, H. P. (1998). *Stewards of Vietnam's upland forests*. Center for south-east Asia Studies. Retrieved September 20, 2008, from http://www.mekonginfo.org/mrc_en/doclib.nsf/0/E5E0A84E9B42A19F80256690003862FB/$FILE/FULLTEXT.html

Pontius, R. G., Huffaker, D., & Denman, K. (2004). Useful techniques of validation for spatially explicit land change models. *Ecological Modelling, 179*, 445–461.

Sader, S. A., & Joyce, A. T. (1998). Deforestation rates and trends in Costa Rica, 1940–1983. *Biotropica, 20*, 11–19.

Sala, O. E., Chapin, F. S., III, Armesto, J. J., Berlow, E., Bloomfield, J., Dirzo, R., et al. (2000). Global biodiversity scenarios for the year 2100. *Science, 287*, 1770–1774.

Sam, D. D. (1994). *Shifting cultivation in Vietnam: Social, economic and environmental values relative to alternative land use* (pp. 3–15). London: International Institute for Environment and Development.

TDMP (Tam Dao National Park and Buffer Zone Management Project). (2005). *Rural household economics baseline survey*. Tam Dao National Park. Retrieved October 15, 2009, from http://tamdaonp.com.vn/

Turner, W., Spector, S., Gardiner, N., Fladeland, M., Sterling, E., & Steininger, M. (2003). Remote sensing for biodiversity science and conservation. *Trends in Ecology and Evolutions, 18*, 306–314.

Vitousek, P. M. (1994). Beyond global warming: Ecology and global change. *Ecology, 75*, 1861–1876.

Walker, R. (2004). Theorizing land-cover and land-use change: The case of tropical deforestation. *International Regional Science Review, 27*, 247–270.

Wilkie, D., Shaw, E., Rotberg, F., Morelli, G., & Auzel, P. (2000). Roads, development, and conservation in the Congo Basin. *Conservation Biology, 14*, 1614–1622.

World Bank. (1992). *World development report 1992: Development and the environment*. Oxford University Press. Retrieved June 12, 2009, from http://econ.worldbank.org/

external/default/main?pagePK=64165259&theSitePK=469372&piPK=64165421&menuPK=64166093&entityID=000178830_9810191106175

Wright, S. J. (2005). Tropical forests in a changing environment. *Trends in Ecology and Evolutions*, *20*, 553–560.

Zhou, J., & Civco, D. (1996). Using genetic learning neural networks for spatial decision making in GIS. *Photogrammetric Engineering and Remote Sensing*, *62*, 1287–1295.

Part IV
Multi-criteria GIS Analysis

Chapter 12
Land Suitability Analysis for Peri-Urban Agriculture

Rajesh B. Thapa, Frederic Borne, and Yuji Murayama

12.1 Introduction

Recently, population and food problems have become most serious global problems. The total population in the world is 6.8 billion and is still increasing at significant rate in the developing countries (United Nations, 2009). It is projected to reach 10 billion by the year 2050. Much of the projected growth in population will be centered on the world's cities. In this respect, the urban population, currently estimated to be about 3.4 billion, is expected to double by 2050. On the other hand, the rate of cultivating area is decreasing because of the increase of industrial land use and the remarkable trend of city sprawl to urban and peri-urban agricultural areas. Part of the reason for the observed growth in urban and peri-urban area is due to its adaptability and mobility compared with rural agriculture. As cities expand physically, the frontiers between urban, peri-urban and rural activity distort and merge, thereby, presenting opportunities for beneficial linkages.

Peri-urban area is considered as a transitional zone between urban and rural areas. The habitat of a diversity of populations, the heterogeneity of land uses, the morphological conditions and densities of the built areas, the complex functional relations and the changing social structure are some of the characteristics of the peri-urban area (Adell, 1999; Allen, 2001; Thapa & Murayama, 2008). These characteristics of peri-urban area will be transformed to the urban system. The transformation process decreases the cultivated area because of significant trend of city sprawl to urban and peri-urban agricultural areas (Zeng, Sui, & Li, 2005). Urban expansion is governed

R.B. Thapa (✉)
Division of Spatial Information Science, Graduate School of Life and Environmental Sciences, University of Tsukuba, Tsukuba, Ibaraki, Japan
e-mail: thaparb@yahoo.com; thaparb@gmail.com

This chapter is improved from "Rajesh Bahadur Thapa and Yuji Murayama (2008), Land evaluation for peri-urban agriculture using analytical hierarchical process and geographic information system techniques: A case study of Hanoi, Land Use Policy, 25, 225–239", Copyright (2010), with permission from Elsevier.

Y. Murayama, R.B. Thapa (eds.), *Spatial Analysis and Modeling in Geographical Transformation Process*, The GeoJournal Library 100, DOI 10.1007/978-94-007-0671-2_12,

by geographic and socio-economic factors such as population growth, policy and economic development (Thapa & Murayama, 2010).

In developing countries, a substantial and growing proportion of the population lives in or around metropolitan areas and large cities including the peri-urban zone, where their livelihoods depend to some extent on natural resources such as land for food, water, natural resources, and space for living (Allen, 2001; Thapa, 2009). Rapid growth at the peri-urban fringe has resulted in increased commercial development along arterial roads connecting cities and the countryside (Sullivan & Lovell, 2006). The sustainability of both urban and rural areas is affected by the dynamic and changing flows of commodities, capital, natural resources, people and pollution at the peri-urban interface (Allen, 2001; Brook & Davila, 2000). Peri-urban agriculture shall provide a solution to ecologically unhealthy development of large urban agglomerations, whereas urbanization is driven by the desire for short run economic growth and wealth in ever growing cities. According to local ecological conditions and habitat, peri-urban agriculture can contribute to preserve natural areas despite the increase of the price of land to favor intensive production of perishable foods like fruits, vegetables, meat and fish, dairy products. The demand of perishable products can be expected to remain high in urban areas because people living in those areas mainly depend on market supplies for their food consumption as compared to rural people (Jansen, Midmore, Binh, Valasayya, & Tru, 1996). Supplying perishable products to the urban dwellers, peri-urban agriculture can also generate formal and informal employment for farmers themselves as well as food processors and distributors. Many urban farmers, especially women are likely to use income earned from farming on food provision for the family (McGee & Robinson, 1995).

Uncontrolled momentum of urban sprawl and land use change raises many issues (Brook & Davila, 2000) which might have both positive and negative impacts in natural, social, and economic environment. Vegetable production in peri-urban areas may face difficulties to survive in the long run due to the scarcity of land and labor resources, unless alternative production technologies become available and the positive externalities generated by the peri-urban agriculture become internalized (Midmore & Jansen, 2003). Better planning and understanding of optimum resource utilization while considering environmental issues, should be given priority for peri-urban areas which may facilitate the criterion of a safe urban environment for human beings. Rapidly increasing population in cities and limited cultivated areas require higher agricultural productivity without polluting soil, air and water. Therefore, urban resources planners in developing countries should give more attention to peri-urban agricultural practices which can assist in improving urban environment, generating employment and reducing future food insecurity.

Policy makers and other stakeholders concerned with regional rural development increasingly face the need for instruments that can improve transparency in the policy debate and that enhance understanding of opportunities for and limitations to development (Jansen, Bouman, Schipper, Hengsdijk, & Nieuwenhuyse, 2005). Understanding the dynamics of complex urban systems and solving the real world problems requires robust methods and technologies for urban resources managers (Thapa & Murayama, 2008; Wilson, 2006). Spatial organization of urban system can be extracted into GIS as a set of points, lines and spheres to describe the socio-economic properties, linkage interactions and extent of the cities, respectively. Remote sensing provides an efficient tool to monitor long term land-cover changes in and around urban areas whereas geographic information systems provide a framework for spatial analysis and modeling based on geographic principles and seeks to integrate the analytical capabilities to broaden the understanding of the real world system (Thapa & Murayama, 2009).

In order to gain the higher productivity and profitability, the competition between demand and supply of land for different activities are playing crucial role in land use morphology that challenges to urban and peri-urban land use planner and managers. Conversion of peri-urban agricultural land into urban uses is a natural economic response to the increasing demand for space in a growing urban economy (Thapa, 2009). The Census Bureau of Vietnam (2001) mentioned that the rate of urbanization in Vietnam has been increased from 15% in 1960 to 25% in 2001 and expected to be doubled (31.6%) in 2015. The speed of urban sprawl and land use change might raise many problems such as inadequate infrastructure, population and employment pressure, overcrowding, slum occurred form low income groups, fresh nutrient rich food insecurity and environmental degradation (natural and social). Hanoi, capital city of Vietnam, is surrounded by the agricultural activities where the solution of traffic congestion is an urgency thereby maintaining the satisfactory transport facilities in the future. Resolution in decentralization of population distribution by establishing small city and satellite towns have to be considered as optimal decisions in strategies for encouraging development at the central city of Hanoi. The competition between demand and supply of land for different activities also plays a critical role in land use morphology of Hanoi urban and peri-urban areas. The peri-urban districts of Hanoi province have to be well planned for regulation of urban development process in the future (HARDD, 2002). The peri-urban districts of Hanoi province have to be reserved as storing land for regulation of sustainable urban development process in future. Therefore, the timely updated urban spatial information, land use information of peri-urban districts and presentation of major linkages between peri-urban and urban areas are needed for sustainable development of urban areas. The main objective of this chapter is to evaluate the land for peri-urban agriculture in the urban fringe area of Hanoi using Geographic Information System (GIS) and Analytical Hierarchical Process (AHP) techniques.

12.2 Study Area: Hanoi Province

A case study was carried out in Hanoi province (Fig. 12.1). Provincial boundaries as administrative units play an important role while preparing local and regional development policies in Vietnam. Therefore, the areas outside the provincial boundary were excluded in the study. The province is located in the Red River delta in northern Vietnam spread over 928 km^2 with flat topography from northwest to southeast. The province comprises of seven inner urban districts, i.e., Hoan Kiem, Ba Dinh, Hai Ba Trung, Dong Da, Tay Ho, Thanh Xuan and Cau Giay, and five peri-urban districts, i.e., Tu Liem in the west, Thanh Tri in the south, Dong Anh and Soc Son in the north and Gia Lam in the east. The peri-urban districts cover 91% (i.e., 845.93 km^2) of the

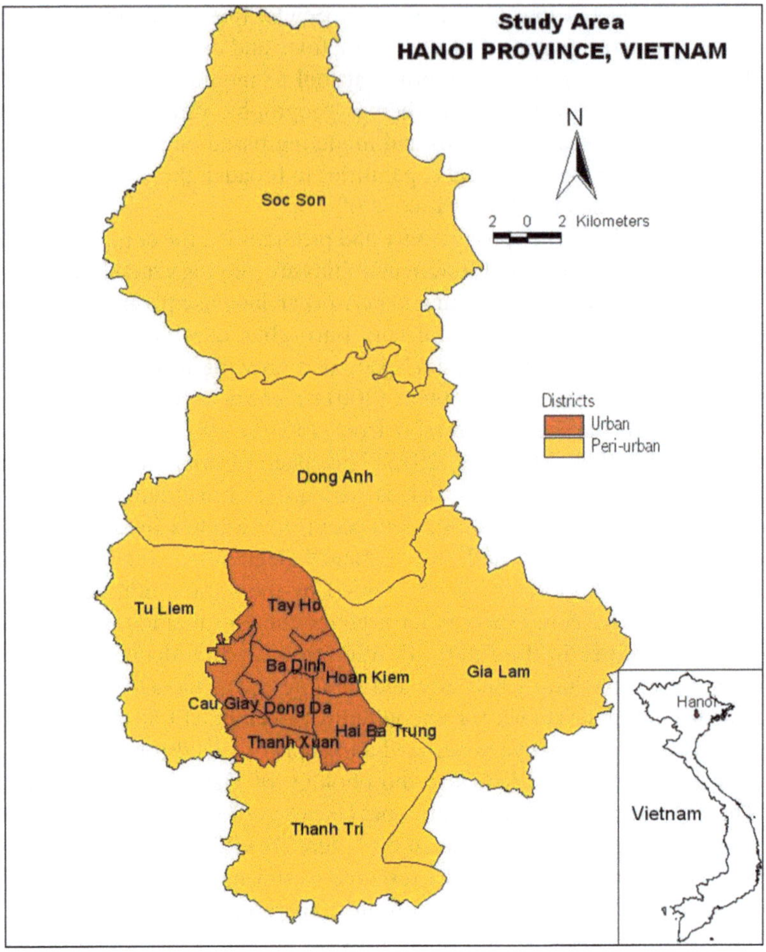

Fig. 12.1 Study area, Hanoi province, Vietnam

total land. All together there are 220 small administrative units (102 in urban and 118 in peri-urban) called commune. There are eight small towns in the peri-urban districts (Thapa, 2009). The topography of Hanoi is flat with a slope gradient of 0.3% from northwest to southeast. A large part of Hanoi spreads with height ranges from 1 to 30 m from mean sea level. However, there are some mountains as much as high of 462 m in far northern part of this province. Because of sunny and tropical weather along with monsoons, the climate is favorable to agricultural development in Hanoi. The average annual rainfall is 1533.3 mm during the period of 1995–2001. There are two seasons, Wet from May to October and Dry from November to April. Normally, 85% of total rainfall occurs in wet season (Thapa, 2009).

In 2001, the population of the province was 2.81 million with an annual 3.2% growth rate where 53.6% population lives in the urban area and the rest lives in

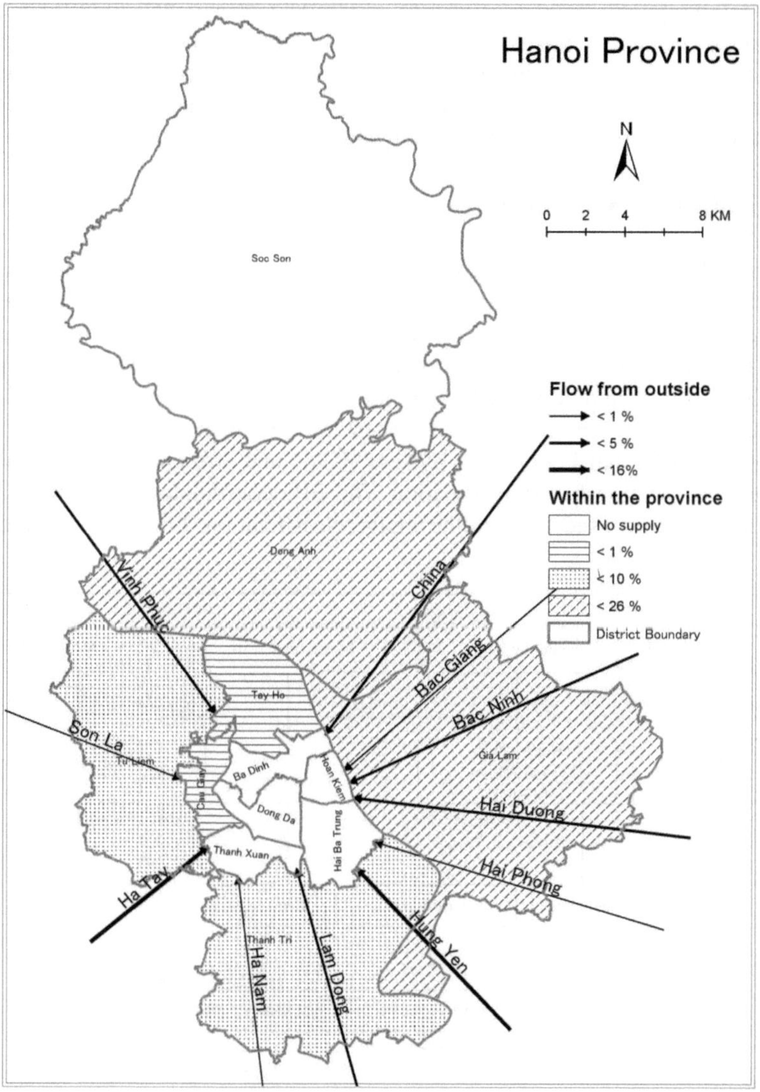

Fig. 12.2 Vegetables input linkages

peri-urban area (CBS, 2001). The population densities were 18.11 and 1.54 thousand people per km^2 in urban and peri-urban areas respectively. It is evident that the population is highly concentrated in the urban districts. Urban population in Hanoi is growing rapidly which fueled the demand for timely supplies of fresh vegetables to the cities. Demand of perishable products through peri-urban areas has significantly increased since the change from centrally planned and collective systems to a market economy (Jansen et al., 1996; Thapa & Murayama, 2008). A map

(Fig. 12.2) prepared based on Baseline Study data shows that quite a significant demand for vegetables from urban dwellers is supplied by the peri-urban districts of Hanoi and from its surrounding provinces. Although it presents only an overview of the vegetable consumption and supply in the province, it does give a picture of the economic interaction between peri-urban and urban areas of the province.

12.3 Database Methodology

GIS and AHP techniques in decision making were integrated while assessing the peri-urban land for agriculture (Fig. 12.3). Understanding the existing land use pattern greatly helps the decision/policy makers for identifying the developable spaces in the region and preparing appropriate zoning of land uses. Therefore, updated land use information and major linkage information between peri-urban and urban areas are necessary for optimum peri-urban resource management. A field survey equipped with Global Positioning System (GPS) and still camera was conducted to collect the location-based information. The GPS coordinates and photographs of each location were used for satellite image classification. Local experts meetings

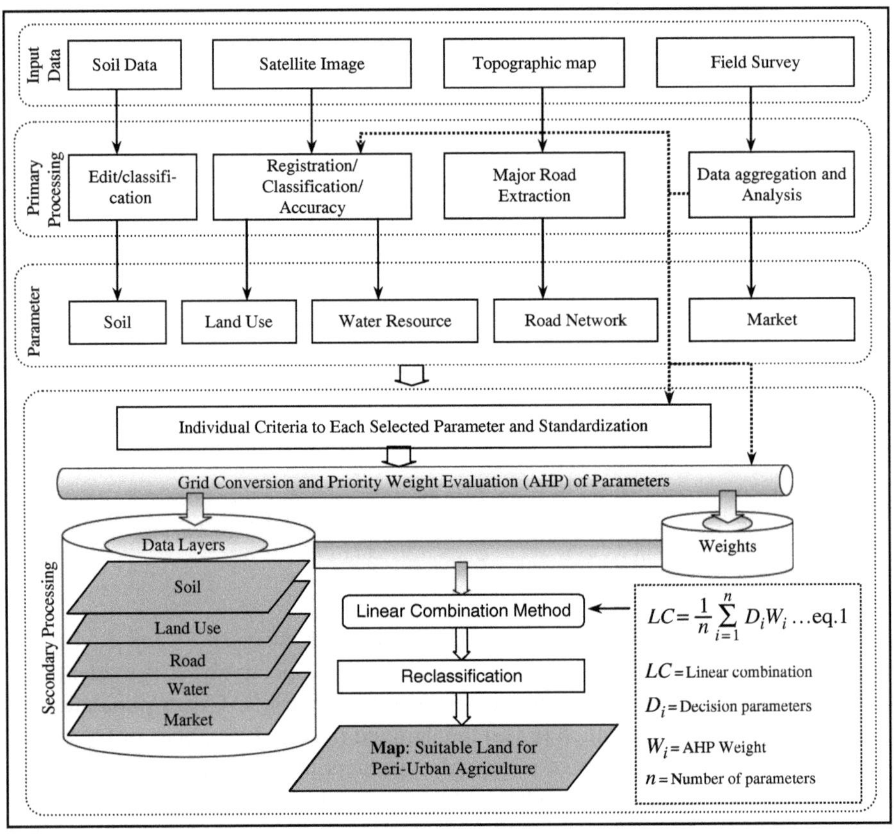

Fig. 12.3 Land assessment method for peri-urban agriculture

and focused group discussions were also accomplished during the field survey period. Based upon intensive discussions made with Hanoi agriculture experts and focused groups (local farmers and community leaders) during field visits, five spatial parameters (land use, soil, water, road, and market) were considered as the most influencing factors in decision making process for peri-urban agriculture in Hanoi. Landsat TM image acquired on November 23, 2001, a soil map, a road network map, a topographic map (1:10,000) and the field survey data were used as spatial data for the study.

The land use information was extracted from the Landsat image. Figure 12.4 shows detailed procedures of land use map preparation. Supervised classification with Bayesian Maximum Likelihood Classifier was applied to identify major land uses. Furthermore, a 3×-majority filter was used to remove salt and pepper noises on the map. The accuracy assessment was done using Kohen′s Kappa index where 89% of accuracy was achieved. However, the land use extraction method was constrained by the spatial resolution of satellite imagery, the classification system used, and expert knowledge of the study area (Thapa, 2009). Rivers and ponds/lakes were extracted from the Landsat satellite based land use map and used as a water resource parameter. Based on the field survey data analysis, the water proximities at aerial distance of 1 km, 2 km and 3 km from rivers and 0.3 km, 0.6 km and 0.9 km from ponds/lakes were computed. Similarly, the aerial distances of 1 km, 2 km and 3 km from road network and 10 km, 20 km and 30 km from market zone (inner urban districts of Hanoi) were assigned while computing the road and market accessibility to peri-urban agriculture areas. These distances reflect relative results: the nearer the objects higher the potential for peri-urban agriculture. The areas covered from the

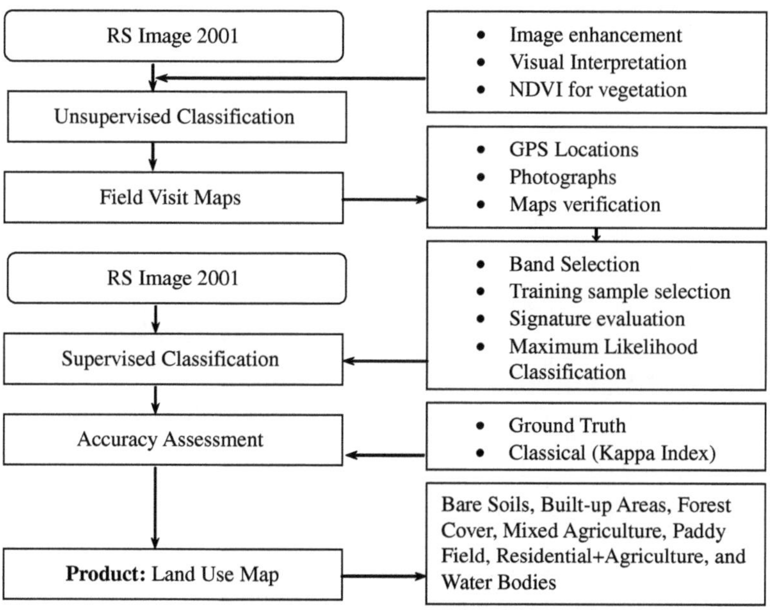

Fig. 12.4 Framework for land use map preparation

chosen distances were subsequently to be considered as suitability classes. The soil map was classified based on Brady (1974) to identify the level of suitability. Each map was further converted into 30 m grid for spatial analysis.

Depending on the location and the study objective, the contribution level of the selected parameters in decision making process can be different. In this research, the AHP method by Saaty (1990) was used to evaluate the decision making parameters for peri-urban agriculture land suitability. An AHP framework was designed (Fig. 12.5). This framework evaluates the consistent weight of each parameter through pair-wise comparison (Thapa & Murayama, 2010; Thapa, 2009). A set of questionnaires within the AHP framework was developed. In the questionnaire, respondents can determine relative importance of each criterion with respect to other, for example, importance of soil with respect to land use, water, road and market and vice versa. Sets of questionnaires were disseminated to 12 key people in the province covering various backgrounds (local farmers/community leaders, agriculture experts, market experts and agriculture planners) during the field trip. These key people were chosen based on focused group consultation.

Separate matrix was prepared for each respondent. All together 12 matrices were developed. As per the AHP rule, reciprocal computation, value normalization, principal vector weights computation, consistency calculations were performed to the individual matrix, respectively. Nine matrices out of 12 were found consistent. The consistent matrices were further linearly combined using average mean to

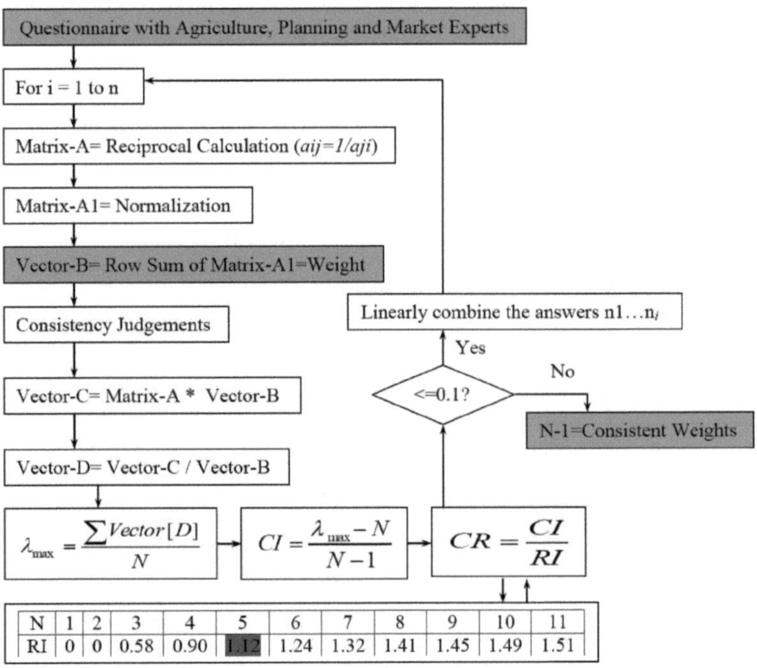

Fig. 12.5 AHP framework

prepare a final matrix. Then, AHP rule was applied to derive the final weightings. The consistency ratio (CR) coefficient for the final weightings obtained was 0.097 which falls in the acceptable range (CR<=0.10) (Canada, Sullivan, & White, 1996; Saaty, 1990). Details of the weight evaluation procedure can be found at Thapa (2009).

Linear combination method (Fig. 12.3; eq. (1)) was used to overlay the raster data and integrate the AHP results. In this study, five raster map results (i.e., soil, land use, road accessibility, water resources accessibility and market accessibility) were multiplied by their corresponding AHP weights individually and then all raster data layers were arithmetically overlaid to derive a final suitability map.

The suitability classification was done based on FAO (1976) guidelines. According to FAO framework for land evaluation, four categories were identified for peri-urban agriculture, i.e., High suitable (S1), Medium suitable (S2), Low suitable (S3) and Unsuitable (U). The values of S1, S2 and S3 were derived based on relative scale in decreasing order. But the values for U were computed from the land parcels that are neither S1, S2, and S3 nor water resources, built-up areas and forest coverage. The following equation was used for identifying the S1, S2, and S3:

$$\frac{Si = Ls_{\max} - Ls_{\min}}{3}$$

where, S_i = Land suitability class, Ls = Land suitability index which is derived from the linear combination method (See, Fig. 12.3). Max and min denote maximum and minimum values in the land suitability index map.

12.4 Results

12.4.1 Soil

Eight different types of soil namely, recent alluvial, sub-recent alluvial, mottle alluvial, hydromorphic, alluvial podzolic, gley, faralite mottle, and bare sandy were found at different locations. Brady (1974) suggested that the recent, sub-recent and mottle alluvial soil contain rich nutrients for crops. Therefore, these kinds of soils are assigned as highly suitable for agriculture. The hydromorphic, alluvial podzolic and gley soils require additional external inputs for crop and vegetable production, therefore these soils are assumed to be of medium suitability. The feralite mottle soil needs very high external inputs to support plants growth, hence, it is considered as of low suitability. The bare sandy soil is unsuitable for agriculture production. The soil suitability map of Hanoi (Fig. 12.6) demonstrates that the highly suitable soil covers 29% of the peri-urban land in the province (Table 12.1). The Red River and the associated streams formed this type of soil. The locations with these soils are considered highly suitable for agriculture. The soils of medium suitability are found slightly over 31% of the peri-urban land in the province. Low suitable soil was found in 9%

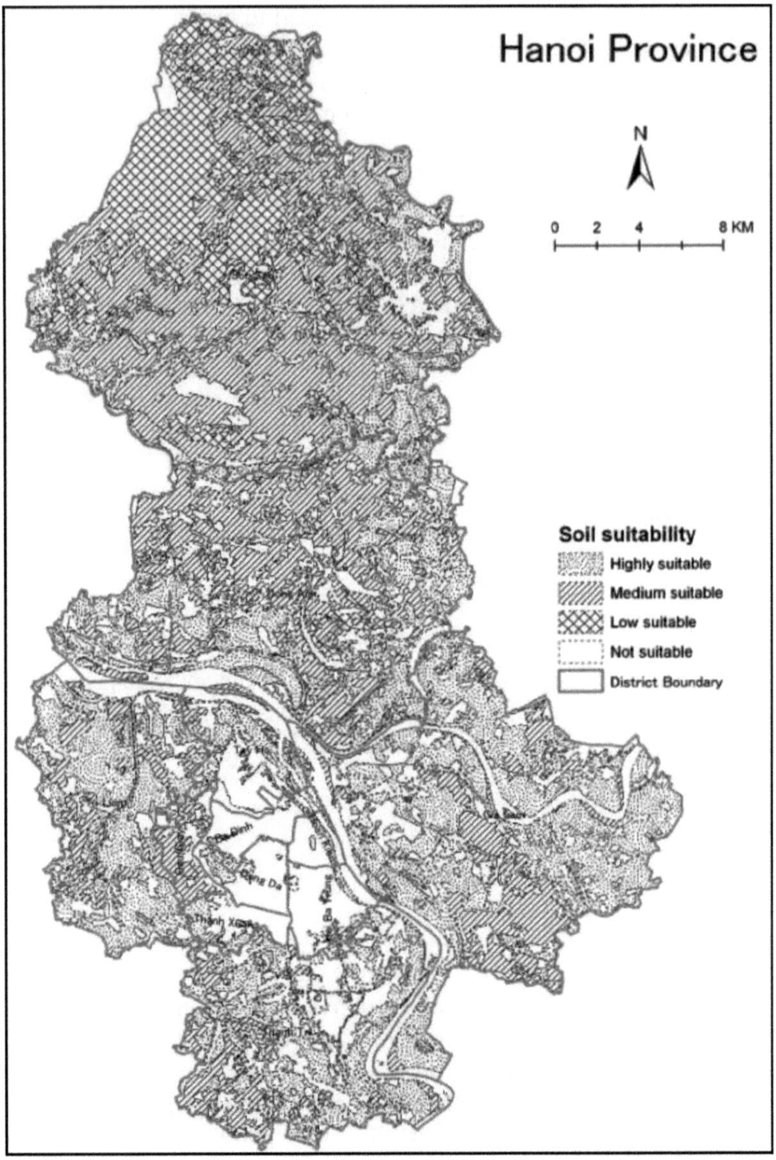

Fig. 12.6 Soil suitability

Table 12.1 Soil suitability classification

District	Soil suitability (in percentage)			
	High	Medium	Low	Not
Soc Son	3.74	16.39	8.62	7.37
Dong Anh	6.38	8.14	0.51	6.93
Gia Lam	9.89	3.57	0.04	7.24
Tu Liem	5.22	1.53	0.01	2.66
Thanh Tri	4.07	1.95	0.00	5.74
Total peri-urban land	29.29	31.58	9.18	29.94

of the area. The other types of parcels in soil map such as water bodies, bare sandy soil, and unidentified parcels cover about 30%, which are considered unsuitable for agriculture.

12.4.2 *Land Use Pattern*

Based on spectral responses from Landsat imagery, seven types of land use, namely, Bare Soils, Built-up Areas, Forest Cover, Mixed Agriculture, Paddy Field, Residential & Agriculture and Water Bodies (Fig. 12.7) were identified. The paddy field is recognized as the largest cover, which covers about 290 km^2 of the total provincial land (Table 12.2). Paddy field and mixed agriculture together occupied about 53% of the peri-urban land in the province and can be considered as pure agricultural land. Residential & Agriculture land use occupies about 11% land of the province. This area might involve agricultural practices within the households in order to support their daily needs for vegetables while at the same time earning cash. Another prominent land use is the built-up area. The built-up area covers slightly over 14% of land which is a combination of urban houses, industrial houses, and roads where these bodies reflect almost the same electromagnetic energy. Water bodies and forest cover extend to about 15 and 5%, respectively. Presence of bare soils was nominal. The bare soils were mostly located in riverside flooded area. The parcels of Paddy field and Mixed agriculture are classified as highly suitable areas for peri-urban agriculture whereas Residential & Agriculture parcels are classified as medium suitable land. Remaining types of land use are considered as unsuitable for agriculture.

12.4.3 *Road Accessibility*

From the field observation, most of the farmers use motorbike and bicycle as daily means of transportation to commute to the markets in urban districts. Some farmers also commute by foot. More often, farmers go to market in early morning to sell their goods and in the evening to acquire market information. In peri-urban areas of Hanoi, roads account for 1030 km (Fig. 12.8, Table 12.3). The existing roads, based

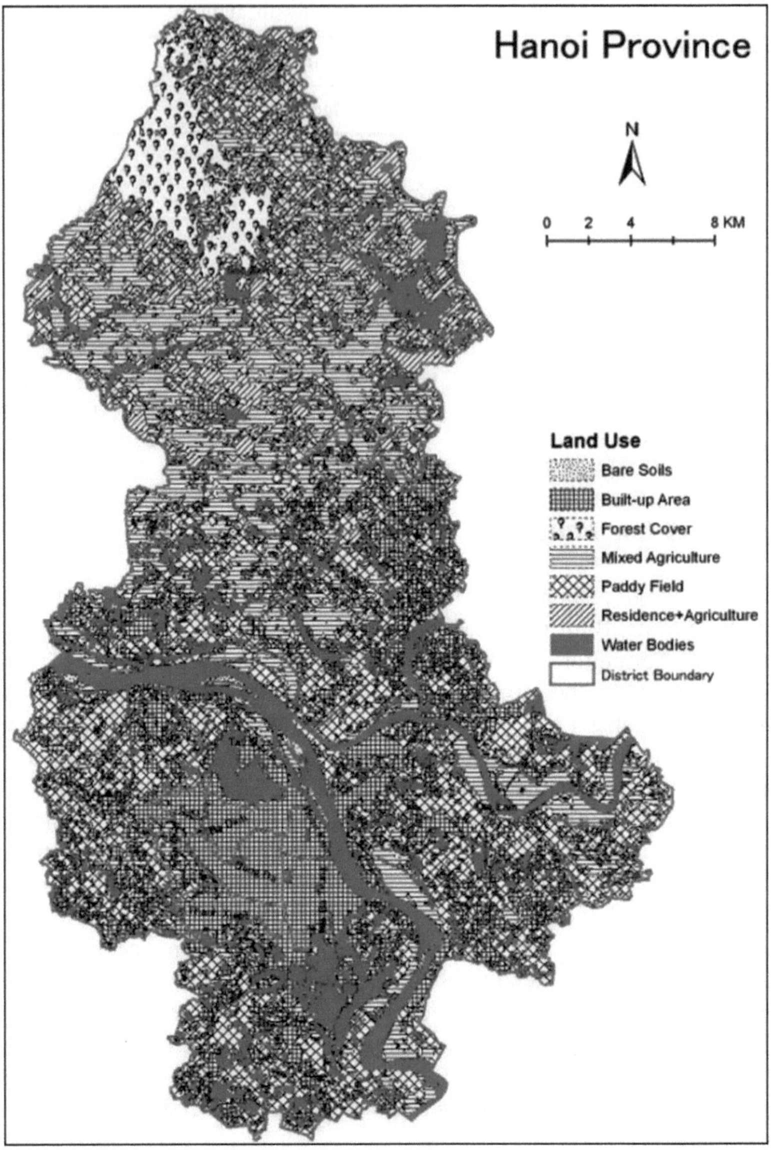

Fig. 12.7 Land use patterns

on their condition, are differentiated into four types, namely, All Weather, Main Earth, Paved Surface and Railroads. These roads play a crucial role in interlinking several communes of the 12 districts of Hanoi as well as other provinces of the country. The accessibility of the physical transportation network shows that a large part of the province is highly accessible. About 89% is covered by the highly accessible zone and 9% falls in between 1 and 2 km aerial distance (medium accessible zone) from the physical transportation network. A few parts, about 2% of the peri-urban

Table 12.2 Land use pattern in Hanoi province

	Land use (in percentage)						
Districts	PF	MA	R&A	FC	BuA	BS	WB
Soc Son	8.63	9.86	8.68	4.99	0.26	–	3.70
Dong Anh	7.68	5.48	1.00	–	4.12	0.03	3.66
Gia Lam	8.18	2.80	0.94	0.06	5.03	0.05	3.67
Tu Liem	4.75	0.56	0.43	–	2.74	0.03	0.91
Thanh Tri	4.19	0.98	0.49	–	2.47	0.01	3.62
Total peri-urban land	33.44	19.68	11.55	5.05	14.61	0.11	15.56
Suitability	High	High	Medium	Not	Not	Not	Not

Note: PF= Paddy Field, MA= Mixed Agriculture, R&A= Residential & Agriculture, FC= Forest Cover, BS= Bare Soils, BuA= Built-up Area and WB= Water Bodies

area are 2 km away from the road network. Small spatial unit of the Soc Son and Than Tri districts were found to be inaccessible because people from that areas have to walk more than 3 km of to reach an accessible road. However, the road linkages in the province are very accessible and facilitate the transportation of rural products to the urban market. Therefore, there is a great potential for peri-urban agriculture. Farmers of those areas can easily market their agricultural products to the city dwellers.

12.4.4 Water Resources Accessibility

The water resources layer was prepared from the Landsat based land use map (Fig. 12.9). The total coverage of water resources in the peri-urban area is 132 km^2 (Table 12.4) where rivers and lakes cover about 48 and 52% area, respectively. Water resources are being used for aquaculture, irrigation, transportation and recreational purposes. Peri-urban agriculture fully depends on these water resources. Traditional and modern techniques have been observed for irrigation systems. Pumping and small canal techniques are used in lakes and ponds. The modern canal systems supply water from the rivers to agriculture land. The accessibility of water resources was measured based on the rivers and lakes. Figure 12.9 shows that a total of 49% of the peri-urban land falls into the high accessibility zone of water resources. About 20 and 9% are found in medium and low accessibility of water resources, respectively. Five percent of the total peri-urban land located in Soc Son district has very difficult access to water resources. Few areas of the Tu Liem and Dong Anh district also face the same problem. Such areas fall into the inaccessible zone of water resources for peri-urban agriculture.

12.4.5 Market Accessibility

Figure 12.2 and Table 12.5 show that most of the vegetable products in Hanoi are sourced from peri-urban districts, i.e., Dong Anh, Gia Lam, Tu Liem and Thanh

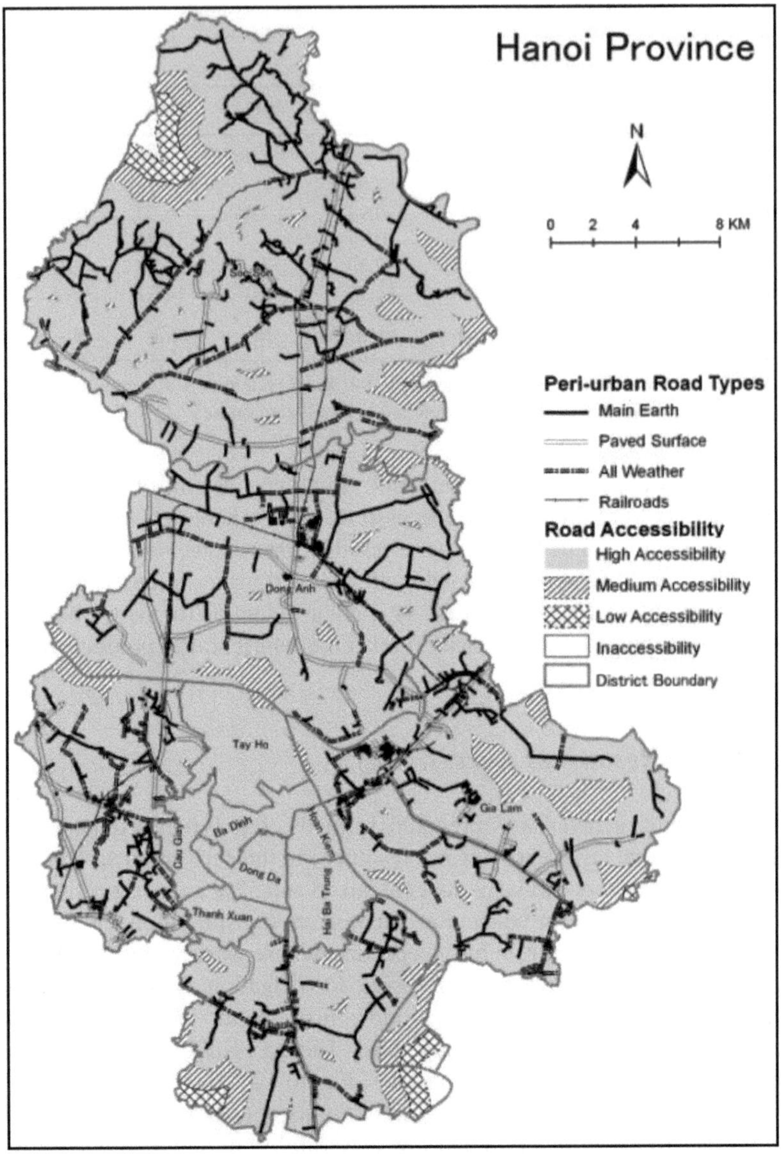

Fig. 12.8 Road accessibility

Tri. Some provinces of Vietnam and China are also providing vegetable products into the urban Hanoi. The map shows that there is a significant linkage between the peri-urban and urban as well as areas outside the Hanoi province in different directions. The market accessibility map (Fig. 12.10) suggests that 56% of the peri-urban district has high accessibility to market, the whole of Thanh Tri and Tu Liem districts are in the high accessibility zone (Table 12.5). Most of the Dong Anh and Gia

Table 12.3 Road network accessibility

		Road accessibility area (in percentage)			
Districts	Road length (km)	High	Medium	Low	Inaccessibility
Soc Son	324.49	31.84	3.27	0.81	0.21
Dong Anh	246.04	20.70	1.26	0.00	0.00
Gia Lam	218.99	17.65	2.68	0.31	0.09
Tu Liem	141.88	9.13	0.29	0.00	0.00
Thanh Tri	98.89	9.70	1.33	0.55	0.18
Total peri-urban land	1030.30	89.02	8.83	1.67	0.48

Lam districts are near to the market center. For the high accessible zone, the farmers can easily secure their market information needs and supply their agricultural products to the urban areas due to proximity to market. They can also access their daily basic needs from the city center. The Soc Son district seems very far from the Hanoi market access. About 25% of the total area of the peri-urban district is found in the medium accessibility zone. Slightly over 1% of the peri-urban land located in the northern part of Soc Son district is unable to access the market center easily. It can be solved through the provision of transportation infrastructure but additional transportation cost may raise the price to the products.

12.4.6 Land Allocation for Peri-Urban Agriculture

Five geographic maps namely, Soil (Fig. 12.6), Land use (Fig. 12.7), Road accessibility (Fig. 12.8), Water accessibility (Fig. 12.9) and Market accessibility (Fig. 12.10) were used as major maps for identifying the suitable land for peri-urban agriculture. The land suitability assessment has been performed in peri-urban districts of Hanoi province. Based on the AHP method, the soil is identified as the most important indicator for agriculture, which has 37 weights (Table 12.6). The second most important indicator is land use assigned a weight of 31. The other indicators road, water resources and market have been assigned a weight of 16, 10 and 6, respectively. All decision parameters with their corresponding weights were linearly combined.

In Hanoi peri-urban area, 52% land is found as arable land (Table 12.7). The resultant map of suitable land for peri-urban agriculture (Fig. 12.11) shows that 36% area of arable land is highly suitable, 14% is medium suitable, and 2% is low suitable. An area of 13% area was found to be unsuitable. It may be due to the lack of fertile soil, water resources accessibility, among others. The rest of the area (35%) of peri-urban is either covered by the water resources, built-up or forest cover. In the highly suitable land, the coverage of 10% is in the Dong Anh district (northern) alone whereas the two districts, namely, Soc Son and Gia Lam have similar area of 9% while the remaining two districts have the area around 4%. About 10% of the

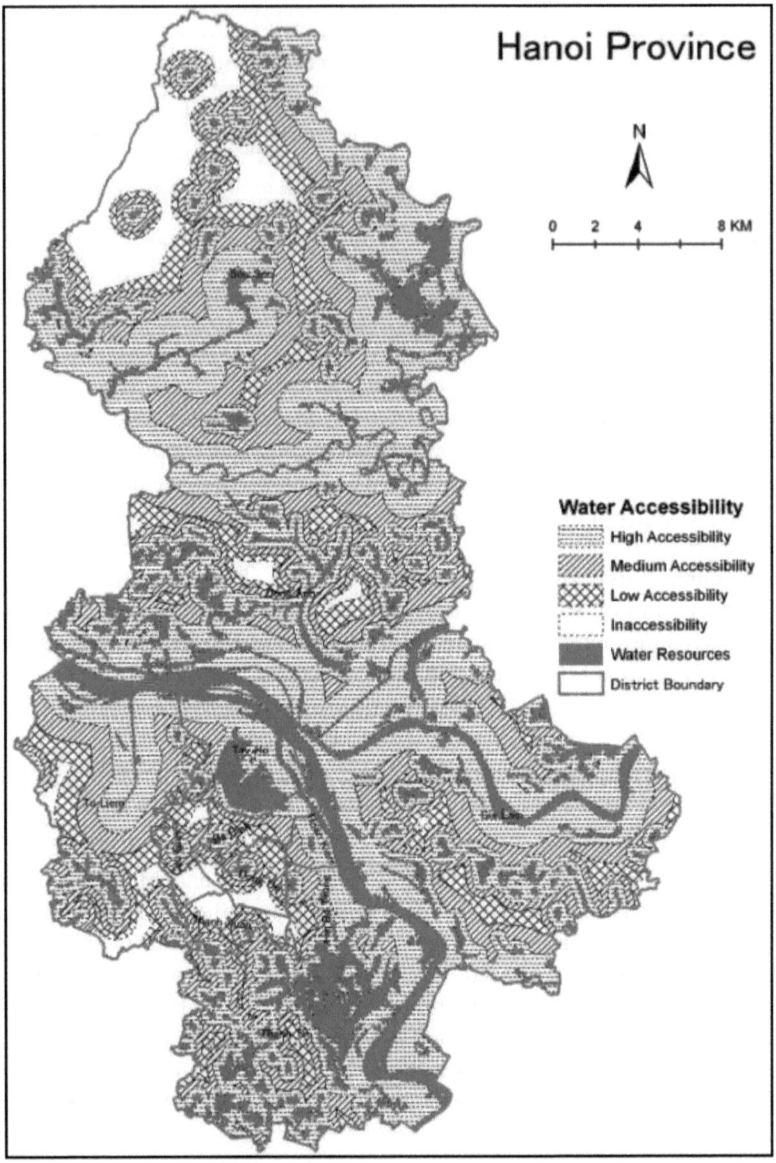

Fig. 12.9 Water resources accessibility

total peri-urban area located in the Soc Son district is accounted as medium suitable land for peri-urban agriculture. The Dong Anh district has only 2% of medium suitable land area. The other three districts with less than 1% fall under the medium unsuitable land. Only 2% is in the low suitable category, mostly located in Soc Son district. An addition, a total of 6% of the total peri-urban land located in the Soc Son district was found as unsuitable for peri-urban agriculture. Some areas comprising

Table 12.4 Water resources and accessibility

Districts	Resources (km^2)		Water resources accessibility area (in percentage)			
	Rivers	Lakes	High	Medium	Low	Inaccessibility
Soc Son	20.98	10.33	16.70	7.09	3.65	4.99
Dong Anh	13.77	17.19	12.04	4.24	1.64	0.39
Gia Lam	18.64	12.40	11.10	4.55	1.33	0.08
Tu Liem	4.69	2.99	3.44	2.82	1.68	0.58
Thanh Tri	5.57	25.08	5.68	1.79	0.62	0.05
Total peri-urban land	63.65	67.99	48.96	20.48	8.92	6.08

Table 12.5 Market accessibility

Districts	Vegetables input %[a]	Accessible area in (in percentage)			
		High	Medium	Low	Inaccessibility
Dong Anh	25.51	18.48	3.49	0.00	0.00
Gia Lam	14.28	16.33	4.41	0.00	0.00
Soc Son	n.a.	0.01	17.71	17.18	1.22
Thanh Tri	8.34	11.76	0.00	0.00	0.00
Tu Liem	9.47	9.42	0.00	0.00	0.00
Total peri-urban land	57.6	55.99	25.61	17.18	1.22

[a]To Hanoi urban area

about 2.3, 2.1, 1.2 and 1.1% of the total peri-urban lands are located Dong Anh, Gia Lam, Tu Liem, and Than Tri, respectively, also fall into the unsuitable category.

12.5 Discussion and Conclusion

The 10-year Hanoi provincial master plan and policy documents about peri-urban agriculture (HARDD, 2002) were also reviewed. Priority is given by these to the development of peri-urban agriculture at institution level with business perspectives that can be sustainable for long periods. For example, transform paddy field to fruits, flowers, fish farm, apply clean and safe vegetables approach, reform agricultural structure, etc. If the agriculture goes in business perspective, definitely it will follow the profit that means the peri-urban agriculture can be sustainable. If it will sustain for a long period, then it will have dual benefits of (a) accelerating the economic growth and employment generation and (b) reducing the environmental degradation adding more greenery as well as food insecurity in some extent. Emphasizing market demand based on agricultural specializations; focus is given to enhance the suburban economic development through peri-urban agricultural

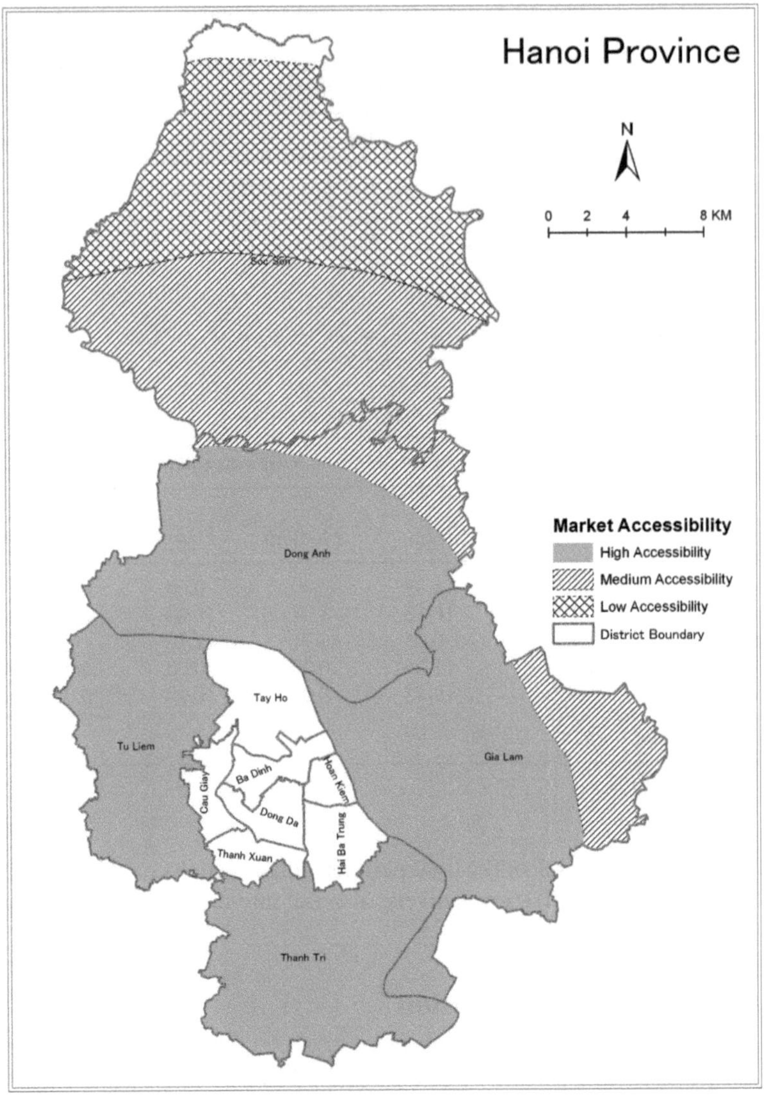

Fig. 12.10 Market accessibility

practices. Identifying the suitable zone for peri-urban agriculture and developing special products areas: clean vegetables, fruits, pork and fish may play an important role for economic growth as well as food security. The integrated technique of GIS and AHP evaluated the land of peri-urban area of Hanoi at different scale of suitability for peri-urban agriculture which may support to improve government plan and land use policy in some extent. Consolidation of local people knowledge and expert's views with modern geographic techniques to evaluate the peri-urban land leads this research results more empirical and original.

Table 12.6 Weights[a] of individual variables

Indicators	Weights
Soil	37
Land use	31
Road	16
Water	10
Market	6

[a]Weights are computed based on AHP guidelines (Saaty, 1990)

Table 12.7 Suitable land for peri-urban agriculture

	Suitability area (in percentage)				
Peri-urban districts	High	Medium	Low	Unsuitable	Excluded area
Soc Son	9.19	10.28	1.68	6.03	8.94
Dong Anh	9.97	1.81	0.02	2.39	7.78
Gia Lam	9.14	0.71	0.00	2.11	8.76
Tu Liem	4.21	0.33	0.00	1.24	3.64
Thanh Tri	3.96	0.51	0.00	1.19	6.09
Total peri-urban land	36.47	13.64	1.71	12.96	35.22

Higher resolution data, for example, ALOS or SPOT at 10 m, 5 m or 2.5 m can enhance land uses in more details and accuracy for peri-urban studies. Inclusion of more parameters such as social, cultural, technological and economical may add robustness of the framework. Some rapid primary survey may be suitable for these kinds of parameter determination. Although the AHP (Saaty, 1990) method is widely accepted in solving multi-criterion problems, the accuracy of the results depends on available sources of spatial data. Moreover, the selection of land suitability assessment parameters, priority weights within the AHP framework are greatly influenced by objectives, location, topography, people involved in discussions and key informants. However, this chapter provides a thematic framework including opportunities and limitations of Hanoi peri-urban area to the Vietnamese provincial planners. Every level of suitability can be utilized for different purposes, for example, the unsuitable land for peri-urban agriculture may be used to establish industrial estates. The low suitable land can be checked for fish farming or dairy farming or other species based on their requirements. Separate assessment is necessary before making appropriate decision for alternative use of land. If the planners and decision makers of the Hanoi province follow such integrated scenario while making decision for allocating economic activity zones in peri-urban areas, the land use might be sustainable. Therefore, government policy plays a crucial role for sustainability of the peri-urban agriculture.

This chapter shows an empirical land assessment technique integrating GIS and AHP which may help the policy makers for rapid assessment. To understand the complexity of the urban fringe, geographically referenced maps and local key

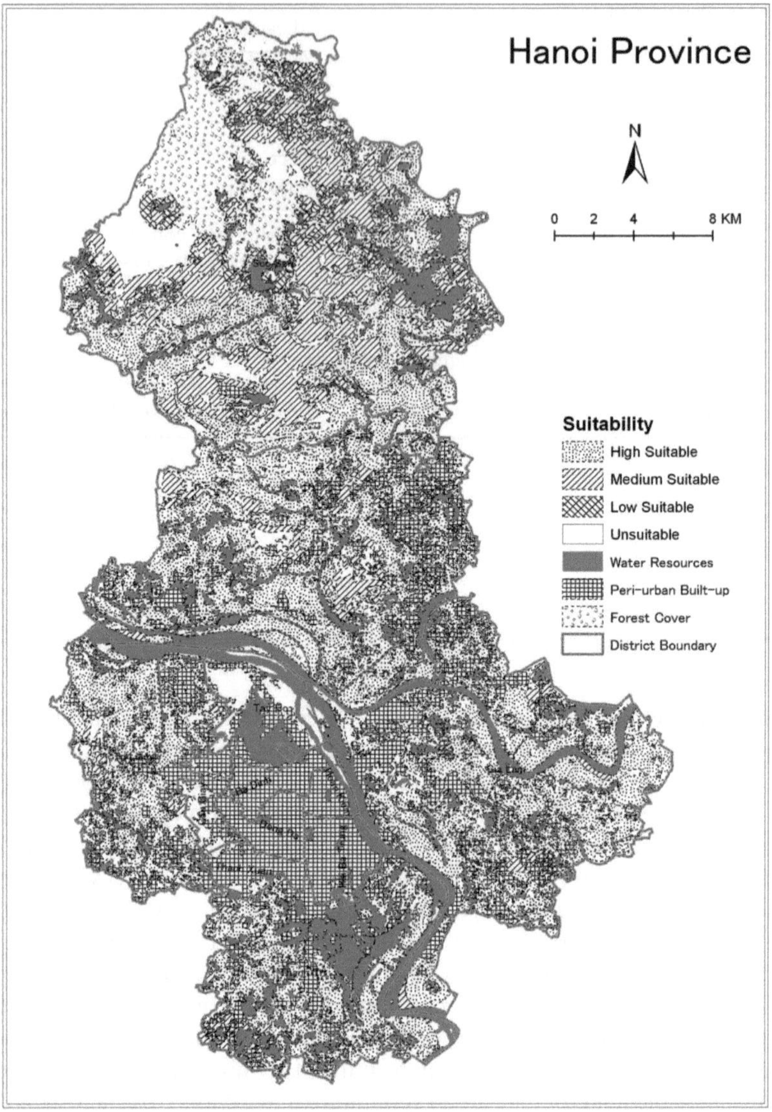

Fig. 12.11 Suitable land for peri-urban agriculture

informants are very important. In this research, the techniques applied in the field survey (location information collection, experts meetings and focused group discussions) have played a very important role in rapid data acquisition at the grassroot level. Remote sensing techniques assessed the recent land use information cost effectively. This empirical method accepts key input from experts and local people through group discussion and questionnaire survey, which significantly enhance the decision making capabilities while preparing the land use plan. The AHP framework provides this platform very efficiently. Integrating local knowledge in the method has shown the people participation in land use planning. The integrated method was able to map the suitable land for a peri-urban agricultural activity precisely. More decision factors can be added if deemed necessary. The selection of decision factors is case dependent because influencing factors in decision can differ from place to place. As for Hanoi peri-urban area, decision factors are selected based on focused group discussion analysis. The government policy is also being focused towards the sustainable peri-urban agriculture. Although this chapter does not discuss some issues such as the urbanization patterns in Hanoi and negative impact on soil environment due to vegetable cultivation malpractices as argued by Midmore, Jansen, and Dumsday (1996), the application of this chapter will be useful to policy makers and urban and regional planners to manage peri-urban resources and provide services to the people living in the rapidly changing environment especially in developing countries. The scenario also provides a framework to the urban researchers for rapid information acquisition within a short duration.

References

Adell, G. (1999). Theories and models of the peri-urban interface: A changing conceptual landscape. Strategic Environmental Planning and Management for the Peri-urban Interface (Research Project Report), Development Planning Unit, University College London, UK.

Allen, A. (2001). Environmental planning and management of the periurban interface. Proceedings of the Conference on Rural-Urban Encounters: Managing the Environment of the Periurban Interface, London, UK.

Brady, N. C. (1974). The nature and properties of soils. London: McMillian.

Brook, R. M., & Davila, J. D. (2000). *The peri-urban interface: A tale of two cities*. London, UK: School of Agricultural and Forest Sciences, University of Wales and Development Planning Unit, University College London.

CBS. (2001). *Hanoi statistical year book*. Vietnam: Central Bureau of Statistics.

Canada, J. R., Sullivan, W. G., & White, J. A. (1996). *Capital investment analysis for engineering and management*. Englewood Cliffs, NJ: Prentice-Hall.

HARDD. (2002). *Ten-year (2001–2010) master plan and policies for peri-urban agriculture*. Hanoi, Vietnam: Agricultural and Rural Development Department.

Jansen, H. G. P., Bouman, B. A. M., Schipper, R. A., Hengsdijk, H., & Nieuwenhuyse, A. (2005). An interdisciplinary approach to regional land use analysis using GIS, with applications to the Atlantic Zone of Costa Rica. *Agricultural Economics*, *32*, 87–104.

Jansen, H. G. P., Midmore, D. J., Binh, P. T., Valasayya, S., & Tru, L. C. (1996). Profitability and sustainability of peri-urban vegetable production system in Vietnam. *Netherlands Journal of Agricultural Science*, *44*, 125–143.

McGee, T. G., & Robinson, I. M. (1995). *The mega-urban regions of Southeast Asia*. Vancouver: UBC Press.

Midmore, D. J., & Jansen, H. G. P. (2003). Supplying vegetables to Asian cities: Is there a case for peri-urban production?. *Food Policy*, *28*, 13–27.

Midmore, D. J., Jansen, H. G. P., & Dumsday, R. G. (1996). Soil erosion and environmental impact of vegetable production in the Cameron Highlands, Malaysia. *Agriculture, Ecosystems and Environment*, *60*, 29–46.

Saaty, T. L. (1990). *Multi criteria decision making - the Analytic Hierarchy Process*. Pittsburgh, PA: RWS Publication.

Sullivan, W. C., & Lovell, S. T. (2006). Improving the visual quality of commercial development at the rural-urban fringe. *Landscape and Urban Planning*, *77*, 152–166.
Thapa, R. B. (2009). *Land assessment framework for peri-urban agriculture*. KG Koln, Germany: Lambert Academic.
Thapa, R. B., & Murayama, Y. (2008). Land evaluation for peri-urban agriculture using analytical hierarchical process and geographic information system techniques: A case study of Hanoi. *Land Use Policy*, *25*, 225–239.
Thapa, R. B., & Murayama, Y. (2009). Examining spatiotemporal urbanization patterns in Kathmandu Valley, Nepal: Remote sensing and spatial metrics approaches. *Remote Sensing*, *1*, 534–556.
Thapa, R. B., & Murayama, Y. (2010). Drivers of urban growth in the Kathmandu Valley, Nepal: Examining the efficacy of the analytic hierarchy process. *Applied Geography*, *30*, 70–83.
United Nations. (2009). World population prospects: The 2008 revision. Department of Economic and Social Affairs, Population Division. Working Paper No. ESA/P/WP.210.
Wilson, A. G. (2006). Ecological and urban systems models: Some explorations of similarities in the context of complexity theory. *Environment and Planning A*, *38*, 633–646.
Zeng, H., Sui, D. Z., & Li, S. (2005). Linking urban field theory with GIS and remote sensing to detect signatures of rapid urbanization on the landscape: Toward a new approach for characterizing urban sprawl. *Urban Geography*, *26*, 416–434.

Chapter 13
Suitability Analysis for Beekeeping Sites Integrating GIS & MCE Techniques

Ronald C. Estoque and Yuji Murayama

13.1 Introduction

13.1.1 Economic Importance of Beekeeping

Beekeeping is the management of honeybees for the production of honey and other by-products, and for the pollination of crops. Technically, the science and art of beekeeping is called apiculture (FAO, 2003). Beekeeping plays an important role by providing the rural poor additional income especially in many developing countries. In a case study conducted in the northern part of the Philippines involving *Apis mellifera* colonies, De Padua (2009) found that the average annual return on investment for 4 years of beekeeping operation was 182.20% equivalent to 62,596.00 Philippine Peso or about US $1,391 annual net income. In Uganda, based on a case study conducted by FAO in 2001, each of the 150 families engaged in beekeeping was able to gain an annual net profit of US $1,400 (FAO, 2003). Aside from this economic incentive, honeybees are very important because they are the key pollinators of about 33% of crop species (Oldroyd & Nanork, 2009). Although the degree of pollination dependency of plants varies significantly, the yield of highly pollinator-dependent crops would decrease up to 100% with the absence of pollinators (Garibaldi, Aizen, Cunningham, & Klein, 2009).

R.C. Estoque (✉)
Division of Spatial Information Science, Graduate School of Life and Environmental Sciences, University of Tsukuba, Tsukuba City, Ibaraki, Japan; Don Mariano Marcos Memorial State University, La Union, Philippines
e-mail: purplebee80@yahoo.co.uk

This chapter is improved from "Ronald C. Estoque and Yuji Murayama (2010), Suitability analysis for beekeeping sites in La Union, Philippines, using GIS and multi-criteria evaluation techniques, Research Journal of Applied Sciences, 5, 242–253".

Y. Murayama, R.B. Thapa (eds.), *Spatial Analysis and Modeling in Geographical Transformation Process*, The GeoJournal Library 100, DOI 10.1007/978-94-007-0671-2_13,

13.1.2 Geographic Information System (GIS) and Multi-criteria Evaluation (MCE) Techniques

The recent developments in computer-based tools like GIS and MCE techniques can assist the analysts and decision-makers deal with the complexity of decision making (Jorein, Theriault, & Musy, 2001). The integration of MCE with GIS is also a powerful tool in environmental monitoring and decision-making (Hubina & Ghribi, 2008) and in land suitability assessments (Jorein, Theriault, & Musy, 2001). Furthermore, MCE techniques have been recognized as decision-support tools in dealing with complex scenarios where technological, economical, ecological and social aspects have to be considered for proper land use planning (Marinoni, 2005). Likewise, computer-based system decision support tools strongly help decision-makers in evaluating criteria and alternatives for a specific objective (Ghribi, 2005). There had been several studies that were able to demonstrate the application of GIS along with MCE techniques on a wide range of area of interests; from the technical aspect of integrating GIS with MCE techniques (Jankowski, 1995; Laaribi, Chevallier, & Martel, 1996), land suitability mapping for industry using raster-based GIS (Eastman, Kyem, & Toledano, 1993) and Analytical Hierarchy Process (AHP) (Saaty, 1980, 1994, 2008), to combining MCE and GIS in evaluating habitat for endangered species (Pereira & Duckstein, 1993).

More specifically, Boonyanuphap, Wattanachaiyingcharoen, and Sakurai (2004) concluded that the integration of GIS with a multifactor spatial analysis made them effectively assess the environmental suitability of an area for banana plantation in Lower Northern Thailand. Roy and Grealish (2004) were able to demonstrate the capability of GIS in mapping arable soils using detailed soil characteristics. Likewise, Thapa and Murayama (2008) successfully applied AHP and GIS in evaluating suitable lands for peri-urban agriculture in Hanoi province, Vietnam. In land suitability assessments like these, rather than isolating the best alternatives, it is helpful to map the suitability index of the whole area being studied (Jorein, Theriault, & Musy, 2001).

Recognizing that the data needed in any multi-criteria assessment usually come from various fields of disciplines, careful evaluation and prioritization based on the actual characteristics and potential use of the land must be done. The determination of priorities and the importance of factors for the efficient use of land may be done using MCE techniques combined with GIS (Tudes & Yigiter, 2010). In this aspect, weighting methods for the different factors and alternatives like the AHP (Saaty, 1980) may be used. In a study conducted by Farajzadeh, Bayati, Rahimi, and Modares (2007), they found that a pairwise comparisons method of the AHP showed better results when compared to the other weighting methods like Boolean, ranking and rating.

13.1.3 Statement of the Problem and Objective of the Study

The economic benefits that can be derived from beekeeping can help lift people's economic status especially the marginalized. However, productive beekeeping depends on good colony management and richness of the foraging areas, and in order to promote it as a profitable agricultural occupation, areas with a good potential for beekeeping must be located and evaluated (FAO, 1987). Furthermore, its sustainability as a livelihood must also be considered to ensure long-term benefits.

Sustainable livelihood concept had first appeared in research literature in the 1980s (Solesbury, 2003). It includes the capabilities and assets, including both material and social resources, for a means of living. A livelihood is said to be sustainable if and when (1) it can cope with and recover from stresses and shocks; and (2) it has the ability to maintain or enhance its capabilities and assets both now and in the future, while not undermining the natural resource base (DFID, 1999). In general, the concept of sustainable livelihood is broad and complex encompassing different approaches across different locations and settings. A comprehensive review of the different approaches used by some leading organizations like the United Nations Development Program (UNDP), CARE of the United States of America (USA), and the Department for International Development (DFID) of United Kingdom (UK) can be found in Krantz (2001). While it is not the main objective of this study neither to evaluate nor to develop approaches for sustainable livelihood, this paper presents how new technological tools like GIS coupled with Remote Sensing (RS) and MCE techniques can be of help in attaining sustainability. It is based mainly on the basic concept of a strong foundation that deals with the understanding of the fundamental requirements of a particular livelihood project. For livelihood projects like beekeeping, it is basically necessary that the area is carefully and scientifically evaluated first for suitability prior to implementation. Suitability assessment is one way to strengthen the foundation of beekeeping, a foundation that could help make this project a sustainable livelihood.

Thus, this study was conceptualized taking into consideration the economic importance of beekeeping, the need for locating and evaluating suitable areas, and the capabilities and potentials of GIS and MCE techniques. Setting in mind sustainability as the goal, this study aims to evaluate the suitability of the province of La Union, Philippines for beekeeping using GIS and MCE techniques. The paper presents an empirical method for suitability analysis involving the participation of stakeholders and experts in the decision-making process.

13.2 Materials and Methods

13.2.1 Study Area: Province of La Union, Philippines

In the province of La Union, the science and art of beekeeping was first implemented and showcased officially to the public when the former Apiculture Training and Development Center (ATDC) of the Don Mariano Marcos Memorial State University (DMMMSU) in the municipality of Bacnotan was established on September 27, 1991. Furthermore, an undergraduate program in agriculture-major in apiculture has also been part of DMMMSU's curricular offerings since the late 1990s. Prior to these, though, local people have been familiar with honey and bees through the wild honeybees present in the locality and also through honey hunting.

On August 10, 2001, considering the economic importance of beekeeping, the Philippine Congress enacted Republic Act 9151, an act to abolish ATDC at DMMMSU, Municipality of Bacnotan, Province of La Union, and create the National Apiculture Research, Training and Development Institute (NARTDI). NARTDI, being a national institute, has been involved in many extension activities throughout the entire country promoting beekeeping as a profitable venture. In La Union alone, as of 2008, NARTDI has been supporting at least 43 successful beekeepers through its extension programs (NARTDI, 2008). There are also other private individuals, trained by NARTDI but have not been directly under its supervision, who have already initiated their beekeeping activities. In 2005, being the host province and one of the primary collaborators of NARTDI and DMMMSU in the promotion of beekeeping, the Provincial Government of La Union has officially adopted honey as its banner agricultural product in response to the One-Town-One-Product (OTOP) Program of the Philippine national government.

Geographically, La Union is situated in the southwestern part of the Ilocos Region between latitudes 16°12′ N and 16°55′ N and longitudes 120°17′ E and 120°35′ E (Fig. 13.1). The capital city, San Fernando, is 273 km north of Manila and 57 km from Baguio City. The climate of La Union is classified based on rainfall and belongs to the Climatic Type 1 of the Philippine Atmospheric Geophysical

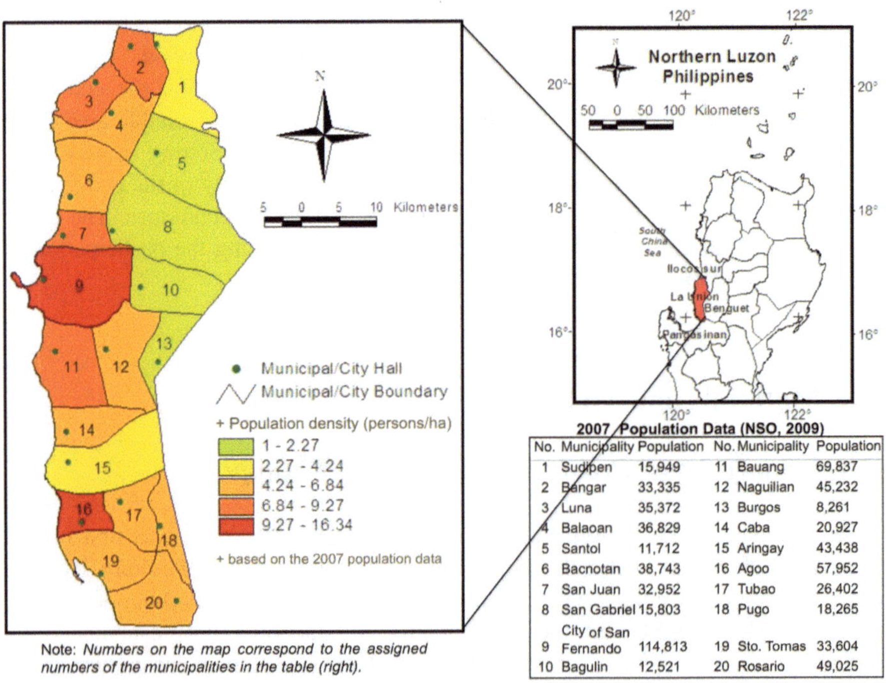

Note: *Numbers on the map correspond to the assigned numbers of the municipalities in the table (right).*

2007 Population Data (NSO, 2009)

No.	Municipality	Population	No.	Municipality	Population
1	Sudipen	15,949	11	Bauang	69,837
2	Bangar	33,335	12	Naguilian	45,232
3	Luna	35,372	13	Burgos	8,261
4	Balaoan	36,829	14	Caba	20,927
5	Santol	11,712	15	Aringay	43,438
6	Bacnotan	38,743	16	Agoo	57,952
7	San Juan	32,952	17	Tubao	26,402
8	San Gabriel	15,803	18	Pugo	18,265
9	City of San Fernando	114,813	19	Sto. Tomas	33,604
10	Bagulin	12,521	20	Rosario	49,025

Fig. 13.1 Location map of La Union, Philippines

Astronomical Services Administration (PAGASA). It is characterized with one distinct wet and one dry season. The dry period is generally lasting from November to April. The temperature is high (above 27°C) or intermediate (26–27°C), seldom or never mild (below 26°C) except during night.

The population of the province has increased from 657,945 in 2002 to 720,972 in 2007 (NSO, 2009). The primary sources of livelihoods in the province are agriculture (53%), services (37%) and industry (10%) (La Union Provincial Government, 2010). The continuous growth of population along with its inherent increasing resources requirement poses a great pressure to the province's land resources

Fig. 13.2 Photographs during the field survey and monitoring of the NARTDI team. (**a**) Landscape of Bacnotan, La Union. (**b**) NARTDI demonstration bee colonies. (**c**) *From left*: Elisio Pera (Cooperator-Balaoan, La Union), Reynaldo Laquidan (NARTDI Extension and Development Services Division Head). (**d**) *From left*: Anicet Derotchers (Canadian Executive Services Organization Volunteer); David Atolba (Cooperator-San Juan, La Union). (**e**) *From left*: Anicet Derotchers (Canadian Executive Services Organization Volunteer); Joseph Panas (NARTDI Staff) and Reynaldo Mostrales (Cooperator-Oya-uy, Bacnotan, La Union). *Source*: Photographs: **c**, **d**, and **e** are contributed by NARTDI

particularly to its arable land. Apparently, additional sources of livelihood are needed to meet the people's increasing needs while rationalizing the escalating pressure to the natural resources. Cognizant to the foregoing and to provide more scientific basis in the promotion of beekeeping not only in the province of La Union but also in other places, this study was framed and carried out. Figure 13.2 shows some photographs taken from the study area.

13.2.2 Determination and Preparation of the Criteria

In this study, criteria are of two types. These are factors and constraints. Generally, the ideal apiary site should be away from playgrounds and busy commercial or industrial areas; near to a fresh water supply: the banks of a river, lake or fish-pond, or even a dripping faucet; near food sources; fairly dry, away from swampy or flooding valley or any bottom land with stagnant water; accessible to good roads; away from danger of vandalism and unfriendly neighbors, and with annual rainfall between 1275 and 1875 mm (FAO, 1990).

In consideration of the above-mentioned criteria and in consultation with experts, the following factors were believed to be the most important features to consider in selecting beekeeping sites in the province of La Union: land use/cover (nectar and pollen sources), distance to river, distance to road, elevation and rainfall. On the other hand, the following were considered constraints because they do not contain any nectar and pollen source plants, and it is not advisable to place the beehives exactly in these locations: built up areas, water body, sand, river wash, and areas within 25 m from the road. Although the average annual rainfall in the province of La Union from 1997 to 2003 (2810.50 mm) (PAGASA Agromet Station, DMMMSU-NLUC, 2004) is beyond the limit set by FAO (1990), beekeeping can still be done in the area since about 70% of the total rainfall usually occur during the months of July-September only. The data used and their sources are presented in Table 13.1.

Table 13.1 Data used and their sources

Data sources	Descriptions
1. Landsat TM image (Source: USGS)	Acquired on 6th December 2004; resolution = 30 m; used to derive the land use/cover of La Union.
2. Topographic map of La Union (Source: NAMRIA, Philippines)	Scale = 1:50,000; used to derive DEM, river and road networks
3. Ground survey data	1st – June 2005 – used in image classification and accuracy assessment; 2nd – Nov. 2008 – visited the beekeeping projects.
4. 12 beekeeping projects	2005 data on honey yield in kg/colony were collected; geographic locations were recorded.
5. Experts	Sources of expert's opinion used in the Analytical Hierarchy Process (AHP)

13.2.2.1 DEM, River and Road Networks

The Digital Elevation Model (DEM) was developed from the digitized contour lines of the topographic map of the province with a scale of 1:50,000 using Vertical Mapper – MapInfo software. The river and road networks were also extracted from the same topographic map.

13.2.2.2 Land Use/Cover Map Development

A detailed vegetation map of the "bee plants" is ideal to represent the source of nectar and pollen. However, mapping individual plant species over a relatively wide area has been a major challenge in the beekeeping industry due to the complexity and limitations. Recognizing this given situation, we believe that, for purposes of identifying sites with high potential as a source of nectar and pollen, a land use/cover map can be used to represent the source of nectar and pollen. It is also one of the specific objectives of this study to contribute potential approach towards addressing this issue by using Remote Sensing (RS) image in mapping the land use/cover of the area under investigation. In this regard, a two-scene Landsat TM image with a resolution of 30 m was used. The derived land use/cover map of La Union (Fig. 13.3), supported by the floral calendar of northern Luzon (ATDC & FAO, 1997) and field observations, was then used to represent the source of nectar and pollen.

During the field survey in June 2005, ground truth points were collected on the different land uses/covers with the aid of a Global Positioning System (GPS) receiver. The vegetation was described and coordinates were recorded. The different cropping systems were known through interview with the local farmers. These points and attribute information were used as a basis in the conduct of a supervised classification on the Landsat TM image.

Generally, the process of supervised classification is divided into two parts: training and classifying. The process of defining the criteria by which patterns are recognized and by which the computer is trained to recognize these patterns is called training (Leica Geosystems, 2005). The training sites were created using the ERDAS Imagine "Signature Editor" with the ground truth points as guide. A maximum likelihood supervised classification method was employed to identify pixels on the image with similar characteristics. The different land use/cover categories (Table 13.2 and Fig. 13.3) were selected and prepared with guidance from the Department of Environment and Natural Resources (DENR), Department of Agriculture (DA) and National Economic and Development Authority (NEDA) in Region 1, Philippines.

Accuracy assessment is a methodical way of evaluating the result of the classification with reference data like, but not limited to, ground truth data, previously tested map and aerial photos (Leica Geosystems, 2005). Furthermore, it is a process of comparing the classified results to any geographical data that are assumed to be true as reference (Richards & Jia, 1999). In this study, since there was no available up-to-date land use/cover map of the study area to be used as a reference

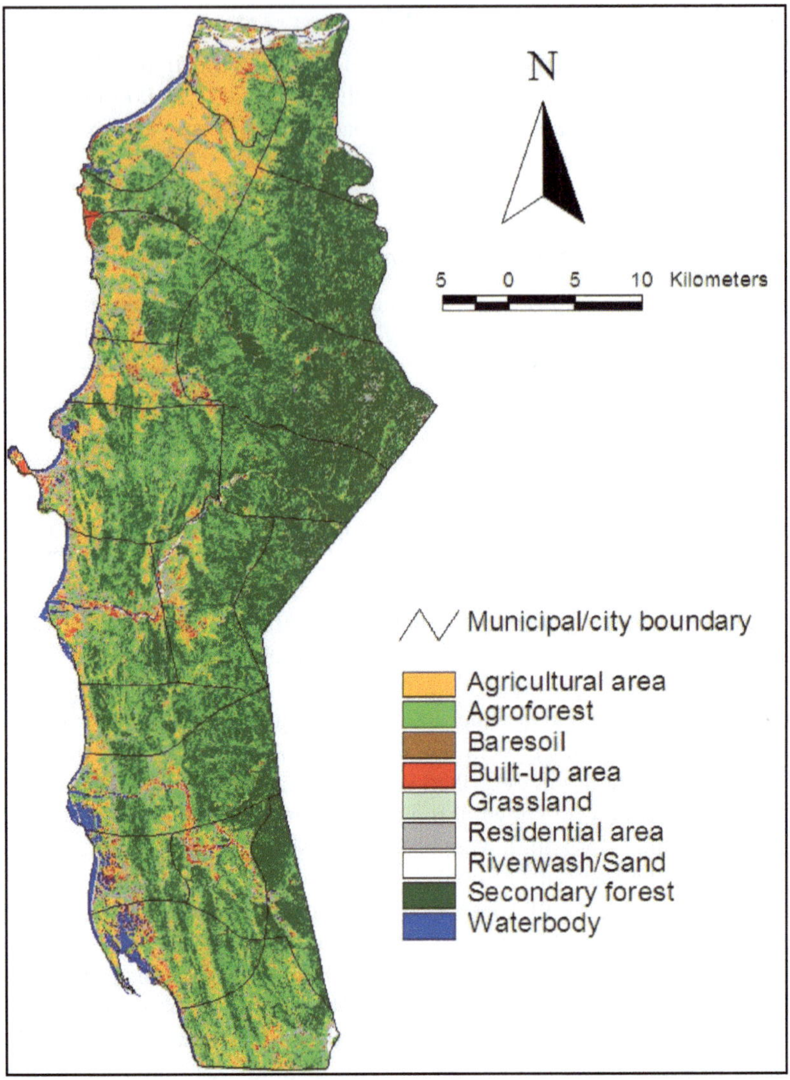

Fig. 13.3 Land use/cover map of La Union, 2004 (Note: Riverwash and sand were merged into one category)

during the accuracy assessment, ground truth points were collected. A total of 300 points (30 points per land use/cover category) were collected through random and purposive sampling techniques. Random sampling technique was necessary to give every location or pixel a chance to become part of the sample while the latter was necessary to ensure that equal samples are collected for each land use/cover category. It was also ensured that the recorded land use/cover in a particular sampling location is the actual land use/cover when the Landsat TM image was acquired.

Table 13.2 La Union land use/cover types and their extent

Land use/cover	Area: Hectare (ha)	Percent (%)	Description
Agricultural area (AG)	25,678.71	17.20	Includes areas cultivated for annual crops such as rice, corn, tobacco, vegetables, legumes, etc.
Agroforest (AF)	30,358.98	20.33	Includes areas with trees and agricultural crops
Baresoil (BS)	2,001.63	1.34	Includes areas without any vegetative cover
Built up area (BU)	2,444.63	1.64	Includes industrial and commercial areas
Grassland (GL)	2,691.17	1.80	Includes grasses and bushes
Residential area (RA)	22,018.31	14.75	Includes housing areas with gardens
Riverwash (RW)	740.93	0.50	Includes soils, sand and rocks/stones deposited along river banks
Sand (SD)	1,077.21	0.72	Includes sand along seashore
Secondary forest (SF)	58,886.91	39.44	Includes second growth forests
Water body (WB)	3,410.72	2.28	Includes sea, ponds and rivers
Total	149,309.19	100.00	

The collected points were encoded in MS excel format and saved as "Windows formatted txt file." Using the "Import user-defined points" option in the "Classifier - Accuracy Assessment" modules in Erdas Imagine software, this file was imported as "ASCII point file." The reference values assigned to each corresponding geographic coordinates were referred to the ground survey data. These values were then compared to the class values of the corresponding locations in the classified image. The error matrix of the classification is presented in Table 13.3. The results showed that there were areas classified incorrectly, for example, RW classified as SD; BS classified as AG; BU classified as RA and SF classified as AF. It was observed that the training sites digitized for each category were the major sources of error because they failed to completely delineate the spectral patterns. Because of this, the training sites were further refined. However, error was not totally eliminated. This is due to the comparable color or spectral patterns of some of the land use/cover categories like RW and SD; SF and AF, and BU and RW.

13.2.3 The Conceptual Model

The conceptual model (Fig. 13.4) shows the integration of GIS and MCE techniques in the suitability analysis for beekeeping sites in the province of La Union, Philippines. It uses information ranging from ground truth to digital data. The model draws three main steps such as database creation and management, spatial multi-criteria analysis, and validation process.

Database creation and management include the collection of both primary and secondary data, and preparation of thematic layers for analysis. Spatial multi-criteria

Table 13.3 Error matrix of the land use/cover supervised classification

Classified data	Reference data											
	BS	GL	RW	SD	SF	WB	AG	RA	AF	BU	Total	UA (%)
BS	23	1	1	0	0	0	1	0	0	0	26	88
GL	1	26	0	0	0	0	1	0	0	0	28	93
RW	1	1	25	2	0	0	1	0	0	1	31	81
SD	2	1	2	26	0	0	0	0	0	0	31	84
SF	0	0	0	0	27	0	0	1	2	0	30	90
WB	0	0	0	0	0	28	1	0	1	0	30	93
AG	1	1	1	1	0	1	26	0	0	0	31	84
RA	1	0	0	0	1	0	0	26	1	2	31	84
AF	0	0	0	0	2	1	0	1	26	0	30	87
BU	1		1	1	0	0	0	2	0	27	32	84
Total	30	30	30	30	30	30	30	30	30	30	300	
Producer's accuracy (%)	77	87	83	87	90	93	87	87	87	90		

Overall classification accuracy = 86.67%; Overall kappa statistics = 0.8519; UA= User's accuracy

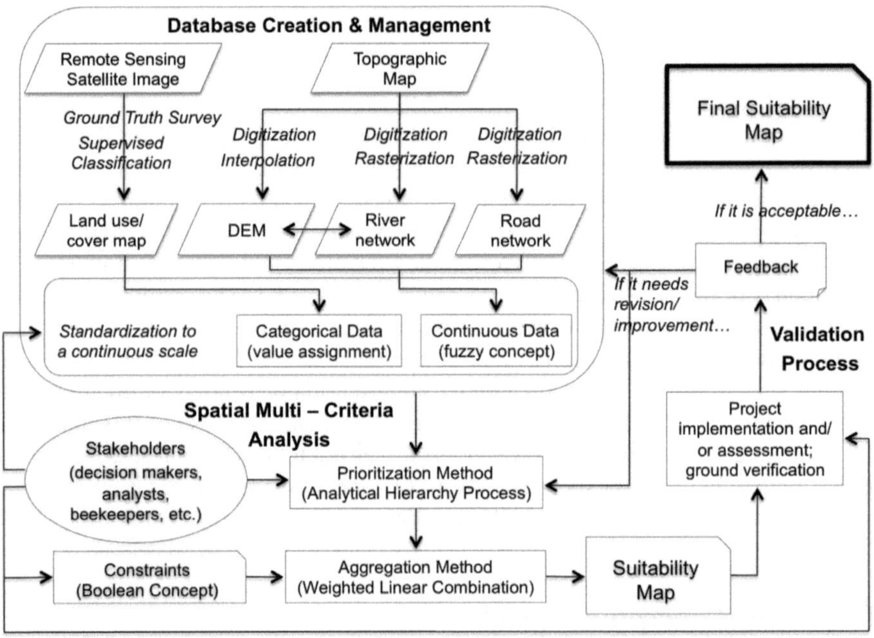

Fig. 13.4 The conceptual model

analysis involves the assessment, prioritization, and weighing of criteria (factors and sub-factors) and constraints. It also includes the determination of the consistency ratio in the AHP based on the experts' knowledge and perceptions. The validation component of the model provides opportunity to determine the reliability of the suitability map based on assessments. Assessments may be done through, but not limited to either one or a combination of the following: (1) project implementation; (2) ground verification; and (3) use of existing projects.

13.2.4 The Standardization of Factors to a Continuous Scale

The "fuzzy" concept was used to standardize the factors to a continuous scale. Standardization was needed "to transform the disparate measurement units of the factor images into comparable suitability values" (Eastman, 2006). The factors were standardized to a byte-level range of 0–255 in order to use "fuzzy" factors with the multi-criteria evaluation with 255 indicating the highest suitability. The experts' knowledge of how suitability changes for each factor was used as a basis in the selection of parameter for this standardization. Specifically, since most of the bee plants currently grown in the province do not grow well in higher elevation (*i.e.* >1000 m asl) where temperature can drop below the minimum temperature for active honeybee foraging, which is approximately 55°F (13°C), the elevation factor was standardized showing an inverse relationship with the suitability values. Good accessibility to water source is important because bees use water to cool their hives during hot weather and to dilute the honey for their own consumption during extreme conditions (BBKA, 2006). Likewise, access to a road is essential for better monitoring and evaluation of the beehives as part of the management routine. Thus, areas closer to these features were considered more suitable.

On the other hand, the land use/cover factor was standardized using a different approach. Since it is a categorical set of data, the suitability of the different land uses/covers was scaled in the range of 0–255 by assigning suitability values based on experts' knowledge, and careful and fair judgment. Agroforest was given a value of 255, while agricultural area, secondary forest, residential area with gardens and grassland were given a value of 200, 150, 100 and 50, respectively. Other land uses/covers considered constraints were masked out. The main reason why agroforest was given the highest value is that it is the only land use that sustains the availability of nectar and pollen almost all year around. Agroforestry is a cropping system of growing agricultural crops and trees, be it forest or fruit trees, simultaneously or sequentially in a given piece of land. Trees during their blooming seasons augment the fallow periods in agriculture and vice versa. Based on the floral calendar of Northern Luzon, Philippines (ATDC & FAO, 1997), not a single bee plant flowers continuously, except for coconut with a relatively short flowering period gap and usually a main component of an agroforestry farm. On the other hand, there are more bee plants in the agricultural area than the rest of the land uses. Some produce flowers in different times of the year, stretching the period of availability of nectar and pollen. Thus, agricultural area was given the second highest suitability value. The suitability values of the rest of the land uses were also based on the experts' knowledge, floral calendar and ground observation. The standardized factors are presented in Fig. 13.5.

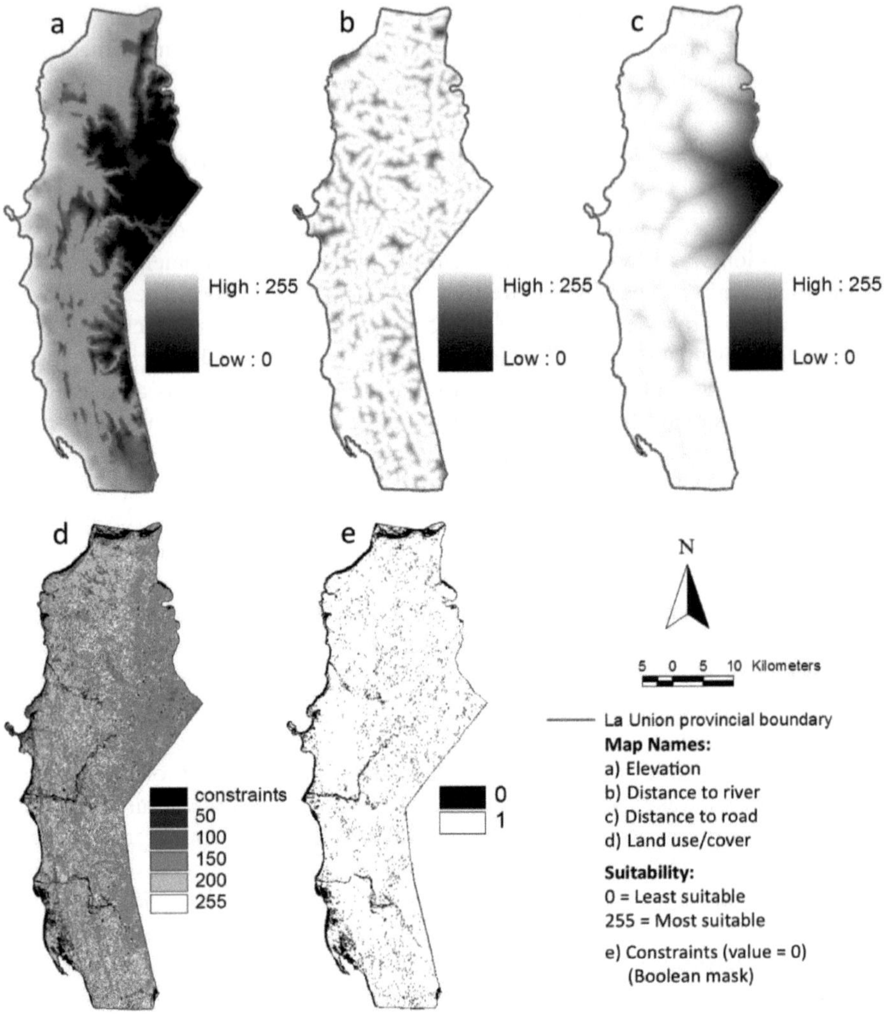

Fig. 13.5 The standardized factors

13.2.5 The Analytical Hierarchy Process (AHP)

The Analytical Hierarchy Process (AHP) "*is a theory of measurement through pairwise comparisons that relies on the judgments of experts to derive priority scales*" (Saaty, 2008). Furthermore, it is a decision-making technique based on the inherent ability of people to make excellent decisions (Saaty, 1994). It allows the analysts and decision-makers to explore all possible options to fully understand the underlying problems before a choice is made.

In this study, the development of a suitability map for beekeeping involved the consideration of a set of criteria that includes the factors, sub-factors and constraints

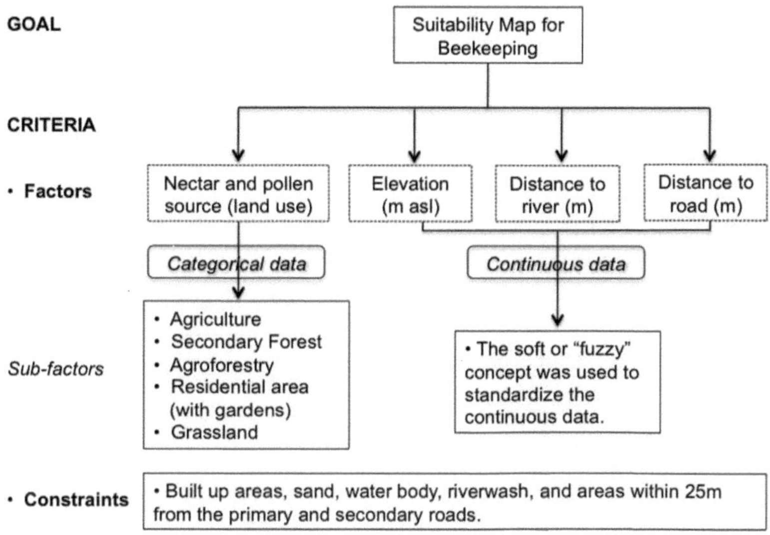

Fig. 13.6 Suitability analysis hierarchy for beekeeping

as presented in Fig. 13.6. AHP was used to determine the respective weight of each factor based on existing knowledge and information, deep understanding on the requirements, and consistency of judgment based on preference. Numerical values, expressing the judgment of the relative importance (or preference) of one factor against another, had to be assigned to each factor according to a comparison scale consisting of values ranging from 1 to 9 which describe the intensity of importance (preference/dominance) as suggested by Saaty and Vargas (1991) and Saaty (2008) as follows: 1 – Equal importance; 3 – Moderate importance of one factor over another; 5 – Strong or essential importance; 7 – Very strong importance; 9 – Extreme importance; and 2, 4, 6, 8 are intermediate values.

The experts' opinions were necessary in assigning the weights of each criterion as well as the suitability values of each of the sub-criteria. Thus, a core group of decision makers, composed of a beekeeping specialist, agriculturist, forester and analyst, was consulted for this purpose to minimize bias. After a thorough deliberation, it was decided to come up with a "consensus" evaluation among the group. The paired comparison matrix including the final weight of each of the criteria is presented in Table 13.4. The suitability map was produced using the standardized criteria and AHP weight following the WLC method as presented in Eq. (13.1). The final suitability map was re-classed into 5 qualitative scales using the equal interval method (very high suitability, high suitability, medium suitability, low suitability and very low suitability) (Fig. 13.7).

$$\mathrm{WLC} = \left(\sum_{i=1}^{n} X_i \times W_i \right) \times C \tag{13.1}$$

Table 13.4 Paired comparison matrix

Factors	Land use	Elevation	Distance to river	Distance to road	Priority vector or weight
Land use	1	7	3	5	0.5650
Elevation	1/7	1	1/5	1/3	0.0553
Distance to river	1/3	5	1	3	0.2622
Distance to road	1/5	3	1/3	1	0.1175

Consistency ratio = 0.04 < 0.10 (acceptable)

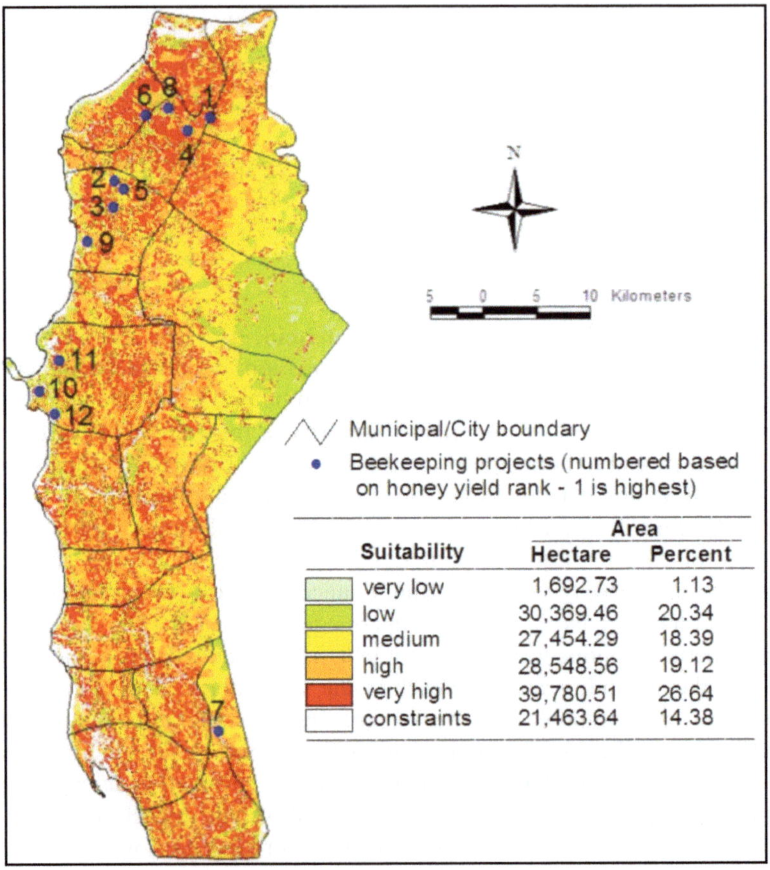

Fig. 13.7 The La Union suitability map for beekeeping

where WLC = Weighted Linear Combination, X_i = Decision parameters (factors and sub-factors), W_i = AHP Weight, n = Number of parameters (factors), C = Constraints.

13.2.6 The Validation Process

The effectiveness of the conceptual model used and the reliability of the output produced was validated through the use and assessment of existing beekeeping projects. As mentioned earlier, the suitability map can be validated in several ways. In this study, the final suitability map was validated through a correlation analysis of the honey yield of the existing beekeeping projects and the calculated suitability values.

In order to obtain a more scientific and reliable assessment, established commercial scale projects should be used as parameters in the analysis. However, since beekeeping industry in the province has just started several years ago, such kind of projects is not yet available especially during the conduct of this study. Thus, the twelve 3–5 years old small-scale projects, raising *Apis mellifera* honeybee species in the study area, were used. In November 2008, a ground survey was conducted to gather information about the geographic locations, physical characteristics, honey yield, number of colonies, and problems encountered of these projects. The survey was aided with GPS receiver. In order to assess the validity of the suitability map, geographic coordinates of each project were loaded into the GIS platform. This made it possible to do some spatial analyses like respective locations of the projects in the suitability map (*i.e.* determining the elevation and land use/cover where the projects are established), proximity analysis (*i.e.* determining the projects' distances to roads and rivers), and correlation analysis between honey yield and suitability index. Since the status of the land use/cover map is as of December 2004, the 2005 set of honey yield data was used in the analysis.

13.3 Results and Discussion

Both GIS and MCE showed their usefulness in dealing with the complexity of suitability analysis for beekeeping sites where a multifaceted set of guidelines had to be considered. By using the empirical conceptual model, the development and validation of the suitability map for beekeeping in the province of La Union, Philippines, was successful (Fig. 13.7 and Table 13.5). The model's inherent opportunities to invoke experts' participation and its cyclical characteristics have demonstrated some strength in achieving agreeable and acceptable results within scientific standards.

The standardization of data to a common 0–255 continuous scale has helped in the determination of suitability values of the areas where existing knowledge and information could not justify any appropriate value or specific weight. This was experienced when the thematic layers of the continuous data like the DEM, distances to river and road were being prepared. The aggregation method used in this study, WLC, allowed retention of the variability from the continuous factors, and made it possible to trade off with each other. For example, in a particular area, a low suitability value in one factor was balanced or counterweighed by a high suitability value in the other factors. Furthermore, detailed examination of the suitability map and the land use/cover map revealed that although agroforestry was given the highest suitability value, this did not guarantee that all agroforestry areas had to have the

Table 13.5 The result of the validation using the 12 existing beekeeping projects

Rank based on production rate	Geographic coordinates (UTM 51 N) *x*	*y*	2005 production rate (kg/colony)	Suitability value where the project is located
1	229,019	1,862,524	37.50	252
2	220,036	1,856,655	33.33	240
3	219,872	1,854,181	21.05	222
4	226,969	1,861,279	20.00	214
5	220,892	1,855,846	19.17	211
6	223,035	1,862,736	14.00	204
7	229,723	1,804,923	13.50	194
8	225,145	1,863,452	12.67	204
9	217,464	1,850,916	11.00	123
10	212,872	1,836,852	10.00	146
11	214,720	1,839,716	8.00	151
12	214,298	1,834,668	5.00	49
Pearson product moment correlation coefficient, r, between the production rate and suitability value				*0.8010*

same high suitability values. Moreover, this did not preclude other land uses like agriculture to be included and become part of the most suitable areas in the final suitability map. It was really a dynamic process since all factors came into play. The extent or degree of trade-off, however, depends on the AHP weights that defined the relative importance of each factor.

Although areas under very high suitability class are the most ideal in the final suitability map, the acceptable limit of suitability still depends on the stakeholders' decision. Nevertheless, the result of the validation process has strengthened the reliability of the suitability map as indicated by the good correlation between the honey yield of the existing projects and the suitability index. It must be pointed out, however, that the projects' yield performance was used directly in the analysis without bringing into the analysis whatever management factors may have influenced each project. Furthermore, it does not necessarily mean that the honeybees have foraged only where the beehives were placed and where the suitability values used in the correlation analysis were recorded. Honeybees might have flown to other places too. However, the distance of flight depends on the spatial variability and availability of nectar and pollen in the foraging landscape. According to Steffan-Dewenter and Kuhn (2003), foraging flight distance depends on the plant and floral density that generally provides the needed nectar and pollen for the honeybees. Thus, flight distance is shorter if there is a high plant and floral density in a given area. BBKA (2006) mentioned, "honeybees mostly forage for both nectar and pollen within a kilometer of their hive and up to about five kilometers for exceptionally rewarding sources." Moreover, if within the foraging area nectar and pollen are scarce, bees are going to find suitable foraging patches

(Naug, 2009). In the simulation study conducted by Ricketts et al. (2008) concerning the visitation rate of *Apis mellifera* to a beeplant, they determined that the rate decreased to 50% of its maximum between 2 and 4 km distances depending on the characteristic of the foraging area. In tropical areas where the landscapes are generally more diverse, visitation rates to farther distances are expected to drop more drastically than in temperate areas (Sande, Crewe, Raina, Nicolson, & Gordon, 2009).

However, although honeybees may fly to other places to forage, other advantage of knowing the location of the most suitable areas, aside from good harvest, is the inherent possibility and opportunity of minimizing the potential risk of absconding. Absconding usually happens when the foraging landscape is fragmented and there is a scarce source of nectar and pollen near the beehives that forces honeybees to increase their flight distance. Efficiency is another advantage. The shorter the distance honeybees fly in collecting nectar and pollen, the more efficient they are. The life expectancy of foraging bees may also be affected by foraging at extreme distances wearing out their wings, and eventually affecting the efficiency of the colony. Thus, in order to really achieve these advantages, spatial contiguity of suitable areas is important in the final decision-making process as to where the beehives shall be placed. In this instance, the suitability map will be of great importance.

13.4 Conclusions

Despite of the limitations encountered particularly on the unavailability of a detailed map of bee plants, this study has successfully evaluated the suitability of the province of La Union for beekeeping. The empirical conceptual model used, which integrates GIS, RS and MCE techniques, has made it possible to attain the objective of this study. The conceptual model did not only invoke experts' knowledge and participation, and enhance the decision-making capabilities, but it has also made the decision-making process more dynamic. The good correlation between the honey yield of the existing projects and the calculated suitability values has indicated that the model used and the output produced are reliable. Thus, this study can provide significant input to those who are interested in beekeeping, GIS-based MCE techniques and suitability analysis and mapping. It can also give insights in the design of a more comprehensive conceptual model in the future, taking into consideration the limitations that were encountered.

Finally, the suitability analysis for beekeeping sites integrating GIS and MCE techniques is particularly helpful in the establishment of a good foundation to make beekeeping a sustainable livelihood. It can help farmers/beekeepers minimize the potential risk of failure especially during the initial stage of their engagement into this activity, where financial capital is usually involved. In the implementation of any livelihood projects, a strong foundation must always be taken into account in order to attain success and sustainability in the future.

Acknowledgments The authors wish to thank the NARTDI staff, Forester Leonard C. Cachero, and Ms. Ria S. Estoque of SRDI, DMMMSU, for their assistance during the fieldwork and data gathering.

References

ATDC & FAO. (1997). *Beekeeping guide for the Philippines*. La Union, Philippines: Don Mariano Marcos Memorial State University.

BBKA. (2006). *Choosing an apiary site*. British Beekeepers Association Advisory Leaflet Number B11. The British Beekeepers' Association, The National Agricultural Centre, Stoneleigh, Warwickshire CV8 2LG.

Boonyanuphap, J., Wattanachaiyingcharoen, D., & Sakurai, K. (2004). GIS-based land suitability assessment for Musa (ABB group) plantation. *Journal of Applied Horticulture*, *6*, 3–10.

DFID. (1999). *Sustainable livelihoods guidance sheets, numbers 1–8*. London: Author.

De Padua, V. M. (2009). *Cost and return analysis for Apis mellifera colonies in the Ilocos Region*. Technical Report. La Union, Philippines: Don Mariano Marcos Memorial State University.

Eastman, J. R., Kyem, P. A. K., & Toledano, J. (1993). A procedure for multi objective decision making in GIS under conditions of conflicting objectives. *Proceedings of the fourth European Conference on Geographic Information Systems*, 29 March-1 April 1993, Genoa, Italy.

Eastman, J. R. (2006). *IDRISI Andes tutorial*. Woroester, MA: Clark University.

FAO. (1987). *Beekeeping in Asia. Agricultural services Bulletin 68/4. Agricultural support system division*. Rome, Italy: FAO.

FAO. (1990). *Beekeeping in Africa. Agricultural services bulletin 68/6. Agricultural support system division*. Rome, Italy: FAO.

FAO. (2003). *Beekeeping and sustainable livelihoods. diversification booklet 1. Agricultural support system division*. Rome, Italy: FAO.

Farajzadeh, M., Bayati, R. M., Rahimi, M., & Modares, T. (2007). *Preparation of saffron cultivation suitability map based on the comparison of different weighting methods in GIS environment*. GIS Development. Retrieved February 23, 2010, from www.gisdevelopment.net

Garibaldi, L. A., Aizen, M. A., Cunningham, S. A., & Klein, A. M. (2009). Pollinator shortage and global crop yield: Looking at the whole spectrum of pollinator dependency. *Communicative & Integrative Biology*, *2*, 37–39.

Ghribi, M. (2005). *GIS applications for monitoring environmental change and supporting decision-making in developing countries*. Italy: ICS-UNIDO.

Hubina, T., & Ghribi, M. (2008). *GIS-based decision support tool for optimal spatial planning of landfill in Minsk region, Belarus*. Proceedings of 11th AGILE International Conference on Geographic Information Science, University of Girona, Spain.

Jankowski, P. (1995). Integrating geographical information systems and multiple criteria decision – making methods. *International Journal of Geographical Information Systems*, *9*, 251–273.

Jorein, F., Theriault, M., & Musy, A. (2001). Using GIS and outranking multi-criteria analysis for land use suitability assessment. *International Journal of Geographical Information Systems*, *15*, 153–174.

Krantz, L. (2001). *The sustainable livelihood approach to poverty reduction: An introduction*. Swedish International Development Agency (SIDA).

La Union Provincial Government. (2010). *La Union profile*. Retrieved January 7, 2010, from http://www.launion.gov.ph

Laaribi, A., Chevallier, J. J., & Martel, J. M. (1996). A spatial decision aid: A multi-criterion evaluation approach. *Computers, Environment and Urban Systems*, *20*, 351–366.

LeicaGeosystems. (2005). *ERDAS field guide*. Norcross, GA: Leica Geosystems Geospatial Imaging, LLC.

Marinoni, O. (2005). A discussion on the computational limitations of outranking methods for land use suitability assessment. *International Journal of Geographic Information Science*, *20*, 69–87.

NARTDI. (2008). *Technical report*. Philippines: Don Mariano Marcos Memorial State University.

National Statistics Office (NSO). (2009). *Population census data*. Republic of the Philippines: National Statistics Office (NSO).

Naug, D. (2009). Nutritional stress due to habitat loss may explain recent honeybee colony collapses. *Biological Conservation, 142*, 2369–2372.

Oldroyd, B. P., & Nanork, P. (2009). Conservation of Asian honey bees – Apidologie. *Bee Conservation, 40*, 296–312.

PAGASA Agromet Station, DMMMSU-NLUC. (2004). *Daily rainfall record*. La Union, Philippines: Don Mariano Marcos Memorial State University.

Pereira, J. M. C., & Duckstein, L. (1993). A multiple criteria decision- making approach to GIS-based land suitability evaluation. *International Journal of Geographical Information Systems, 7*, 407–424.

Richards, J. A., & Jia, X. (1999). *Remote sensing digital image analysis*. Berlin: Springer.

Ricketts, T. H., Regetz, J., Stephan-Dewenter, I Cunningham, S. A., Kremen, C., Bogdanski, A., Gemmill-Herren, B., Greenleaf, S. S., Klein, A. M., Mayfield, M. M., Morandin, L. A., Ochieng, A., & Viana, B. F. (2008). Landscape effects on crop pollination services: Are there general patterns? *Ecology Letters, 11*, 499–515.

Roy, W., & Grealish, G. (2004). Mapping arable soils using GIS-based soil information database in Kuwait. *Management of Environmental Quality: An International Journal, 15*, 229–237.

Saaty, T. L. (1980). *The analytic hierarchy process*. New York: McGraw Hill.

Saaty, T. L. (1994). How to make a decision: The analytic hierarchy process. *Interfaces, 24*, 19–43.

Saaty, T. L. (2008). Decision making with the analytic hierarchy process. *International Journal of Services Sciences, 1*, 83–98.

Saaty, T. L., & Vargas, L. G. (1991). *Prediction, projection and forecasting*. Norwell: Kluwer.

Sande, S. O., Crewe, R. M., Raina, S. K., Nicolson, S. W., & Gordon, I. (2009). Proximity to a forest leads to higher honey yield: Another reason to conserve. *Biological Conservation, 142*, 2703–2709.

Solesbury, W. (2003). *Sustainable livelihoods: A case study of the evolution of DFID policy. Working paper 217*. London, UK: Overseas Development Institute.

Steffan-Dewenter, I., & Kuhn, A. (2003). Honeybee foraging in differentially structured landscapes. *Proceedings of the Royal Society of London, 270*, 569–575.

Thapa, R. B., & Murayama, Y. (2008). Land evaluation for peri-urban agriculture using analytical hierarchy process and geographic information system techniques. *Land Use Policy, 25*, 225–239.

Tudes, S., & Yigiter, N. D. (2010). Preparation of land use planning model using GIS based on AHP: Case study Adana-Turkey. *Bulletin of Engineering Geology and the Environment, 69*, 235–245.

Chapter 14
Spatial Allocation of the Best Shipping Canal in South Thailand

Rajesh B. Thapa, Michiro Kusanagi, Akira Kitazumi, and Yuji Murayama

14.1 Introduction

The mobility of people and freight from one place to another and information communication are becoming basic needs of human beings. Contemporary economic processes have been accompanied by a significant growth in mobility and higher levels of accessibility. Now, societies have been increasingly dependent on their transport systems to support a wide variety of activities ranging from travelling, supplying energy needs to distributing products (Rodrigue, 2005; Thapa, Kusanagi, Kitazumi, & Murayama, 2007). Globalization processes, economic integration and expansion of international trade have extended considerably the need for international transportation. International transportation is concerned with the highest scale in the mobility of freight and passengers with intercontinental and inter-regional movements by air, land and ocean using modern means of transportation. World output grew in 2007 by 3.8% over 2006. Developed economies were able to grow at 2.5%, whereas developing countries recorded an average growth of 7.3%. World merchandise exports grew by 5.5% (UNCTAD, 2008). Furthermore, world seaborne trade recorded 8.02 billion tons in 2007 for an annual growth rate of 4.8%.

The international shipping industry is responsible for the carriage of 90% of world trade. Shipping is the backbone of the global economy. Intercontinental trade, the bulk transport of raw materials, and the import/export of affordable food and manufactured goods are dependent on shipping (ShippingFact, 2005). For instance, consider the oil movement; a large volume of oil consumption occurs mainly in the industrialized countries, while oil production takes place largely in the Middle

R.B. Thapa (✉)
Division of Spatial Information Science, Graduate School of Life and Environmental Sciences, University of Tsukuba, Tsukuba, Ibaraki, Japan
e-mail: thaparb@yahoo.com; thaparb@gmail.com

This chapter is improved from "Rajesh Bahadur Thapa, Michiro Kusanagi, Akira Kitazumi, and Yuji Murayama (2007), Sea navigation, challenges and potentials in South East Asia: An assessment of suitable sites for shipping canal in South Thai Isthmus, GeoJournal, 70, 161–172", with permission from Springer.

Y. Murayama, R.B. Thapa (eds.), *Spatial Analysis and Modeling in Geographical Transformation Process*, The GeoJournal Library 100, DOI 10.1007/978-94-007-0671-2_14,

East, former Soviet Union, West Africa, and South America. The overall volume of seaborne crude oil trade increased by 4.5% in 2005. Average freight indices for VLCC/ULCC, Suezmax and Aframax tonnage increased by 86, 40 and 27% respectively (UNCTAD, 2006). A significant volume of oil is traded internationally of which 66% of the world's oil trade (crude and refined oil) moves by VLCC's tanker carrying 2 million barrels of oil per voyage. Tankers have made global (intercontinental) transport of oil possible because of low cost, efficiency, and extreme flexibility as compared to other means of transportation (EIA, 2008; Thapa et al., 2007).

The world trading fleet was made up of 46,222 ships, with a combined tonnage of 598 million gross tons (ShippingFact, 2005). Transportation by sea generally follows a fixed set of maritime routes (Fig. 14.1). Along the way, it encounters several narrow channels, such as the Strait of Malacca linking the Indian Ocean and the Strait of Hormuz leading out of the Persian Gulf. Other important maritime passage-ways include the Bab el-Mandab passage from the Arabian Sea to the Red Sea; the Panama Canal connecting the Pacific and Atlantic Oceans; the Suez Canal connecting the Red Sea and Mediterranean Sea; and the Turkish Straits linking the Black Sea to the Mediterranean Sea. Such narrow points are critically important to world trade because physically they are narrow and theoretically could be blocked temporarily and are more susceptible to pirate attacks and shipping accidents.

UNCTAD (2004) reported the growth rates for merchandise trade measured in value for most of the Asian countries were impressive. The average export and import growth rates for 40 selected economies reached 14.8% in 2003. Asian countries are major players in world maritime transport, with sizeable shares in several activities. These countries accounted for 35.8% of containership ownership, 45.7% of containership operation, 60.4% of seamen, 62.3% of container port throughput, 64.7% of container port operators, 83.2% of containership shipbuilding and 99% of

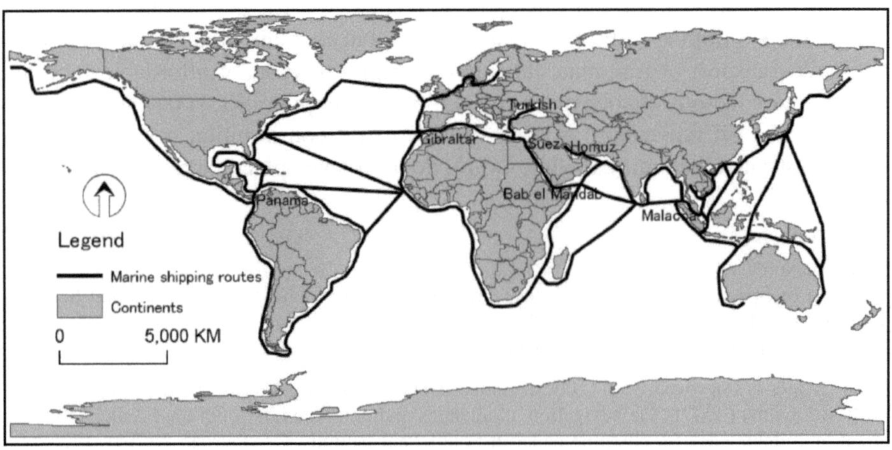

Fig. 14.1 World trading – marine shipping routes

ship demolition. In addition to being one of the focuses of the main east-west shipping routes articulated around world port leaders such as Hong Kong (China) and Singapore, it is also the focus of an intensive and significant intra-Asian shipping trade.

14.2 Sea Navigation in the Asia Pacific Region

At present, three routes (Malacca, Sunda and Lambok straits) are in operation (Thapa et al., 2007) for sea navigation between the Indian Ocean and the South China Sea. The Sunda route is very long as compared to Malacca for the world's giant traders like China, Japan, South Korea and the Middle East countries. The Malacca route is most common and very busy for east-west trade. Malacca is a passage between Sumatra and the Malay Peninsula covering 900 km long distance with a width of 3–350 km. The shallowest depth of the strait is reported to be 25 m (Kuppuswamy, 2004). By the Malacca route, a mass of refined and crude oil, iron ore, raw materials as well as readymade products pass west to east and vice versa on a daily basis. VLCC/DEPP Draft CR, Tanker Vessel, LNG/LPG Carrier, Cargo Vessel, Container Vessel, and Bulk Carrier are the major means of shipping transportation in the strait (Table 14.1). Passenger vessels, government/navy vessels, fishing vessels also use the route. Container vessels and tanker vessels are more prominent users of the strait. The sea traffic passing through the Malacca Strait was counted as 43.9 thousand vessels in the year 1999 with an average of 120 vessels per day. The traffic greatly increased setting a record of 194 vessels per day in 2007. Traffic has been growing gradually in the Malacca Strait with the

Table 14.1 Shipping traffic in the Malacca Strait (1999–2007)

Traffic type	1999	2001	2003	2005	2007
VLCC/DEPP draft CR	2027	3303	3487	3788	3753
Tanker vessel	11,474	14,276	15,667	14,759	14,931
LNG/LPG carrier	2473	3086	3277	3099	3413
Cargo vessel	5674	6476	6193	6340	8467
Container Vessel	14,521	20,101	19,575	20,818	23,736
Bulk carrier	3438	5370	6256	7394	9684
RORO/car carrier	1229	1764	2182	2515	3137
Passenger vessel	1919	3151	3033	2299	1870
Others[a]	1210	1787	2664	1609	1727
Total	43,965	59,314	62,334	62,621	70,718
Average traffic per day	*120*	*163*	*171*	*172*	*194*

Source: MDM (2008)
STRAITREP ship reporting system in the Strait of Malacca and Singapore has been operating since December 1998. The statistics cover only the Malacca Strait from One Fathom Bank to Iyu Kechil
[a]Includes livestock carriers, TUG/TOW vessels, government/navy vessels, fishing vessels, etc

exceptions of slight slowdown in 2005. Reduction in tanker vessels and the passenger ships caused the slow growth in 2005. However, continues growth in all types of traffic except a sharp declining of passenger vessels after 2001, made the strait more crowded and vulnerable in recent years.

14.3 Environmental and Social Consequences

Mitropoulos (2004) reported that ship movements carrying one quarter of the world's commerce and half the world's oil pass through the Straits of Malacca each year. It shows that traffic is increasing heavily in the Malacca Strait and Singapore seaport, which will exceed beyond all capacity in the next few decades if alternative setups are not introduced for sea navigation. At present, due to heavy traffic the strait is fascinating for its environmental problems, ship building and demolition, and oil spillages for example. In 2007, VLCC/DEPP Draft CR and Tankers accounted for 10 and 41 vessels per day on an average, respectively. These types of shipping are considered as source of oil spills. Collision and oil spillage are major environmental threats in the Strait of Malacca. Oil spills happen more frequently in certain parts of the world. The Oil Spill Intelligence Report 1999 listed Malaysia and Singapore as hot oil spill spots from vessels accounting, for 39 cases since 1960 (UNEP, 2005).

In 1993, the 255,312-ton Singapore-registered tanker *Maersk Navigator* collided with the empty tanker *Sanko Honour* in the Andaman Sea en route from Oman to Japan. It was carrying a cargo of nearly 2 million barrels of oil. The ruptured tanker leaked burning oil and spread a slick up to 35 miles (56 km) long off Sumatra, drifting towards India's Nicobar Islands (Mariner Group, 2005). Oil spills on water devastates the marine environment. The *Treasure* accident can be taken as an example of an oil spill and its impact on marine environment. In the early morning of 23 June 2000, the bulk ore carrier MV *Treasure* sank off western South Africa between Dassen and Robben islands, both are important for seabird colonies. About 19,000 oiled penguins were collected for cleaning, 150 died on the spot and about 2000 died within the first month. In addition to the African Penguin, other endangered seabirds were at risk from the *Treasure* spill (Crawford et al., 2000).

In addition to oil, hazardous and poisonous chemicals are also transported via Malacca. It could be an environmental problem if a collision or sinking took place. An Indonesian tanker laden with a toxic chemical capsized off Malaysia's southern Johor state, just across from Singapore in 2001. MV *Endah Lestari* was on its way to East Kalimantan in Indonesia with some 600 tones of the poisonous industrial chemical phenol, and 18 tones of diesel. The toxic spill killed thousands of fish and cockles reared in 85 offshore cages, and the Singapore authorities warned its citizens to stay away from nearby waters. It was tough to mop up the phenol, as it is soluble in water (Mariner Group, 2005).

Leaking oil in the sea and some growing social problems like robberies are some prominent examples of socio-environmental degradation in the area. In recent years, pirate attacks seem to be more violent and sophisticated in nature (Mukundan, 2008). Although, piracy and armed robbery cases rose by 10% world-wide in 2007 (ICC-IMB, 2008), significant improvement in the decline of cases have been

Table 14.2 Piracy incidents in Indonesia, Malaysia, Singapore and Malacca Straits

Places/year	2001[a]	2002[a]	2003	2004	2005	2006	2007
Indonesia	72	103	121	94	79	50	43
Malaysia	9	14	5	9	3	10	9
Singapore	4	5	2	8	7	5	3
Malacca Straits	11	16	28	38	12	11	7
Total	96	138	156	149	101	76	62

Source: ICC-IMB (2008)
[a]Compiled based on the piracy maps available online http://www.icc-ccs.org/prc/piracyreport.php

observed in the Malacca Strait (Table 14.2). The number of reported incidents in Indonesia, Malaysia, Singapore and Strait of Malacca has fallen to 62 in 2007, a steady and yearly decline from 156 in 2003. Attacks in the Malacca Strait, previously known as a hotspot, as well as in Singapore, have continued to drop since 2004. The littoral states (Malaysia, Indonesia and Singapore) started a coordinated patrolling in the middle of 2004. This might be a cause for the fall of piracy and armed robbery cases in the strait (Permal, 2006).

However, sea piracies, robberies and hijackings are kinds of social illnesses which are difficult to predict in the current situation of growing worldwide terrorism. The security concern should always be given high priority in such a bottleneck and economically strategic waterway. As more than 70 thousand ships sail this waterway each year, where approximately one third of the world's trade and half of the world's oil passes to countries such as China, Korea and Japan. Terrorist attacks in such places would have the potential to cause large-scale economic impact, not just regionally but on a global scale.

Considering heavy traffic and its pressure on the environment in the coming decades, research and negotiations for seeking an alternative to reduce the volume of traffic in Malacca Strait are underway in Thailand and Malaysia. The Land Bridge Project and Kra-Canal Project in the South Thai Isthmus and the Trans-Peninsula Pipeline Project in Northern Malaysia for oil transportation (Permatasari, 2007) are some of the prominent alternatives being debated in the South East Asian region. The Land Bridge (Krabi-Khanom) project consisting of a highway, railway and oil pipeline was approved for constructed in 1993 by the Thai government. The project was initially started but has now completely stopped because of some environmental concerns raised in the South Thai territory. Similarly, a recent announcement of an ambitious plan for constructing a pipeline in Northern Malaysia is being debated in Malaysia. The proposed Trans-Peninsula Pipeline will provide a shortcut oil shipment route for the exporting countries in the Middle East to importing East Asian nations, bypassing the Malacca Straits (Khalid, 2007; Permatasari, 2007). It will reduce a significant number of VLCC and oil tankers currently passing through Malacca and transportation cost as well, if the project is implemented. But it requires a detailed feasibility study that clearly addresses several issues, for example, environmental, socio-economic, investment, security and geopolitical. These processes will certainly take considerable time for making decisions on the fate of the pipeline project in the peninsula. A potential shipping route passing through the Indo-Chinese Peninsula in the south Thai Isthmus as an alternative to lower the volume of traffic in the Strait of Malacca has been studied since the 19th century (VIC, 2002).

14.4 Potential Sites Assessment for Shipping Canal

14.4.1 Recent Status in Thailand

Some historical efforts seeking an alternative shipping route for Malacca Strait were found in Thailand. The idea of constructing a shipping canal in Southern Thailand to link the Pacific and Indian Oceans was started, 300 years ago during the reign of King Narai (ES, 2002). Lots of studies on the project have been done since then. It has been considered many times under various Thai cabinets. For instance, King Narai was the first to suggest building a canal across the Kra Isthmus, requesting the French engineer De Lamar to conduct a survey to test the feasibility of connecting the existing waterway out of Songkla in the Gulf of Thailand to Marid (Myanmar). The Kra Isthmus, southern Thailand is narrower Isthmus in Thailand, which separates the Indian and the Pacific oceans. About 300 year old dream to build a canal across this isthmus has been revived many times, as in the mid-1950s, early 1970s, 1990s and 2000s (Pasertsri, 2005). However, the project was dropped but it has been surfaced periodically ever since.

After comparing different routes of the canal, the TAMS (1973) suggested that the Satun-Songkhla route was the most suitable, which is similar to the most recent research of the Mitsubishi Research Institute and Global Infrastructure Fund (MRI-GIF), Japan (TED, 2005). The MRI-GIF study proposed a king-sized canal to run from Satun to Songkhla near the Thai-Malaysian border capable of handling 150,000 dwt vessels going in both directions and the construction of seaports and major industrial facilities at both ends of the canal (VIC, 2002). The engineering suitability highlighted that the size and depth of canal should be 380 and 23 m, respectively. The study concluded that the canal project is feasible both in terms of economics and engineering, but it needs a detailed study on canal alignment, environment, economics, engineering aspects, including a public hearing, to identify the most suitable route.

The aim of this paper is to analyze the sea transportation situation in the Strait of Malacca and evaluate suitable sites in the South Thai Isthmus (Kra Isthmus) for a potential shipping canal using physiographic spatial data.

14.4.2 Study Area, Database and Method

The southern region of Thailand covering 14 administrative provinces was selected for assessing the potential sites of shipping route (Fig. 14.2). Malaysian and

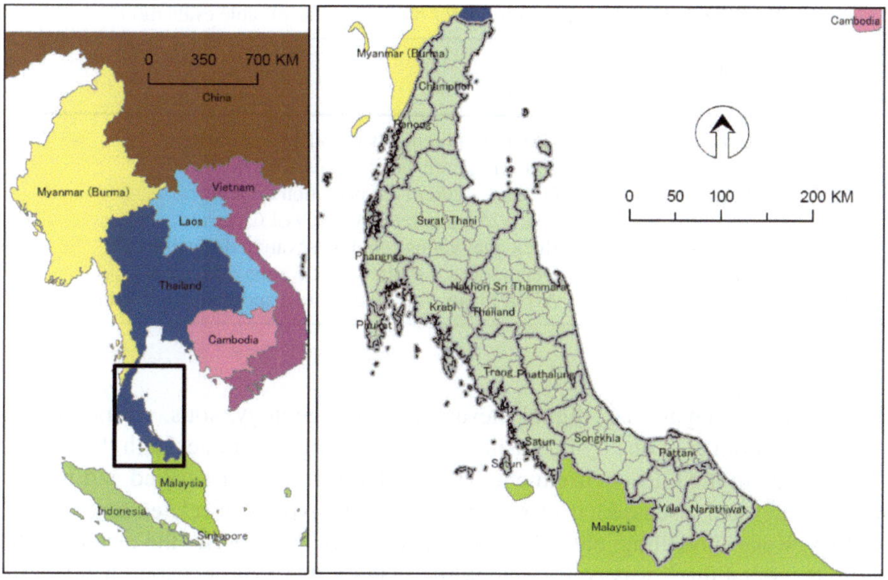

Fig. 14.2 Study area – South Thailand (Kra Isthmus)

Burmese (Myanmar) spaces were excluded in this study. The region consists of a narrow isthmus located in the peninsula running roughly north to south. The isthmus connects Myanmar and Central Thailand to the north and Malaysia to the south. It has a varied topography including basins, hills, mountains and coast between the Andaman Sea of the Indian Ocean to the west and the South China Sea of the Pacific Ocean to the east. This region spans about 700 km long from north to south and ranges from 40 to 200 km in width from east to west. The study area covers about 71,000 km^2 of land surfaces with the highest elevation of 1800 m. In recent years, the isthmus is called Kra Isthmus referring to whole region of the south Thailand, although the name is used locally to describe the narrowest part of the region in Ranong and Chumphon provinces.

Geographically referenced data, namely contour (20 m), ocean charts of the Andaman Sea and Gulf of Thailand, geology, soils and river systems were considered as preliminary physiographic parameters while assessing the suitable site. Several visits were made to different locations of the study area to observe and verify the research results. A digital elevation model was developed using contour data. Hardness of rock, depth of soils and flow of major rivers play vital roles while digging the land surfaces for the potential canal. Rock age, often reflecting the gradient of the hardness of rock, was analyzed using a geological map. Deeper soil is easier to excavate than shallow soils because shallow soils often cover hard rock materials. Therefore, a soil depth map was prepared for analyzing the distribution of soil depth. Similarly, a map of the major river systems was prepared considering major alterations in the agriculture, ecosystem and sediment loadings. Altogether,

Table 14.3 Decision criteria setup for physiographic variable evaluation

SN	Spatial variables	Criteria
1.	Canal size	400 m (requirements for VLLC carrier)
2.	Canal depth	25 m (requirements for VLLC carrier)
3.	Elevation	Ground: <200 m (construction difficulties) Sea: >25 (requirements for VLLC carrier)
4.	Geology	Rock age: young ~ old (excavation difficulties)
5.	Soils	Depth: deep ~ shallow (excavation difficulties)
6.	Rivers	Avoid major rivers (ecosystem and sediment loadings)

five geographic layers, i.e. digital elevation model, geology, soils, rivers and sea charts were evaluated one by one using map overlaying techniques with the criteria (Table 14.3) in the GIS environment. The results were further verified and refined several times and discussed at several meetings with experts of related fields (soils, geology, water modelling, etc.). The rule-of-thumb of "the shorter the distances the better the suitability" was considered while making the final decision on selecting suitable sites for the potential canal route. Figure 14.3 shows a general flow chart applied while assessing the land.

Spatial variables of evaluation criterions (Table 14.3) were setup based upon the requirement of shipping, excavation difficulties and sedimentation problems. All the maps with the selected criterions were overlaid in a GIS environment and carefully evaluated to derive the potential sites for the construction of the shipping canal. After evaluating the potential areas, we computed the distance of the potential canal over land as well as at sea.

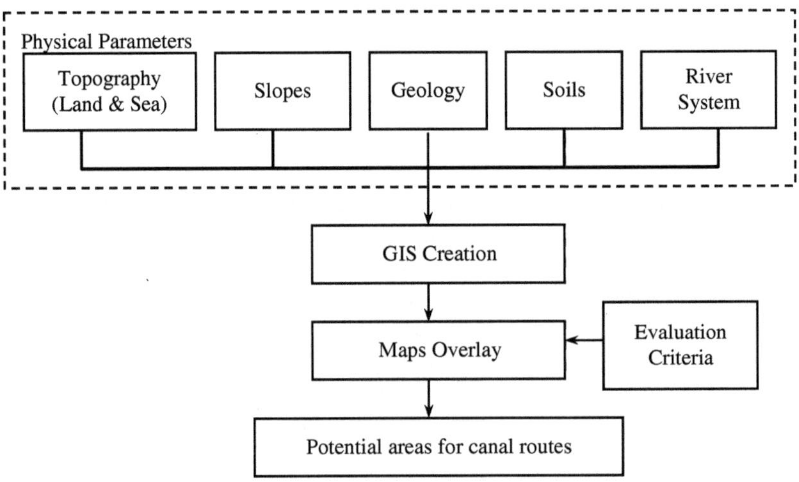

Fig. 14.3 A logical framework for potential sites assessment

14.4.3 Results and Discussion

Figure 14.4 demonstrates the general relief patterns of the Isthmus, whereas Fig. 14.5 shows sample reference maps (Andaman Sea, in the west, and Gulf of Thailand, in the east) of the study area used while assessing the site. About two thirds of the land in the isthmus is found to be less than 100 m in elevation. A

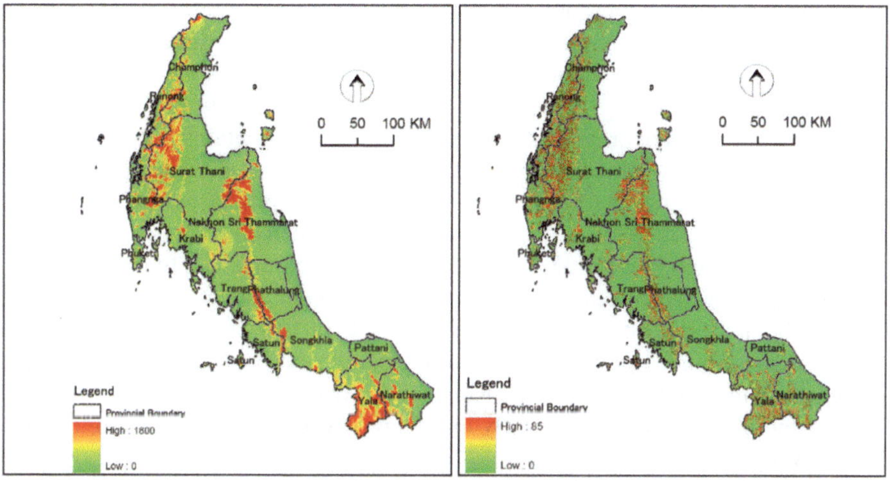

Fig. 14.4 Relief pattern – elevation in meters (*left*) and slopes – in degree (*right*)

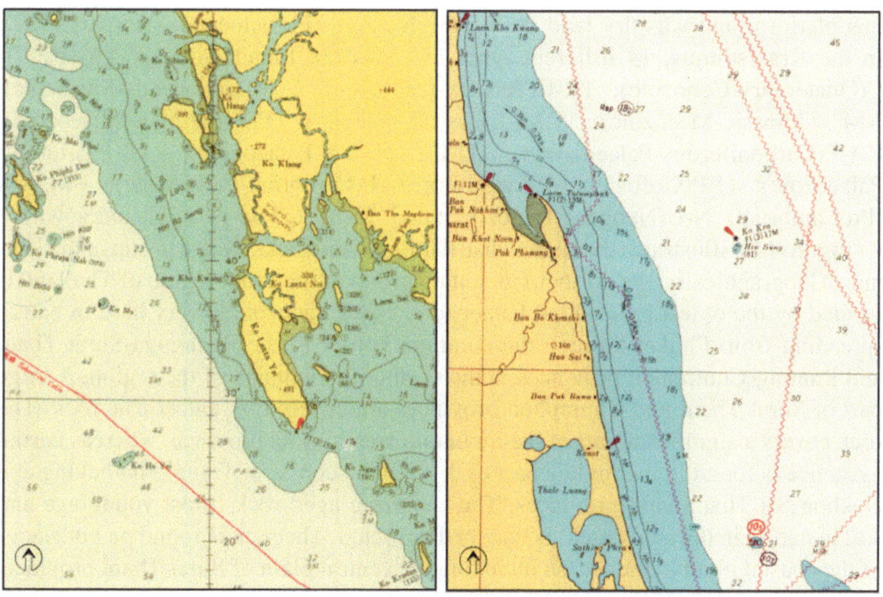

Fig. 14.5 Sea chart as a sample reference – Andaman Sea (*left*) and Gulf of Thailand (*right*)

large proportion of low land lies in the east of Chumphon connected to Surat Thani-Krabi-Trang. In the east, undulating terrain down to a flat coastal plain is in between Nakhon Sri Thammarat, Phathalung and Songkhla. Low-lying, gently undulating terrain lies between and alongside the mountain chains over nearly one third of the region.

The mountains form a series of separate ridges aligned at a slight angle to the region rather than a single backbone. Three major mountain ranges with different shapes and aspects are observed in the isthmus. The northern mountain range runs from Ranong-Chumphon to Phangnga-Phuket, which has small, narrow ridge with a low land valley in the northern part, in between the border of Ranong and Chumphon provinces. This group completely disappears in the Krabi and central plain of the Surat Thani provinces. The middle range of mountains runs as a narrow, hilly chain from the south-east of Surat Thani province along the middle of Nakhon Sri Thammarat, Trang-Phattalung and Satun-Songkhla provinces. This mountain range has a large area of lowland basin as a costal plain on both east and west sides. Some low land areas like a neck (less than 100 m) in the range, are found in Nakhon Sri Thammarat province near the Trang provincial border and Satun-Songkhla joint border. The highest peak in southern Thailand called Khao Luang (1835 m), is also located in the mountain range of Nakhon Sri Thammarat province. Another mountain chain in the southern part of the isthmus runs from the south-west of Songkhla close to the Malaysian border, mostly in Yala province following to the line of the Narathiwat province.

Geologically, South Thailand is part of an old stable landmass known as the Sunda Platform that extends south to east through Malaysia to the Indonesian island and Borneo (CCOP EPF, 2002). The shallow South China Sea is part of this platform and was dry land during the Ice Age (Plesistocene) (Harper, 1999). In the Kra Isthmus, 14 different types of rocks are found with age groups: Q-C(Quaternary-Cenozoic); TR-C(Tertiary-Cenozoic); C-M(Cretaceous-Mesozoic); J-M (Jurassic-Mesozoic); T-M(Triassic-Mesozoic); P-P(Permian-Palaeozoic); CA-P(Carboniferous-Palaeozoic); D-P(Devonian-Palaeozoic); S-P(Silurian-Palaeozoic); O-P(Ordovician-Palaeozoic); C-P(Cambrian-Palaeozoic); PRCAM (PreCambrian); N-C(Neogene-Cenozoic); and -M(Mesozoic-period is unknown).

Figure 14.6 illustrates the spatial distribution of the rocks by age group in the isthmus. Geographically, the northern mountain group of the south region of Thailand is formed by the oldest rock called Palaeozoic. The Palaeozoic is very hard in nature spreading from Phuket through the common border of the Phangnga-Surat Thani and Ranong-Chumphon provinces in the northern mountains of the region. A large part of Surat Thani and Chumphon provinces are covered by Palaeozoic rock. The rock covers a significant part of the mountain area in Yala province, whereas partial existence is found in the middle part of the region such as in Trang, Phathalung and Nakhon Sri Thammarat provinces. The Cenozoic aged rock (most young age and soft in nature in the study area) is widely distributed. The rock is found on both sides of the coastal plain in the region including the central plain of Surat Thani province. Most areas of Surat Thani and Nakhon Sri Thammarat provinces are formed by Cenozoic-age rock. This rock is also found with Mesozoic-age rock in some areas

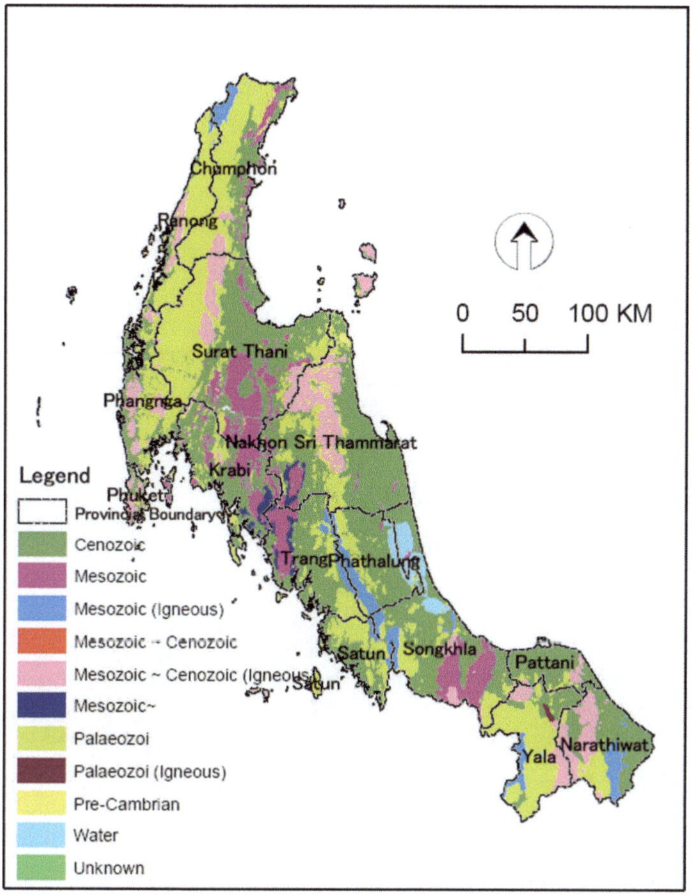

Fig. 14.6 Geology – distribution of rock age

of the southern provinces. A small portion of the area is covered by Mesozoic rock. The rock is also found as igneous in some places. The Mesozoic age rock expands in Surat Thani, Krabi, Trang and Songkhla provinces. Very few areas of Nakhon Sri Thammarat province are found to have this rock. The Mesozoic igneous rock spreads linearly as a strip along the common border of Trang-Phathalung and Satun-Songkhla. Mixture of Mesozoic-Cenozoic (igneous) rock have been found in the Nakhon Sri Thammarat mountains and some eastern landscapes of the northern and southern mountains of the region at relatively lower elevations. Other rock, such as Pre Cambrian, is nominally found in Nakhon Sri Thamarat province. Some land of Songkhala province is covered by water.

The Fig. 14.7 illustrates the soil depth distribution in the south region of Thailand. Four types of depths, deep, medium deep, shallow and very shallow, are identified. The distribution map shows more than half of the region has deep soil that spreads over both side of the coastal plain and lower mountain basin in the whole region.

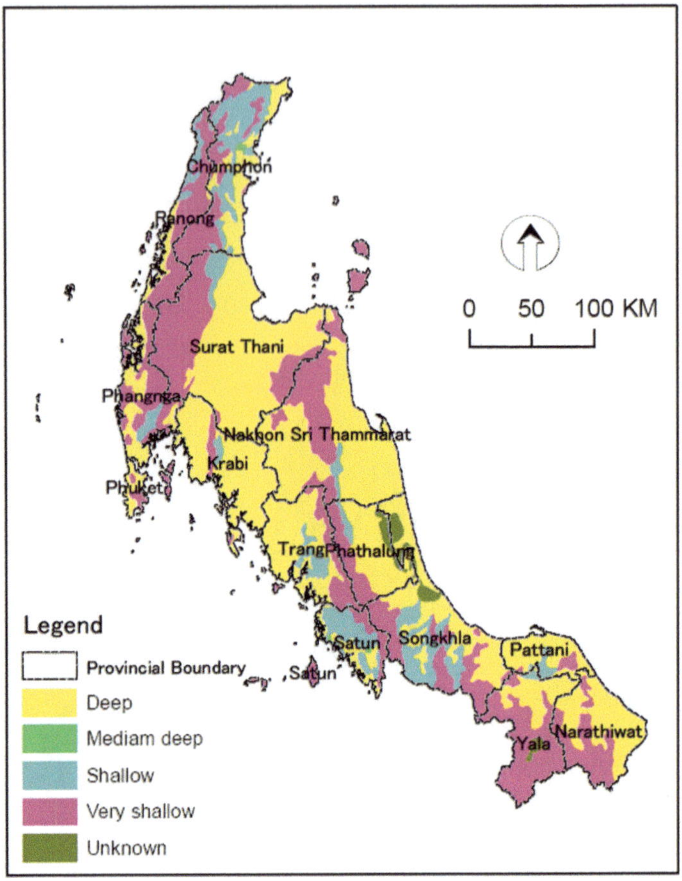

Fig. 14.7 Soils – distribution of soil depth

The major concentration of deep soil is observed over the central plain of Surathani, Krabi, Trang provinces and along the eastern coastal plain started from Nakhon Sri Thammarat to Narathiwat. Despite coastal plains, a huge coverage of deep soil runs continuously over the eastern Surat Thani coastal plain to the western coastal plain of Krabi and Trang provinces. Medium deep soil is found in very few areas such as near the north-east coast of Chumphon and the middle part of Satun provinces. Shallow depth soil covers the lower upland basin of the northern and middle mountain ranges of the region, mostly in Chomphon and Satun provinces, respectively. All the mountain ranges in the region have very shallow soil on top and palaeozoic rock material as an inner formation. The soil depth of the lagoon of Songkhla Lake is unknown.

Figure 14.8 illustrates the rivers in the Isthmus. The largest rivers in the region are the Tha Tapho and Lang Suan in the north, Ta Pi, Phum Duang and Trang in the centre and Pattani, Sai Buri, Bang Nara and Ko Lok in the southern part of the

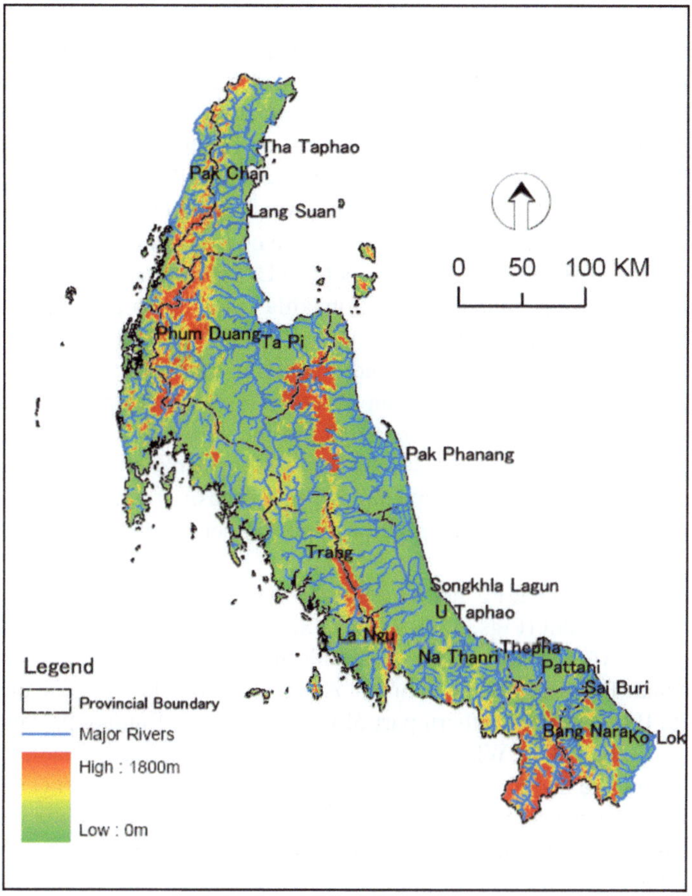

Fig. 14.8 River system with DEM background

region. Most other rivers are small, especially on the west coast where mountains are close to the sea and gradient is very steep. Many intermittent streams in the region originate from upland and create flash floods to the lowland basin during the rainy season. The river Pakchan flows along the border of Myanmar and northwest of Chumphon and Ranong provinces and joins the Andaman Sea. Similarly, Tha Taphao river and Lang Suang river drain the eastern part of the northern hillocks and are submerged to the east coast of Chomphon province (Gulf of Thailand).

The big river called Phum Duang unites with the Ta Pi river, runs from the western mountains of Surat Tahani province irrigating the central lowlands towards the east coast. Originally, it was the result of the unification of four outlets of water from east to south-headed mountain watersheds, namely the Yan, Sang, Sok and Phanom in Surat Thani province. The Ta Pi river drains the middle western mountains of Nakhon Sri Thammarat province and flows through the central plain of Surat Thani. The river amalgamates with the Phum Duang before few kilometers to the east coast of Surat Thani province. These rivers have significantly contributed to the agriculture ecosystem in Surat Thani. Pak Phanang, perennial in nature, is the only river in the Thammarat lowland basin that provides water requirement for inhabitants and agriculture activities. Some other small short streams also exist near to the east coast of the Nakhon Sri Thamarat province. Trang River originates from the western Thammarat mountains and supples the water to the Trang costal plain. The river flows a north-south direction and ends at the Andaman Sea nearby the middle-west coast of the Trang province. On the eastern coastal plain of Songkhla and Pak Phanang provinces, several rivers contribute to U Taphao River and the Na Thanri flows into the complex channels of the Songkhla Lake. Satun province has La Ngu as its main river. The Thepha, Pattani, Sai Buri, Bang Nara and Ko Lok rivers drain the southern mountains of the region and flow to the east coast.

An interesting relationship has been observed in the three spatial variables, soil depth, elevation and rock age. For example, in the mountain range in the Isthmus higher elevation produces the oldest rock material (palaeozoic) which is covered by soils ranging from shallow to very shallow. In the basin, rocks are young, having low elevation and deep soils. The relationship is higher the elevation, older the rock age and lower the soil depth.

After carefully evaluating spatial data of topography, sea charts, rock age, soil depth and river system (Figs. 14.4, 14.5, 14.6, 14.7, and 14.8) using multi-criteria decision (Table 14.3) analysis, we were able to derive five potential sites for the shipping canal (Fig. 14.9) in the Isthmus. As compared to the total distance of Sites A to E, Site B located in southern part of Chomphon and Ranong provinces, is the shortest one (Table 14.4). Whereas Site C passing through Surat Thani, sharing the border of Phangnga and Krabi and exiting at Phuket waters, is the longest. As to the considered land surfaces, Site A, located in northern part of Chomphon and Ranong provinces, is the shortest one, whereas Site D passes through Nakhon Sri Thammarat and Trang becomes the longest compared to other sites. Site A has to share the border with Myanmar (Burma) in the west which posess a long dredge because of the Pakhan river and data unavailability in the Myanmar area. As compared to the total dredge, the dredge of Site E (east coast of Songkhla and west coast of Satun) and Site C (east coast of Surat Thani and south coast of Phangnga) are found to be very long. Due to different construction requirements for dredge and land surfaces, the length of dredges are notable. Some longer time dredging may become costly for excavation and recycling of wastes than on the lands.

Although Sites C and D seem very long, variables such as topography and soils, are favorable because of the low elevation with deep soils which may be easier for excavation. But at site C, a large river (Phum Dung) becomes a major barrier. Mixing the river into the canal may cause sediment loading problems in canal which can raise the daily operating cost of the canal. It may also cause serious problems to the ecosystem in the surrounding area and irrigated land. However, tunneling of river may be possible regardless of cost. The site B looks the shortest one but the existences of hard rock at higher elevations may make excavation more difficult as

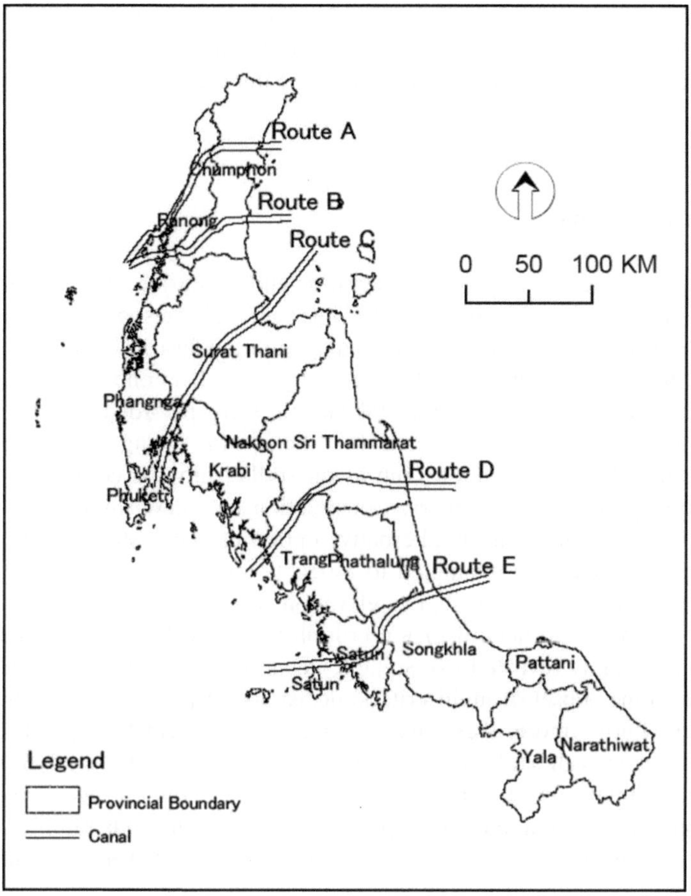

Fig. 14.9 Potential sites for shipping canal

Table 14.4 Length of canal in potential sites

		Dredge			
Site	Land	West	East	Total	Total length
A	94.56	63.18	15.25	78.43	172.99
B	76.09	31.38	35.50	66.88	142.97
C	126.19	35.34	73.10	108.44	234.63
D	141.87	21.24	38.51	59.75	201.62
E	107.12	57.77	42.15	99.92	207.04

Note: Units in kilometers, West (Andaman Sea) and East (Thai Gulf) well as the costs of recycling the wastes. More zigzag courses are observed in Sites A and B than in the other sites. It is because of the mountainous topography. Such snaking may result more risky navigation as compared to a straight and smooth route.

14.5 Conclusions

Globalization processes, economic integration and expansion of international trade have extended considerably the need for international transportation. A big share of world trade depends on the international shipping industries. Half of the world's oil passes through the Strait of Malacca every year. It is one of the busiest narrow channels in the world, where 194 vessels were recorded per day in 2007. Because of continues growth of sea traffic, the strait is facing many challenges in environment and safe navigation such as oil spills, shipping collisions, air pollution and piracy and armed robbery. Construction of a new shipping canal in the South Thai (Kra) Isthmus as an alternative option has been studied since the 19th century. In this paper, we attempted to assess potential sites for a shipping canal based on geophysical parameters, i.e., elevation, sea charts, geology, soils and river systems. Several bottle neck decision criteria were checked while evaluating the sites. The case study revealed five suitable geographic sites including the required length within the subset of multi-criteria decisions for potential shipping canal on the Isthmus. Furthermore, these five potential sites present the right length for a shipping canal on both land and sea. However, a detail feasibility study is necessary into environmental impact, safety measures, socio-economic aspects, and international geopolitics. A detailed topographic survey, rocks and soils samples, land use, including socioeconomic variables are sought to be included in next phase of the study.

Acknowledgements We appreciate the members of the National Kra Committee, Thailand, Prof. Tawatchai Tingsanchali, Asian Institute of Technology, Dr. Sompoch Puntavoungkor, Royal Thai Survey and Mr. Jun Nogami, Asian Institute of Technology for their enthusiastic supports in this research.

References

CCOP EPF. (2002). *Thailand: Geology, petroleum and potentials*. Retrieved January 7, 2007, from http://www.ccop.or.th/epf/thailand/thailand_petroleum.html#5

Crawford, R. J. M., Davis, S. A., Harding, R., Jackson, L. F., Leshoro, T. M., Meÿer, M. A., et al. (2000). *Initial effects of the treasure oil spill on seabirds off Western South Africa*. Avian Demography Unit Department of Statistical Sciences, University of Cape Town. Retrieved 27, 2006, from http://web.uct.ac.za/depts/stats/adu/oilspill/oilspill.htm

EIA. (2008). World oil transit checkpoints, country analysis brief. Energy Information Administration. Official Energy Statistics, the US Government. Retrieved December 26, 2010, from http://www.eia.doe.gov/cabs/World_Oil_Transit_Chokepoints/Full.html

ES. (2002). The executive summary of the Kra Canal project identification. Subcommittee Kra Canal Project, Defence Commission, Thai Parliament.

Harper, S. B. (1999, October). *Morphology of tower Karst in Krabi, Southern Thailand.* Proceedings of the annual meeting of Geological Society of America, Denver, CO.
ICC-IMB. (2008). Piracy and armed robbery against ships. London: Annual Report, ICC IMB.
Khalid, N. (2007). *The trans-peninsula pipeline: A maritime perspective.* Maritime Institute of Malaysia. Retrieved February 3, 2008, from http://www.mima.gov.my/
Kuppuswamy, C. S. (2004). Straits of Malacca: Security implications. South Asia analysis group (SAAG) (paper no 1033).
Mariner Group. (2005). *The mariner group – oil spill history.* Retrieved May 1, 2005, from http://www.marinergroup.com/oil-spill-history.htm
MDM. (2008). *Vessel Report to Klang VTS (1999–2007).* Statistics, Marine Department Malaysia. Retrieved February 2, 2008, from http://www.marine.gov.my/index.html
Mitropoulos, E. (2004). *Special lecture, Japan international transport institute.* Retrieved May 1, 2005 from http://www.imo.org/Pages/home.aspx
Mukundan, P. (2008). *Reported piracy incidents rise sharply in 2007.* International chamber of commerce – commercial crime services. Retrieved February 4, 2008, from http://www.icc-ccs.org/
Pasertsri, N. (2005). *Kra Isthmus history (The Thai-Canal).* Retrieved January 7, 2007, from http://www.thai-canal.org/hist%20E.htm
Permal, S. (2006). *Indonesia's efforts in combating piracy and armed robbery in the Straits of Malacca.* Maritime Institute of Malaysia. Retrieved February 3, 2008, from http://www.mima.gov.my/
Permatasari, S. (2007). Malaysian pipeline developer in stake sale talks. Bloomberg News, International Herald Tribune (published in 21 June 2007)
Rodrigue, J. P. (2005). Transportation geography on the Web. Department of Economics & Geography, Hofstra University, New York.
ShippingFact. (2005). *Overview of the international shipping industry.* London: International Chamber of Shipping and International Shipping Federation.
TAMS. (1973). Preliminary survey report – Kra Canal Complex. TAMS & PRNA, September 1.
TED. (2005). *Canal construction on Thai Isthmus (CANALTH). Trade and environment database – Asia case.* Retrieved May 1, 2005, from http://www.american.edu/TED/canalth.htm
Thapa, R. B., Kusanagi, M., Kitazumi, A., & Murayama, Y. (2007). Sea navigation, challenges and potentials in South East Asia: an assessment of suitable sites for a shipping canal in the South Thai Isthmus. *GeoJournal, 70*, 161–172.
UNCTAD. (2004). *Review of maritime transport 2004.* Geneva: United Nations Conference on Trade and Development.
UNCTAD. (2006). *Review of maritime transport 2006.* Geneva: United Nations Conference on Trade and Development.
UNCTAD. (2008). *Review of maritime transport 2008.* Geneva: United Nations Conference on Trade and Development.
UNEP. (2005). *Accidental discharges of oil. Global marine oil pollution information gateway.* Retrieved July 15, 2006, from http://oils.gpa.unep.org/facts/oilspills.htm#intelligence
VIC. (2002). *Thailand's Kra Canal–special report.* Virtual Information Center. Retrieved May 20, 2004, from http://www.vic-info.org/

Part V
Socio-environmental Applications

Chapter 15
Spatiotemporal Patterns of Urbanization: Mapping, Measurement, and Analysis

Rajesh B. Thapa and Yuji Murayama

15.1 Introduction

As the result of population growth and migration from rural to urban areas, urbanization has been recognized as a critical socioeconomic process in metropolitan areas of Nepal (Thapa & Murayama, 2009a). Socioeconomic processes such as migration, urban sprawl, agriculture, and forest patterns also often contribute to landscape changes. As a city grows, the increasing concentration of population and economic activities demands that more land be developed for public infrastructure (roads, water facilities, and utilities), housing, and industrial and commercial uses. Therefore, the urbanization can be considered as the observable transformation of the spatial pattern of land use and land cover, such as the transformation of agricultural and forest land uses into built-up area or the gradual transformation of rural landscape into urban forms. The transformation of rural landscape to urban landscape has caused various impacts on ecosystem structure, function, and dynamics. Persistent dynamic land use change processes are expected to accelerate in the next several decades. Worsening conditions of crowding, housing shortages, insufficient infrastructure, and increasing urban climatological and ecological problems require consistent monitoring of urban regions.

However, monitoring of land use changes is needed to understand and predict the dynamic process of land use patterns at different times. This was traditionally limited due to labor intensive fieldwork that is often unable to reveal the spatial pattern of landscape changes and environmental consequences that occurred in a given time period. Significant technological advancement in data acquisition and analysis techniques in recent years has made it easier to analyze the spatiotemporal

R.B. Thapa (✉)
Division of Spatial Information Science, Graduate School of Life and Environmental Sciences, University of Tsukuba, Tsukuba, Ibaraki, Japan
e-mail: thaparb@yahoo.com; thaparb@gmail.com

This chapter is improved from "Rajesh Bahadur Thapa and Yuji Murayama (2009), Examining spatiotemporal urbanization patterns in Kathmandu valley, Nepal: Remote sensing and spatial metrics approaches. Remote Sensing, 1, 534–556".

Y. Murayama, R.B. Thapa (eds.), *Spatial Analysis and Modeling in Geographical Transformation Process*, The GeoJournal Library 100, DOI 10.1007/978-94-007-0671-2_15,

dynamics of landscape changes. Remotely sensed images from airborne and satellite sensors (Thapa & Murayama, 2010a) provide a large amount of cost-effective, multi-spectral, and multi-temporal data to monitor landscape processes and estimate biophysical characteristics of land surfaces (Herold et al., 2003; Thapa & Murayama, 2011).

Remote sensing has long been used to map urban growth and urban morphology, and implies the mapping of the form, land uses, and density of urban areas, each having an associated shape, configuration, structure, pattern, and organization of land use (Lillesand et al., 2008). Satellite imagery has the unique ability to provide synoptic views of large areas at a given time, which are not possible using conventional survey methods. A wide range of urban remote sensing applications from both sensors (active and passive) is currently available. These include quantifying urban growth and land use dynamics, landscape pattern analysis, urbanization, socioeconomic applications, life quality improvement, urban infrastructure characterization, microclimate and hydrology, and topographic mapping (Thapa & Murayama, 2009a).

Kathmandu, a bowl shaped valley, which is the most populous metropolitan region in Nepal, is an interesting case to study as it imposes topographic constraints for horizontal urban expansion but faces rapid urbanization, having an annual urban population growth rate of 5.2% (Sharma, 2003). It is the main political

Fig. 15.1 A perspective view of urban landscape development in Kathmandu Valley (Photo by author, fieldwork, 2010)

and administrative center, a major tourist gateway, and an economically strategic location in the country. High population growth, dramatic land use changes, and socioeconomic transformations have brought the paradox of rapid urbanization and environmental consequences to the valley (Thapa et al., 2008). Along with new developments within the city fringes and rural villages, shifts in the natural environment and newly developed socioeconomic strains between residents are emerging. Such rapid demographic and environmental changes and weak land use planning practices in the past decades have resulted in environmental deterioration, haphazard landscape development (Fig. 15.1), and stress on the ecosystem structure. Consequently, more and more agricultural lands and forest lands have been converted into urban areas and human settlements over the past few decades. Therefore, quantifying land use patterns and analyzing the changes over time are essential for monitoring the urbanization and environmental consequences in the valley. The main objective of this chapter is to examine the spatiotemporal pattern of urbanization in the valley using remote sensing and spatial metrics techniques. This is possible due to advancement of remote sensing techniques and multi-temporal satellite data availability while no other data are available for the study.

15.2 Research Methodology

15.2.1 Study Area

Physiographic boundaries formed by the complex topography play an important role in allocating development resources. Therefore, the valley as a study area in this research is delineated based on the watershed boundaries, which were derived from 20-m digital elevation point data (Fig. 15.2). The elevation in the valley ranges from 1100 to 2700 m above the sea level, and forms complex topography within a small geographic area. Half of the study area has slopes of less than 5°, while more than 20% of the land has slopes greater than 20°. Geographically, the valley is situated between 27°31′55″and 27°48′56″ North latitude and 85°11′11″ to 85°31′52″ East longitude. The valley is drained by the Bagmati river system. The river system is the main source of water for drinking and irrigation in the valley. Politically, the valley is composed of five municipal urban centers (Kathmandu, Lalitpur, Bhaktapur, Kirtipur, and Madhyapur Thimi), in addition to 97 surrounding villages. The study area covers 684 km^2, and the urban centers make up only 14% of the land.

15.2.2 Data Sources

Remote sensing provides spatially consistent data sets that cover large areas with both high spatial details and high temporal frequency (Jensen, 2005). Dating back to the 1960s, remote sensing can also provide consistent historical time series data. Because of the lack of temporal and spatially consistent datasets in other forms for

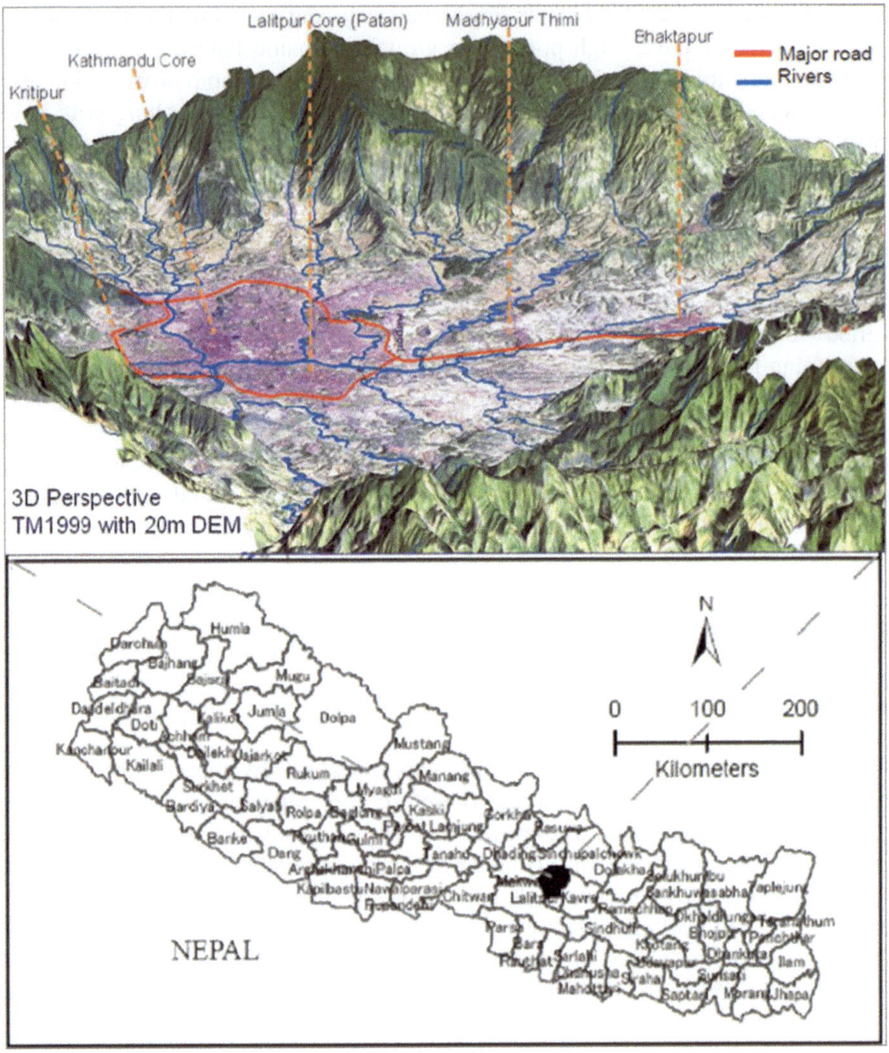

Fig. 15.2 Study area – Kathmandu Valley

the valley, multi-temporal satellite images with high resolution (CORONA, SPIN, and IKONOS) to moderate resolution (LANDSAT MSS and TM) were processed to identify the temporal changes in landscape patterns since the 1960s. The high resolution images are only available for limited areas in the valley; therefore, most care was given to the LANDSAT MSS and TM images. Because of the mountainous terrain and topographic complexity in the study area, the elevation data was also considered as an important source of information. Fieldwork was conducted in December 2007 to acquire first hand data required for the research. A detailed list of the data used in this study is shown in Table 15.1.

Table 15.1 List of databases used in this research

Data types	Year	Resolution/scale	Sources
Satellite imageries:			
CORONA	1967.02.05	1 m	USGS
LANDSAT MSS	1976.10.28	57 m	University of Maryland
LANDSAT TM	1989.10.31	30 m	University of Maryland
SPIN-2	1991[a]	2 m	USGS
LANDSAT TM	1999.11.04	30 m	University of Maryland
IKONOS	2000[a]	1 m	GeoEye, Space Imaging
QuickBird	2007[a]	0.6 m	Google Earth
Aerial photographs	1979, 1981, 1992, 1998	–	Survey Department, Nepal
Vector layers:			
Spot height (points)	1995	20 m	(ICIMOD/UNEP, 2001)
Land cover map	1978	–	
Land use map	1995	1:25,000	
Road map	2000	–	
Field survey	2007.12	–	Kathmandu fieldwork

[a]The data acquisition date (month and day) is missing

15.2.3 Mapping of Spatial Patterns

The satellite data are in image form and contain many details, but not in an objective thematic setting. Image analysis techniques are evolving rapidly, but many operational and applied remote sensing analyses still require extracting discrete thematic land surface information from satellite imagery using classification-based techniques. Thapa and Murayama (Chapter 9) argued that the heterogeneity and complexity of the landscape in urban regions, for example, suburban residential areas forming a complex mosaic of trees, lawns, roofs, concrete, and asphalt roadways, require land use and land cover classification techniques that combine more than one classification procedure to improve remote sensing-based mapping accuracies. Therefore, a series of processing steps (Fig. 15.3) is followed to transform those data into meaningful thematic information.

The geometric rectification process was carried out for all satellite images using a road network map in the local projection system (i.e., UTM WGS 1984). Image enhancement, contrast stretching, and false color composites were created to improve the visual interpretability of the image by increasing the apparent distinctions between the features. Knowledge-based visual interpretation, texture, and association analysis were performed at the preliminary stage. Furthermore, field survey data, aerial photographs, high resolution satellite images, and city planning documents were carefully analyzed while preparing the land use classes. After analyzing all the information collected so far, only 12 types of land uses were considered for mapping, i.e., agricultural areas, forest, shrubs, open space, water, built-up areas, industrial areas, roads, airport, institutional areas, government secretariat area, and royal palace. The last six land uses cover small spaces in the

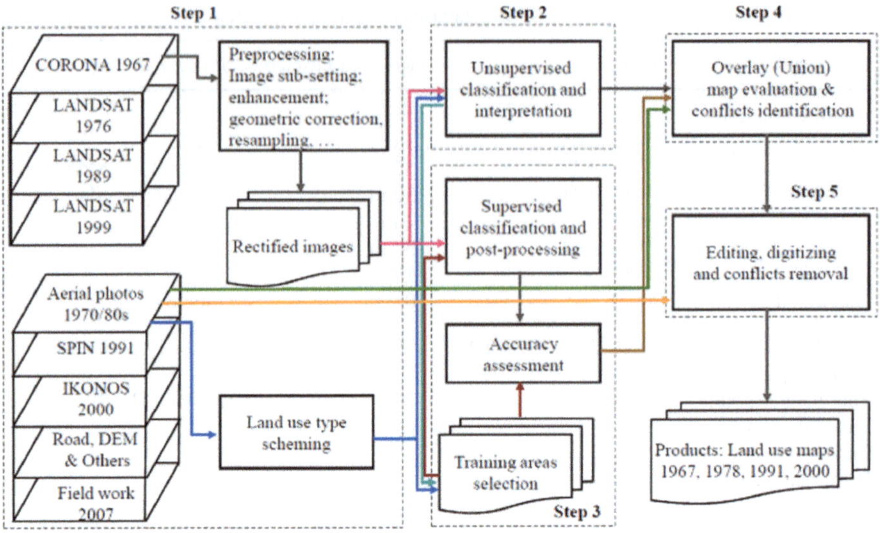

Fig. 15.3 Land use mapping scenario for remote sensing images

valley; therefore, these land uses were merged into an urban/built-up area category for detailed quantitative assessment purposes. However, all twelve legend units were listed in the land use maps.

CORONA image was resampled using a nearest neighbor resampling technique to match the data resolution of LANDSAT (30-m) for maintaining spatial resolution consistency in the data sources. Resampling the CORONA image (1-m) into 30-m may introduce modifiable areal unit problem (MAUP) due to generalization of continuous geographical phenomenon as discussed by Openshaw and Alvanides (1999). When values are averaged over the process of aggregation, variability in the CORONA image is lost which may result the variation in spatial patterns arising from the statistics computed at coarser resolution. This problem is also known as scale effect in MAUP. However, after following the steps shown in Fig. 15.3, such problem is removed in resulting map.

An unsupervised approach with the ISODATA clustering technique available in Erdas Imagine 9.0 was applied to obtain different land use clusters of similar spectral pixels in the Corona, MSS and TM images. This preliminary interpretation reduced the artificial errors and selected the most appropriate clusters for further processing. Then, the supervised approach with the maximum likelihood parameter was run to improve the accuracy of the land use classification for the images for all three dates (1967, 1976, 1989, and 1999). Aggregating the detailed remotely sensed surface characteristics into thematic information always contains some degree of errors, hence, an accuracy assessment should be performed (Thapa & Murayama, Chapter 9, this volume). The image classification accuracy was performed by evaluating the overall classification accuracy using geographically referenced vector datasets. The classification accuracies of 83.66, 80.66, 84.44, and 83.33% were achieved for the years 1967, 1976, 1989, and 1999, respectively.

Because of the complex topography in the valley, land use types are closely related to altitude and slope and have some specific distribution rules. Confusing areas were detected mostly between the water areas and shadows of mountain areas; bare lands, brick factories, and construction sites; and golf courses and shrub lands. The areas of confusion were further verified with DEMs and slope data, road data, the 1978 land cover map, high resolution imagery, including CORONA (1967), SPIN (1991), IKONOS (2000), aerial photographs acquired at different time periods, and fieldwork information to determine the appropriate land use type. Editing and digitizing were carried out to resolve all the confusion and conflicts that occurred in each map. This process helped to improve the accuracy of the mapping. After updating the maps, the reference year of each map was fixed to 1967, 1978, 1991, and 2000, respectively.

15.2.4 Analysis of Spatial Patterns

The land use transition matrix is a useful tool that has been widely accepted in land use change analysis. Three land use transition map layers for the years 1967–1978, 1978–1991, and 1991–2000 were prepared for the detailed land use change pattern analysis in the valley. Empirical studies have substantiated the use of both spatial metrics and remote sensing in urban modeling (Herold et al., 2003; Thapa & Murayama, 2008, 2009b; Zhao & Murayama, Chapter 10). The use of spatial metrics has provided a new platform for describing the spatial land use and land cover heterogeneity and morphological characteristics within the urban environment. Spatial metrics are already commonly used to quantify the shape and pattern of landscapes (O'Neill et al., 1988; McGarigal et al., 2002). Recently, there has been an increasing interest in applying spatial metric techniques in an urban environment to link land use heterogeneity to structures and dynamic changes in urban land uses.

A set of spatial metrics were selected to measure and monitor the landscape fragmentation, land use complexity, proximity, dominancy, and diversity (Table 15.2). The selected metrics are patch density (PD), largest patch index (LPI), edge density (ED), area weighted mean patch fractal dimension (AWMPFD), Euclidian nearest neighbor distance mean (ENNMN), cohesion (COHESION), contagion (CONTAG), and Shannon's diversity index (SHDI). These metrics describe the composition and configuration of landscape pattern changes in the valley.

The metrics were computed for each land use map at the class and landscape levels. Metrics at the class level are helpful for understanding landscape development, while those at the landscape level provide relatively general information on the assessment. All these metrics were calculated using the FRAGSTAT software (McGarigal et al., 2002) while Erdas Imagine and ArcGIS software were used for image analysis and GIS data processing.

Table 15.2 Description of spatial metrics used in this study (Compiled from McGarigal et al., 2002)

Metrics	Description	Units	Measure of
PD	PD equals the number of patches of a specific land cover class divided by total landscape area.	No./100 ha	Fragmentation
ED	The sum of the lengths of all edge segments involving a specific class, divided by the total landscape area multiplied by 10,000.	Meters/ ha	Fragmentation
LPI	The area of largest patch of the corresponding class divided by total area covered by that class, multiplied by 100.	Percent	Dominance
ENNMN	The distance mean value of all patches of a land use to the nearest neighbor patch of the land use based on shortest edge-to-edge distance from cell centre to cell centre.	Meters	Isolation/proximity
AWMPFD	It describes the complexity and fragmentation of a patch by a perimeter-area ratio. Lower values indicate compact form of a patch. If the patches are more complex and fragmented, the perimeter increases representing higher values.	None, range: 1–2	Fragmentation and complexity
COHESION	Approaches 0 as the portion of the landscape comprised of the focal class decreases and becomes increasingly subdivided and less physically connected.	None, range: 0–100	Physical connectedness
CONTAG[a]	Contagion index describes the fragmentation of a landscape by the random and conditional probabilities that a pixel of patch class is adjacent to another patch class. It measures to what extent landscapes are aggregated or clumped.	None, range: 1–100	Fragmentation and the degree of aggregation
SHDI[a]	Shannon's diversity index quantifies the diversity of the landscape based on two components: the number of different patch types and the proportional area distribution among patch types.	Information	Patch diversity

Note: PD: patch density; LPI: largest patch index; ED: edge density; AWMPFD: area weighted mean patch fractal dimension; ENNMN: Euclidian nearest neighbor distance mean; CONTAG: contagion; SHDI: Shannon's diversity index.
[a]Metric can be applied only for landscape level assessment.

15.3 Results

15.3.1 Spatial Patterns of Land Use

Land use statistics and transition matrices are important information to analyze the temporal and spatial changes of land use, and examine the driving forces behind those changes. Figure 15.4 shows the four land use maps for the years 1967, 1978, 1991, and 2000. The urban/built-up areas in the valley had a noticeable increase, from 3% (2010 ha) of the total land in 1967 to 14% (9717 ha) in 2000, showing spatial patterns of urbanization with consistent (5%) growth in 1991 and 2000 (Table 15.3). However, an opposite trend in shrubs (13,563 ha) and forest (15,800 ha) lands was observed, where half of the shrubs land changed to another class by 2000, including a significant loss of area between 1978 and 1991. Similarly, the forest lost 4% of its land over the three and a half decades. However, after gaining a small area in 1978, the forest lost land area in later years.

Agricultural lands still cover half of the valley, but their area has increased and decreased in different time intervals. The agricultural land decreased slightly by 1978, but increased to 56% of the total land by 1991, and had again decreased to 54% in 2000. Only 2% of the total land in the valley is covered by water. A few manmade ponds and the Bagmati river system are the major components of the water coverage. Because water covers a small area, only a small pattern of change

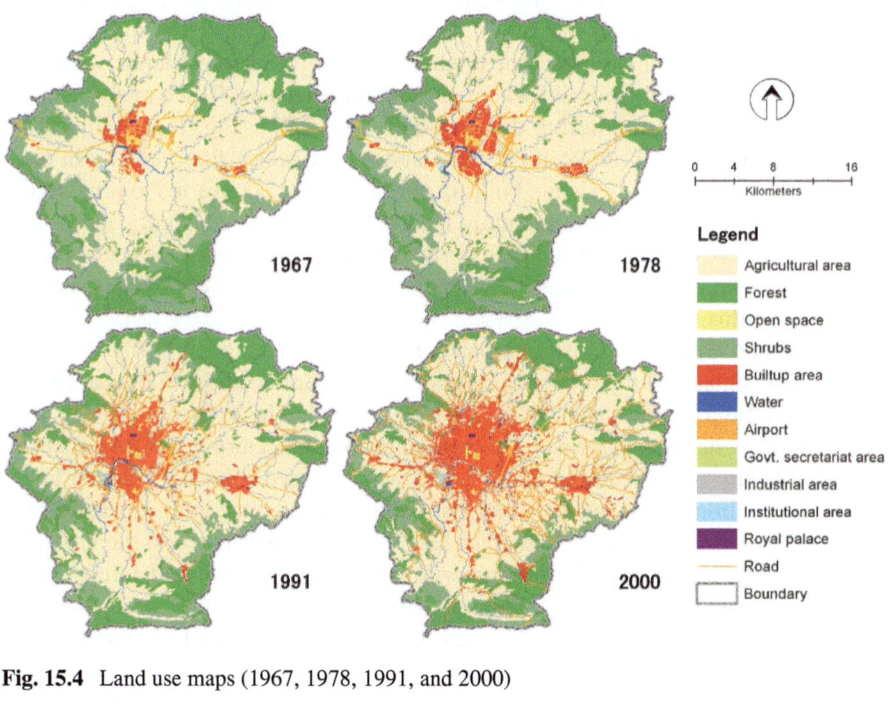

Fig. 15.4 Land use maps (1967, 1978, 1991, and 2000)

Table 15.3 Land use statistics

Year	1967		1978		1991		2000	
Land use type	Hectare	%	Hectare	%	Hectare	%	Hectare	%
Shrubs	13,563	19.81	12,124	17.71	8129	11.87	7150	10.44
Forest	15,800	23.08	16,311	23.83	13,887	20.29	13,301	19.43
Water	1337	1.95	1380	2.02	1341	1.96	1266	1.85
Urban/builtup area[a]	2010	2.94	3362	4.91	6313	9.22	9717	14.19
Open space	100	0.15	95	0.14	135	0.20	171	0.25
Agricultural area	35,649	52.07	35,186	51.40	38,652	56.46	36,854	53.83
Total	68,458	100.00	68,458	100.00	68,458	100.00	68,458	100.00

[a] Includes builtup areas, industrial areas, roads, airport, institutional areas, government secretariat area, and royal palace (Fig. 15.4)

was found for water. Similarly, the availability of open spaces for recreation and sports purposes in the valley is observed to be very low.

15.3.2 Spatial Patterns Changes

Figure 15.5 shows the landscape transition maps for the three time periods. The maps demonstrated significant landscape transitions during the study period. Most of the agricultural lands in the valley floor and near existing built-up areas were transformed into urban/built-up lands, whereas shrubs and forest lands were converted into agricultural lands elsewhere in the rural periphery. Three major land use transitions were observed during the period of 1967–1978 (Fig. 15.5, Table 15.4). The agricultural lands (1.96% of the total land) in the valley floor, mostly in close proximity to the road and existing built-up periphery, were converted to urban/built-up areas. In the southwestern mountain landscape, much of the shrubs lands were converted to forest lands, while in the northeastern area the forest areas changed to agricultural lands. The transitions between the other land uses were found to be very small during this period.

A ring road around the existing urban core was built during the 1970s. This road significantly enhanced the urbanization process in later decades, which can be easily discerned during the years 1978–1991 (Fig. 15.5). The agricultural lands near the road began to be transformed to urban/built-up areas. During this period, a significant amount of agricultural land (3.9%) was changed to urban built-up lands, with urbanization following the road networks and existing built-up peripheries (Table 15.5). In the meantime, the other land uses also contributed to the urbanization process at lower rates. Large proportions of shrubs (5.3%) and forest (3.8%) lands were transformed into agricultural land in the surrounding rural mountain areas in the valley. This can be observed mostly in the northeastern border of the valley, and may be due to conversion of agricultural lands to built-up areas in the urban fringes, which forced the farmers to migrate in vicinities in one hand. On the

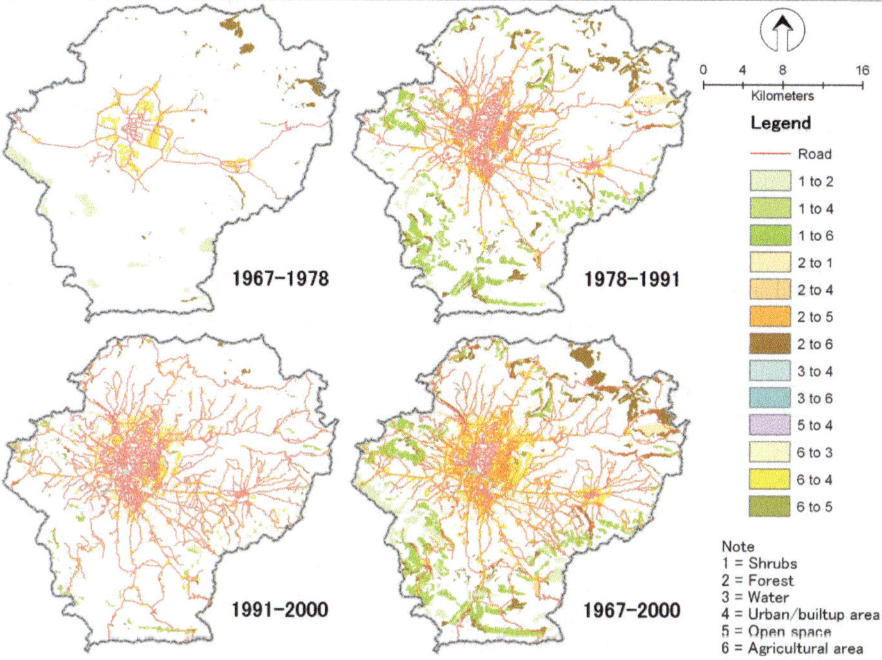

Fig. 15.5 Land use change in different time periods (1967–1978, 1978–1991, 1991–2000, and 1967–2000)

other hand, due to road expansion and market accessibility to rural areas, farmers were encouraged to develop agricultural activities in the rural hills and had spread on the nearby shrubs and forest lands. The land use transition continued in 1991–2000 (Fig. 15.5). A different phenomenon of land conversion is observed in this

Table 15.4 Land use change in percentage (1967–1978)

	1978						
1967	Shrubs	Forest	Water	Urban/builtup area	Open space	Agricultural area	Total
Shrubs	17.68	2.13	0.00	0.00	0.00	0.01	19.81
Forest	0.03	21.70	0.01	0.01	0.00	1.33	23.08
Water	0.00	0.00	1.95	0.00	0.00	0.00	1.95
Urban/builtup area	0.00	0.00	0.00	2.93	0.00	0.00	2.94
Open space	0.00	0.00	0.00	0.01	0.14	0.00	0.15
Agricultural area	0.00	0.00	0.06	1.96	0.00	50.05	52.07
Total	17.71	23.83	2.02	4.91	0.14	51.40	100.00

Table 15.5 Land use change in percentage (1978–1991)

1978	1991 Shrubs	Forest	Water	Urban/builtup area	Open space	Agricultural area	Total
Shrubs	11.41	0.89	0.00	0.13	0.00	5.27	17.71
Forest	0.46	19.32	0.00	0.21	0.03	3.81	23.83
Water	0.00	0.00	1.92	0.04	0.00	0.06	2.02
Urban/builtup area	0.00	0.00	0.00	4.90	0.00	0.01	4.91
Open space	0.00	0.00	0.00	0.03	0.10	0.00	0.14
Agricultural area	0.00	0.07	0.04	3.90	0.07	47.32	51.40
Total	11.87	20.28	1.96	9.22	0.20	56.46	100.00

Table 15.6 Land use change in percentage (1991–2000)

1991	2000 Shrubs	Forest	Water	Urban/builtup area	Open space	Agricultural area	Total
Shrubs	10.35	0.12	0.00	0.19	0.00	1.21	11.87
Forest	0.08	19.23	0.00	0.36	0.04	0.57	20.29
Water	0.00	0.00	1.84	0.03	0.00	0.09	1.96
Urban/builtup area	0.00	0.01	0.00	9.19	0.01	0.01	9.22
Open space	0.00	0.00	0.00	0.02	0.18	0.00	0.20
Agricultural area	0.01	0.07	0.01	4.40	0.02	51.95	56.46
Total	10.44	19.43	1.85	14.19	0.25	53.83	100.00

period as compared to the earlier time period. The transformation of agricultural land into urban/built-up areas was increased (4.4%), but the transformation of the other land uses into agricultural lands remarkably decreased (Table 15.6). However, agricultural encroachment on shrubs lands still continued at slow rate as compared to earlier. Forest (0.36%) and shrubs (0.19%) lands were also changed to built-up areas as a result of the expansion of rural roads in the 1990s.

15.3.3 Landscape Fragmentation and Heterogeneity Analysis

Planners and policy makers are normally concerned about the negative effects of landscape fragmentation and heterogeneity development. There are two processes that can result in these effects, namely the reduction of the total amount of land with a specific land use (decrease in size), and the breaking up of land use into smaller patches (increase in isolation of the land use patches). This process will also be followed by an increase in the total amount of edges in some cases. Agricultural expansion in the rural areas, urban development, and transportation infrastructure development in the valley floor over time are regarded as the main processes that influence landscape fragmentation and heterogeneity development.

Landscape level analysis: At the landscape level, the patch density (PD) in the valley increased from 0.86 in 1967 to 1.99 in 2000 (Fig. 15.6). The number of new

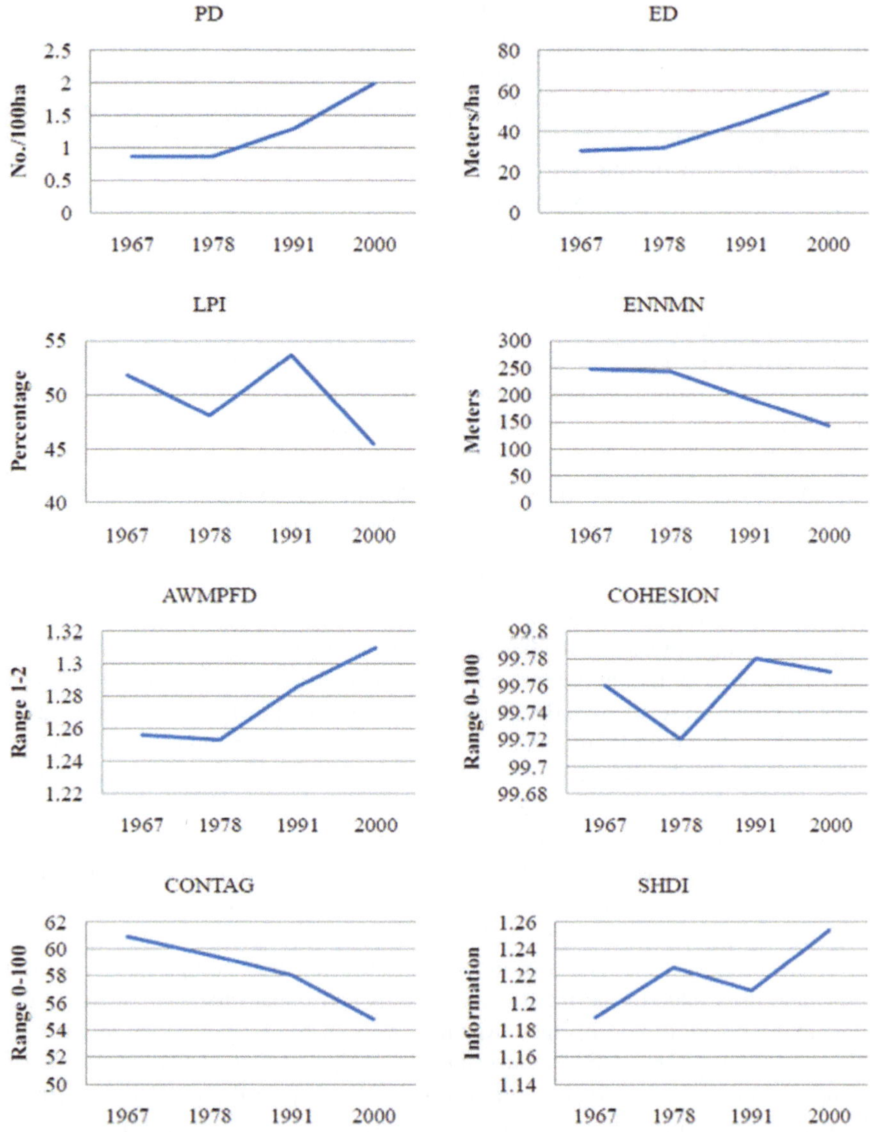

Fig. 15.6 Spatial metrics at landscape level

patches in the landscape significantly increased since 1991. The increase of PD often leads to increased edge density (ED), as it creates new edge segments in the patch. The ED almost doubled in the whole study period, increasing steadily in the 1980s and 1990s after a small increase in the 1970s. Edge density often increased while land use fragmentations occurred due to land use changes. In the valley, it is observed mainly in the urban built-up areas, and the agricultural area particularly increased in the 1980s and 1990s, as shown by the figures of PD and ED. However, the dominance index (LPI) slightly increased in 1991 but decreased in 2000. The land use compactness of the urban/built-up areas in the valley floor and the expansion of agricultural activities into shrubs and forest lands, forming larger patches in the landscape in the 1980s and 1990s (Fig. 15.4), can be commonly observed.

A significant amount of agricultural lands, i.e., 4.4%, was transformed into urban/built-up area during the period 1991–2000 (Table 15.6, Fig. 15.5), which reduces the proximity (ENNMN) of the neighboring land use patches. The ENNMN decreases across the whole study period. The expansion of urban/built-up areas in the existing built-up periphery in the valley floor and agricultural encroachment near shrubs and forest lands in the rural areas could be the main causes of the ENNMN decline in the 1990s.

The AWMPFD increased slightly, reporting the degree that the shapes of the patches became more complex in later years. This may be due to road network expansion towards the rural areas in the 1980s and 1990s. A decreasing trend is observed in CONTAG, which shows the growing dispersion and fragmentation of the urban/rural landscape in the valley. The physical connectedness of the land use in 2000 increased as compared to 1967, but it faced some increases and decreases in the 1970s and 1980s, as shown by the COHESION index.

The diversity of land uses in the valley also increased. Construction of additional bridges over the rivers and the expansion of urban/built-up areas over the agricultural land in fringe areas and the beyond in the valley played key roles in increasing the heterogeneity of the landscape. However, the construction of commercial complexes and planned residential developments, represented by the built-up area in the maps, in the valley floor increased in recent decades, but their impacts on improving homogeneity at the landscape level are small. A temporal reduction of the contagion (CONTAG) and an increase of the patch diversity (SHDI) presented a clear picture of increasing the landscape heterogeneity in the valley.

Class (land use) level analysis: At the class level metrics, changes in the PD of all land use classes are observed, except in the forest land cover (Fig. 15.7). The PD of urban/built-up and water areas had remarkable changes in the 1980s and 1990s. The proportions of these land uses are small in the valley as compared to the other major land uses. Due to the inconsistent widths and linear character of the rivers, construction of additional bridges over the rivers in the later decades might have influenced the increase in the PD of water areas. The increase of PD in urban/built-up areas is a result of the urbanization process that occurred rapidly in the 1990s. The population influxes into the valley floor increased the demand for new housing, subsequently creating new built-up areas in the city fringes that segmented the existing agricultural lands in the later years. This process also increased the PD and ED of urban/built-up areas, including agricultural areas. The ED of agricultural land remained higher than that of other land uses. However, this dominancy could cause higher ED values, as shown by the LPI. A noticeable change is observed in the ED and PD of agricultural and urban/built-up land uses correlating each other in the 1980s and 1990s. The trend of unordered individual housing development in the valley floor, especially in the fringe areas in the corresponding decades enhanced the fragmentation and the heterogeneous landscape development.

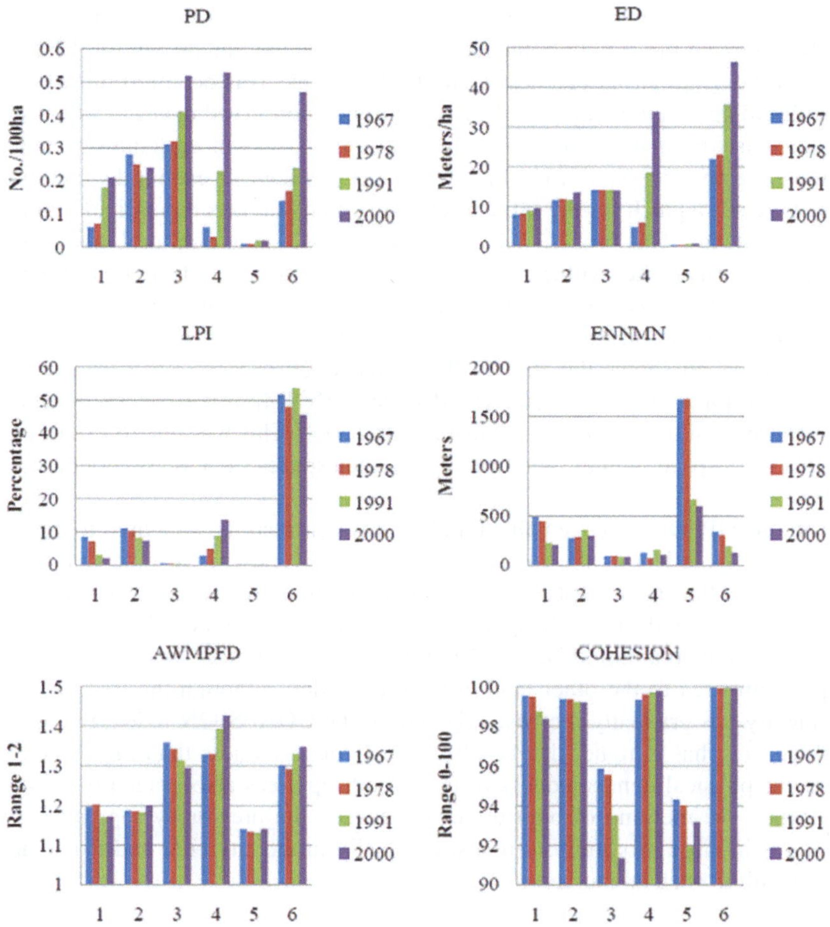

The numbers in X-axis represent as: 1. Shrubs; 2. Forest; 3. Water; 4. Urban/builtup area; 5. Open space; 6. Agricultural area

Fig. 15.7 Spatial metrics at class level

A gradual increase in LPI is found for the urban/built-up land use type. The increasing intensity of buildings and other infrastructure in the core area in the 1990s, for example, increased the LPI of the urban built-up area. In the same time-frame, the LPI of agricultural land decreased significantly, showing a tremendous urbanization pressure over the agricultural lands, as observed in Fig. 15.5. The urbanization process over agricultural land is a common trend in the valley. However, the LPI of other land uses decreased. The expansion of urban infrastructure abutting the land between the built-up areas and surrounding areas in the valley floor in the later decades is the main cause of the improvement of the dominance index of the urban/built-up areas.

Adding new urban structures in agricultural spaces creates new patches, eventually fragmenting the land in both categories and reducing the proximity (ENNMN) between the neighboring patches of similar land uses. This occurred mostly in the agricultural land in the valley. A decreasing trend is observed over the whole study period. It is true that when urbanization activities occur in agricultural land, for example the development of road networks, open spaces, or even individual buildings, it creates a patch for the infrastructure itself and divides the agricultural lands into two patches. This process decreases the ENNMN of agricultural land patches, creating several patches within close proximity. Interestingly, a decrease of ENNMN for urban built-up areas in the 1970s shows that the urbanization process was confined mostly in the margins of the existing built-up areas. In the 1980s, ENNMN increased, which shows the start of the urbanization process away from the city core, i.e., rural areas (Fig. 15.5). The distance (ENNMN) between the patches of open space significantly decreased in 1991 and 2000. This was due to expansion of existing open spaces and the addition of new open spaces such as golf courses in the valley.

The AWMPFD always remained higher in water areas as compared to the other land use types, reflecting the complex shape of the rivers. This is somewhat natural, as rivers pass through complex mountain topography in the valley. The shape complexity of urban/built-up lands significantly increased in later decades. Expansion of roads into rural areas could be a cause of this. The agricultural land still has simple shapes compared to the other land uses in the valley, although the trend of shape complexity has gradually increased. Based on the COHESION index, the agricultural land use has a higher degree of physical connectedness than the others. The degree of physical connectedness of urban/built-up areas also gradually increased over time. The merging of built-up areas between the previously separated urban structures in close proximity to roads, especially in the city core and fringe areas, could result in such an increase.

15.4 Discussion

The spatiotemporal analyses clearly showed that the intensity of human induced activities (Fig. 15.1) affected the landscape pattern of the study area. However, the other important finding was that it also influenced the dynamic changes in the land use types. The rapid urbanization process over productive agricultural lands in close proximity to the road network in the valley is apparent. The built-up area increased by 5% in each decade of the 1980s, and 1990s. This rate of increase is much higher than in the United States, Canada, and New Zealand (Li & Yeh, 2004).

A large share of agricultural space was transformed to urban/built-up areas in different time periods, and this mainly occurred in the valley floors and neighboring villages. A proportional transition of other land uses to the urban/built-up areas was also found to be consistently increasing. Interestingly, a significant level of agricultural encroachment over the shrubs and forest lands was found in the 1980s. The urbanization process resulted in the increase of agricultural activities in rural areas in the valley. This is a common phenomenon near metropolitan regions of developing countries (Abdullah & Nakagoshi, 2006).

The urbanization process in the valley has caused fragmentation of the landscape and heterogeneous land use development. Haphazard land development practices and uncontrolled urban growth enhance those processes and generate a range of environmental problems, water shortage, air pollutions in the lack of green space, for example, affecting human health and welfare. The demand for infrastructure and services has caused the emergence of a number of urban environmental problems in Kathmandu (Thapa et al., 2008). However, the heterogeneous landscape development will continue into the next few decades as the built-up areas in the valley floor have already started to agglomerate. Planned residential developments are emerging recently in the fringe and rural areas. Furthermore, the decreasing trend of nearest neighborhood distance is an indication that homogeneity will increase by refill type development after a certain period of time.

A significant extension of the road network is observed during the 1980s and 1990s. The development of road networks to the rural periphery in the valley reduced the travel time for commuters living in the villages. It made it easier for rural farmers, the sole producers of the perishable goods for urban dwellers in the valley, to commute to the city for urban services and markets. This process enhanced rural prosperity and attracted migrants from neighboring districts. The rural villages gradually become urban frontiers. Along with the increased income, the agricultural landscapes in close proximity to the roads were gradually converted to urban/built-up areas, stretching the urbanization influences to rural areas. Many individual developments occurred nearby in later years, showing that travel time was probably a major driver of change. Usually, the influence of roads is not linear. Bruijn (1991) indicated that the factors influencing land development probabilities often have a strong "distance decay function". For example, the influence of a road or existing built-up areas on land development decreases quickly with distance.

The land use change maps show that the most productive agricultural lands are now being converted to urban uses, particularly in the valley floor and nearby hills. This pattern is driven, in part, by migration (Thapa & Murayama, 2010b). Migrants tend to move to areas in close proximity to economic opportunities and urban services in the valley, which is usual in most developing countries in Asia (Seto & Kaufmann, 2003). Consequently, urban centers emerge from areas where people first moved to the land with highest productivity. From the field observations, most of the lands in the fringes and nearby villages are growing high value agricultural products (i.e., vegetables, flowers, and fruits), where they used to be rice paddies a few years ago. Some of the paddy terraces in the hills of the city fringes, most with easy access to water and roads, were converted to perishable agricultural production land. Many rice paddies were converted to other types of agricultural land use for better revenue under the influences of market mechanisms. Most of the urban structures in the valley are built by bricks (Fig. 15.1). Due to high demand of bricks, the rice paddies having suitable soils for brick production in the fringe and rural areas have also been converted to brick kiln areas in the dry season and back to rice paddies in the rainy season.

Industrialization also contributed to the urbanization of the valley, particularly in the 1970s and 1980s (Industrial Development Management, 2007). Three major industrial estates, Balaju, Patan, and Bhaktapur, and several small scale industries scattered around the ring road and along major highways were established in the valley during these decades (Thapa et al., 2008). The carpet industry was one of the industries that flourished in the valley. High worldwide demand for Nepalese carpet in the 1980s and early 1990s encouraged entrepreneurs to make investments in this sector. By that time, 5000 carpet factories were established in the valley, producing several thousands of jobs (Eastman et al., 1998), which enhanced the population pressure in the valley. This fact shows that the carpet industry may have been a major driving force of urbanization in the 1980s. At present, however, due to the unfavorable environmental consequences including water and air pollution, enforcement of environmental laws and public awareness in the valley has caused the decline of the carpet industry by 50% (Gautam et al., 2008).

All these processes changed the spatial structure of the urban form in the valley floor, which is observed as a multiple-nuclei pattern. Furthermore, the urbanization process enhanced the fragmentation and heterogeneous landscape development in the valley as evident by the spatial metrics. The valley experienced high population influx and a heterogeneous land use environment caused by unrestrained urban development, which may pose serious threats to the inhabitants especially creating water shortages, human wastages, and more squatters (Thapa et al., 2008).

15.5 Conclusions

In this study, we have investigated the spatiotemporal patterns of urbanization in the Kathmandu valley using remote sensing and spatial metrics techniques. The landscape of the valley is diverse and complex and comprises both homogeneous and heterogeneous surface features, which causes problems with the spectral variability in the satellite image data. The increasing population pressure caused the spatial pattern of urbanization to be highly dynamic. The predominantly agricultural landscape gradually changed to an urban landscape with increasing human settlement in the 1960s and 1970s. The changing process has escalated since the 1980s. It has proved to be very high in the urban fringe area. Spatial diffusion of urban/built-up areas has spread outward from the city core and along the major roadways.

The urban built-up area has increased by four times in the last four decades. Almost half of the shrubs land in the valley has disappeared during this period. Similarly, forest land decreased dramatically in the 1980s. Shrubs and forest landscape in rural areas of the valley mostly changed to agricultural areas. Half of the land in the valley is still agricultural area, but it has faced changing circumstances in different time period. Agricultural encroachment in rural hills and mountain peripheries and urbanization in the valley floor area are identified as the most common phenomenon in the study period. A small land use transition between the other land uses was also noticed.

The sparsely developed built-up area with individual unordered housing practices in the fringe areas indicated a complex urbanization process in the valley. An increasing trend of land use diversity is explored using the spatial metrics analysis. The land use patch density significantly increased during the period. Such urbanization process and scattered individual developments created fragmentation and a heterogeneous landscape that gradually increased in the 1980s and 1990s. These processes are mostly observed in the city fringes and adjacent villages in the valley. However, the overall nearest neighbor distance between the similar land use patches in the valley has decreased over the last two decades. The city core area seems to be agglomerated, which eventually creates homogeneity by the refilling type of development in future.

The study presented a consolidated approach of remote sensing and spatial metrics that allowed a separation of land use categories and descriptions of its temporal changes. It has provided a robust quantitative measure of land use dynamics using remotely sensed data for the last four decades of the twentieth century, which will help urban planners and researchers to make assessments of landscape development and change.

References

Abdullah, S. A., & Nakagoshi, N. (2006). Changes in landscape spatial pattern in the highly developing state of Selangor, Peninsular Malaysia. *Landscape and Urban Planning, 77*, 263–275.

Bruijn, C. A. D. (1991). Spatial factors in urban growth: Toward GIS models for cities in development countries. *ITC Journal, 4*, 221–231.

Eastman, J. R., Jiang, H., & Toledano, J. (1998). Multi-criteria and multi-objective decision making for land allocation using GIS. In E. Beinat & P. Nijkamp (Eds.), *Multicriteria analysis for land-use management* (pp. 227–252). Kluwer: Dordrecht

Gautam, R., Baral, S., & Herat, S. (2008). Opportunities and challenges in implementing pollution prevention strategies to help revive the ailing carpet manufacturing sector of Nepal. *Resources, Conservation and Recycling, 52*, 920–930.

Herold, M., Goldstein, N. C., & Clarke, K. C. (2003). The spatiotemporal form of urban growth: Measurement, analysis and modelling. *Remote Sensing of Environment, 86*, 286–302.

ICIMOD/UNEP. (2001). *Kathmandu valley GIS database, CD-ROM*. ICIMOD: Kathmandu.

Industrial Development Management. (2007). *Present status*. Retrieved September 12, 2008, from http://www.idm.com.np/presentstatus.php

Jensen, J. R. (2005). *Introductory digital image processing: A remote sensing perspective*. Upper Saddle River, NJ: Prentice Hall.

Li, X., & Yeh, A. G.-O. (2004). Analyzing spatial restructuring of land use patterns in a fast growing region using remote sensing and GIS. *Landscape and Urban Planning, 69*, 335–354.

Lillesand, T. M., Kiefer, R. W., & Chipman, J. W. (2008). *Remote sensing and image interpretation*. New York: Wiley.

McGarigal, K., Cushman, S. A., Neel, M. C., & Ene, E. (2002). FRAGSTATS: Spatial pattern analysis program for categorical maps. University of Massachusetts Amherst: Amherst, MA. Retrieved February 12, 2008, from http://www.umass.edu/landeco/research/fragstats/fragstats.html

Openshaw, S., & Alvandies, S. (1999). Applying geocomputation to the analysis of spatial distributions. In P. Longley, M. Goodchild, D. Maguire, & D. Rhind (Eds.), *Geograhpic information systems: Principles and technical issues* (pp. 267–282). Wiley: New York.

O'Neill, R. V., Krummel, J. R., Gardner, R. H., Sugihara, G., Jackson, B., DeAngelis, D. L. et al. (1988). Indices of landscape pattern. *Landscape Ecology*, *1*, 153–162.

Seto, K. C., & Kaufmann, R. K. (2003). Modeling the drivers of urban land use change in the Perl River Delta, China: Integrating remote sensing with socioeconomic data. *Land Economics*, *79*, 106–121.

Sharma, P. (2003). Urbanization and development. In *Population monograph of Nepal* (pp. 375–412). Kathmandu: Central Bureau of Statistics.

Thapa, R. B., & Murayama, Y. (2008). Spatial structure of land use dynamics in Kathmandu Valley. In J. Chen, J. Jiang, & A. Peled (Eds.), *The international archives of the photogrammetry, remote sensing and spatial information sciences* (Vol. XXXVII, (Part-B8), pp. 11–16). ISPRS: Beijing.

Thapa, R. B., & Murayama, Y. (2009a). Urban mapping, accuracy, & image classification: A comparison of multiple approaches in Tsukuba City, Japan. *Applied Geography*, *29*, 135–144.

Thapa, R. B., & Murayama, Y. (2009b). Examining spatiotemporal urbanization patterns in Kathmandu valley, Nepal: Remote sensing and spatial metrics approaches. *Remote Sensing*, *1*, 534–556.

Thapa, R. B., & Murayama, Y. (2010a). Remote sensing: Platforms and sensors. In B. Warf (Ed.), *Encyclopedia of geography* (pp. 2420–2424). Thousand Oaks, CA: SAGE

Thapa, R. B., & Murayama, Y. (2010b). Drivers of urban growth in the Kathmandu valley, Nepal: Examining the efficacy of the analytic hierarchy process. *Applied Geography*, *30*, 70–83.

Thapa, R. B., Murayama, Y., & Ale, S. (2008). Kathmandu. *Cities*, *25*, 45–57.

Thapa, R. B., & Murayama, Y. (2011). Urban growth modeling of Kathmandu metropolitan region, Nepal. *Computers, Environment and Urban Systems*, *35*, 25–34.

Chapter 16
Spatial Determinants of Poverty Using GIS-Based Mapping

Brandon Manalo Vista and Yuji Murayama

16.1 Introduction

In many countries, poverty has a geographic dimension (Bigman & Deichmann, 2000). Geography, particularly the physical environment, plays a significant role in the incidence of poverty in every local communities, especially in developing nations (Bigman & Fofack, 2000). Empirical evidences suggest a strong relationship between geography and poverty. Previous mapping studies in Vietnam (Minot, Bob, & Epprecht, 2006), in Uganda (Rogers, Emwanu, & Robinson, 2006), in Nigeria (Legg et al., 2005), in Bangladesh (Kam, Hossain, Bose, & Villano, 2005), in Kenya (Kristjanson, Radeny, Baltenweck, Ogutu, & Notenbaert, 2005), in Malawi (Benson, Chamberlin, & Rhinehart, 2005), in Sri Lanka (Amarasinghe, Samad, & Anputhas, 2005), among others, have shown that significant geographic variation in the incidence of poverty may be due to a variety of geographic factors. These include natural resource endowments and agro-climatic conditions, accessibility and proximity to markets, access to land, as well as aspects of public policy.

On this regard, geography cannot be regarded as a factor affecting the rate of poverty because it has a strong impact on the living standards of people living in the community. Poverty is not merely a socio-economic problem. Equally important, as Glasmeier (2002, p. 158), puts it "poverty is an inherently spatial problem ..." However, debates on poverty and its causes more often rest on the socio-economic realm, not until recently when the role of geography in understanding and analyzing poverty has been increasingly recognized by scholars and development practitioners (Hyman, Larrea, & Farrow, 2005). The passage of the Millennium Development Goals (MDGs) which advocates for poverty reduction triggered initiatives in the international arena to have greater interest in the geographic dimensions of poverty (Hyman et al., 2005). Correspondingly, the development of geographic information systems (GIS) together with advances in remote sensing has led to incorporate spatial data and satellite imageries as methods for poverty mapping and analysis

B.M. Vista (✉)
Department of Geography, University of Otago, P.O. Box 56 Dunedin, New Zealand 9054
e-mail: brandonvista@yahoo.com

Y. Murayama, R.B. Thapa (eds.), *Spatial Analysis and Modeling in Geographical Transformation Process*, The GeoJournal Library 100, DOI 10.1007/978-94-007-0671-2_16,

(Deichmann, 1999; Bigman & Deichmann, 2000; Davis, 2003; Hyman et al., 2005). As such, "Poverty mapping provides a means for integrating biophysical information with socioeconomic indicators to provide a more systematic and analytical picture of human wellbeing and equity" (Henninger & Snel, 2002).

Despite recent advancement in poverty mapping brought about by GIS and remote sensing, little has been done to study the spatial aspects of poverty in the Philippines. More often than not, poverty studies in the country still resides within the realm of economics and public policy which do not consider so much on its spatial dimension (Albert & Collado, 2004; Balisacan & Fuwa, 2004; Estudillo, Sawada, & Hossain, 2005; Fuwa, 2007). Although there were a few poverty mapping initiatives over the past few years (Domingo, 2003), GIS was mostly used for the production of various poverty maps by plotting various socio-economic indicators and not for spatial analysis (Balisacan, Edillon, & Ducanes, 2002; NSCB, 2005). In fact, the Philippines' National Statistics Coordination Board (NSCB) in collaboration with The World Bank commissioned a poverty mapping study to estimate local poverty, i.e. municipal and city level, using socio-economic and census data, and thereafter mapping the results using GIS (NSCB, 2005). However, the incorporation of geographical factors was not considered in many these studies; as such, the geographic dimension of poverty is more often neglected.

Henceforth, in an attempt to underscore the importance of geography in analyzing poverty in the Philippines, this study tries to explore the spatial patterns and the possible underlying determinants affecting poverty using GIS. Specifically, this study aims to determine the key factors that may have an influence on the incidence of poverty.

16.2 Methods

16.2.1 Study Area

To undertake this objective, two adjacent provinces in Bicol, one of the poorest regions in the Philippines, were selected as pilot site. These were Albay and Camarines Sur (see Fig. 16.1). The two provinces have a total population of 2.64 million comprising of 496,812 households as of calendar year (CY) 2000 (NSO, 2000a). Population density is about 328 persons per square kilometer, slightly higher than the regional population density of 265. This means that the Bicol regional population is highly concentrated in the two provinces. The average annual family income in this region is Php 89,277 which is far below the Php 144,039 national average (NSO, 2000b).

The study site is endowed with rich reserve of natural resources for agricultural and economic activities. Albay and Camarines Sur have a land area of 2,565.8 and 5,481.6 km^2, respectively (NSCB, 2006). These land area is 45.6% of the total area of the Bicol Region. A strip of mountain-volcanoes cuts across the two provinces while rolling hills and terrain comprised the eastern and western part. The midsection of the study site forms an elongated valley consisting of flat alluvial lands.

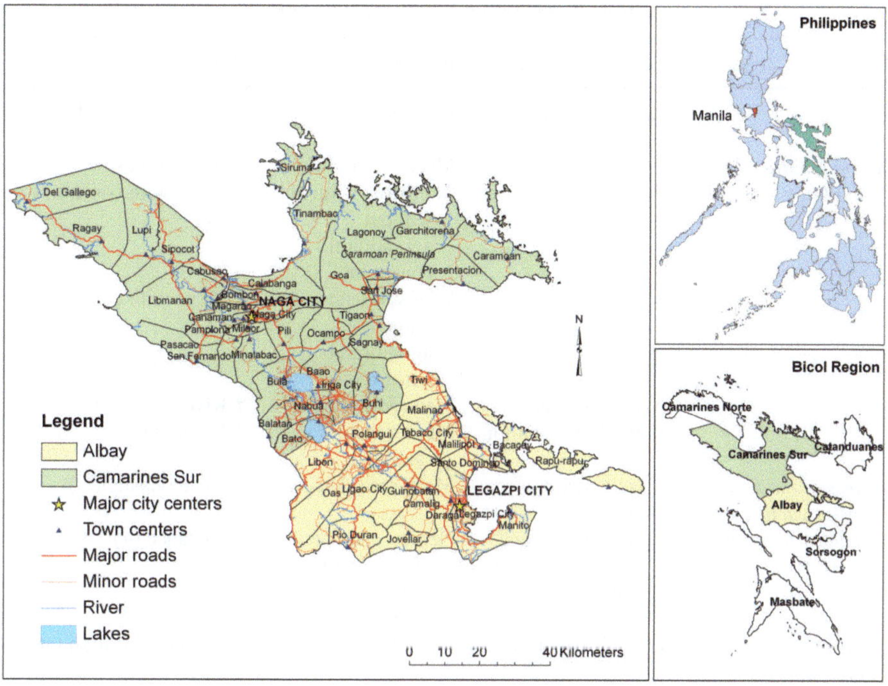

Fig. 16.1 Location and basic geographical features of the pilot site

Network of rivers, streams and creeks traversed the Bicol plain which serve as a natural drainage and are adjoined by three fresh water lakes. Water from these lakes, which flows across the plain, formed the so-called Bicol River Basin which comprise fertile agricultural lands.

These two provinces composed of 55 local government units (LGUs). Two of these LGUs are classified as major cities – Naga City and Legaspi City, with a population of 137,810 and 157,010, respectively (NSO, 2000a). For the purpose of this research and also considering the limited data available at disaggregated level, the unit of analysis would be the LGU or municipal level.

16.2.2 Data Collection

Poverty, which is the independent variable in this study, adopts the traditional measurement or indicator – Poverty Incidence – which is the percentage of the people living below the poverty threshold (Coudouel, Hentschel, & Wodon, 2002). The higher the incidence of poverty in an area, the poorer it is relative to another locality. In view of this, poverty incidence of each municipality/city within the study site was obtained from the local poverty estimates in CY 2000 published by NSCB based from its World Bank – Asia-Europe Meeting (WB-ASEM) poverty mapping project mentioned above (NSCB, 2005). Meanwhile, the administrative boundary of every

Table 16.1 Matrix of variables and hypotheses

Primary variables	Secondary variables	Indicators	Hypothesis
Poverty	Local poverty level	Poverty incidence (PI)	Independent variable
Agro-climatic condition	Elevation	Mean elevation	↑ Elevation (+)
	Slope	% of slope with and without agricultural constraints	↑ Slope with constraint (+)
	Soil texture	% of soil with and without agricultural constraints	↑ Slope with constraint (+)
	Access to water	Mean river density	↑ River density (−)
	Rainfall distribution	Mean annual rainfall	↑ Rainfall (−)
Access to road infrastructure	Road network	Mean road density	↑ Road density (−)
Proximity to major market	Road network and location of town centers	Nearest estimated distance of town centers to major cities	↑ Distance (+)
Influence of public programs and policy	Land distribution	Land distribution rate	↑ Land distribution rate (−)
	Fiscal decentralization	% of IRA share	↑ IRA share (−)
Influence of demographic characteristics	Population growth	Population growth rate	↑ Population growth rate (+)

Note: (↑) Arrow means "higher" e.g. higher elevation would have higher poverty incidence; (+) positive means direct relationship, e.g. higher elevation increases incidence; (−) negative means inverse relationship, e.g. higher elevation decreases incidence

city/municipality was taken from a digital data owned by Cybersoft Informatics, Inc., a private GIS consultancy company engaged in mapping and geospatial data building activities in the Philippines. These boundaries served as the spatial unit or modifiable areal unit of analysis for this study.

To find out the various determinants affecting poverty, the following variables were considered in this study (see Table 16.1), which were argued by different studies to have possible influence on poverty:

First and foremost, agro-climatic condition may have an effect on the level of poverty of every locality, especially in an agriculture-based economy such as the Philippines. This includes the topography, soil texture, rainfall and access to river within the study area. Unfavorable topography such as high altitude and steep slopes or mountainous areas is generally perceived to increase the incidence of poverty because of its agricultural limitations (Minot et al., 2006). Similarly, lack of access to river and less rainfall which normally serve as a source of water for crop irrigation or livestock propagation is thought to have adverse effect on poverty (Kristjanson et al., 2005). Rainfall is important especially in rainfed agriculture. Ecologically poor areas which maybe characterized by drought due to lack of access to water have high poverty rate because these places are not quite suitable for productive farming or livestock. In Vietnam, for instance, low rainfall is associated with high poverty rate (Minot et al., 2006).

In view of these variables, elevation values were collected from satellite imageries of the Shuttle Radar Topographic Mission (SRTM) version 3 with a 90-m resolution (Jarvis, Reuter, & Guevarra, 2006). Soil texture data was obtained from a digitized map produced by the Philippines' Bureau of Soil and Water Management (BSWM). River network was taken from Cybersoft while average annual rainfall values in CY 2000 were based from the data provided by the Philippine Atmospheric, Geophysical and Astronomical Services Administration (PAGASA).

Secondly, the study also considers the likely influence of access to rural infrastructures such as roads, and proximity to major markets to the state of poverty. Road network was also obtained from Cybersoft, on this regard. High access to road is perceived to lower the rate poverty (Minot et al., 2006; Kristjanson et al., 2005) because road is a basic rural infrastructure for agriculture and economic development. Similarly, poor areas are often far from major markets, particularly the major cities. Since major cities serve as centers of trade, commerce and industry, towns far from the major cities tend to have high rate of poverty because major cities have an influence on the economic activities and level of development of towns, especially within its periphery. Minot et al. (2006), for instance, found out that distance to cities was a significant predictor to local level poverty incidence in Vietnam.

Thirdly, the study also takes into account the potential impact of government policies on the condition of poverty. Accordingly, two major policies of the Philippine government were identified – the Comprehensive Agrarian Reform Program (CARP) which aims to distribute lands to uplift the lives of landless farmers nationwide from the shackles of poverty; and the fiscal decentralization policy which is a strategy of the national government to shift development away from major cities such as Manila towards the regions and the countryside through funding support for local development. On these regards, CY 2000 land distribution data in every municipality in the study area was acquired from the Department of Agrarian Reform Bicol Regional Office whereas the amount of Internal Revenue Allotment (IRA) in CY 2000 for every locality was provided by the Department of Budget and Management Regional Operations and Coordination Service. Presumably, higher rate of land distribution and fiscal funding support would alleviate poverty, hence decrease its incidence.

Lastly, the study regards the likely effect of population growth to the condition of poverty because it is normally viewed as contributory to the persistence of poverty in the Philippines (Balisacan, 2007). As such, high population growth rate would further increase poverty incidence. The study used data based from the growth of the regional population between CY 1990 and 2000 (NSCB, 2004).

16.2.3 Pre-processing and Database Built-Up

Appropriate pre-processing techniques were done to the abovementioned datasets in order to perform various spatial and statistical analyses. Administrative boundary, river and road network datasets were compared with the Landsat imageries (NASA Landsat Program, 2003) to assess their quality and reliability. Boundaries of the inland lakes were also digitized to delineate them from land borders. All of these, including the SRTM imageries, were projected into Universal Transverse Mercator (UTM) Zone 51 North to allow the conduct of various spatial analyses. Meanwhile, the percentage of land distribution versus the working scope and the percentage of allotment against the total IRA allocation for the two provinces were calculated in each municipality within the pilot site. Together with poverty incidence and population growth rate datasets, these were encoded in MS access, and spatially joined with administrative boundary layer using ArcGIS.

16.2.4 Spatial Analysis

The study exploited the capability of Spatial Analyst and Network Analyst extension in ArcGIS 9.1 to employ several spatial analyses on each of the spatial variables in this study – elevation, soil, river, road and rainfall – in order to generate auxiliary datasets that could be disaggregated by municipality and be analyzed in relation to local poverty incidence.

Firstly, to represent elevation of every locality, mean elevation values by municipality were extracted from SRTM by performing Zonal Analysis, with administrative boundary shapefile serving as zonal border. Meanwhile, the slope gradient in percent was extracted using the same elevation values from SRTM. Subsequently, the generated slope values were grouped into two categories to delineate those which have agricultural limitations based from the criteria set by the Global Agro-ecological Zone (GAEZ) mapping classification of the Food and Agricultural Organization of the United Nations (FAO) (Fischer, van Velthuizen, Nachtergaele, & Medow, 2000). Those lands with a slope of 0–8% were classified as slope with no agricultural constraints because they are most likely suitable for farming and possess no irrigational restriction whereas those areas with a slope higher than 8% possessed agricultural limitations (Fischer et al., 2000). Finally, a Zonal Analysis was executed to compute for the percentage of slope with and without agricultural constraints in every municipality within the pilot site.

Similarly, the study identified soil textures possessing no agricultural limitations based from the GAEZ criteria (Fischer et al., 2000). According to Fischer et al. (2000), soil with medium and fine textures have no constraints and are suitable for farming activities. These include: (1) Clay; (2) Silty Clay Loam; (3) Clay Loam; (4) Sandy Clay Loam; (5) Loam; (6) Silt Loam. Those lands which do not have any of these soil properties were classified as soil with agricultural constraints. Thereafter, the percentages of soil with, and with no agricultural limitations were computed by municipality using Zonal Statistics in ArcGIS.

Thirdly, the study estimated the annual rainfall values across the pilot site based from existing annual rainfall data in 13 weather stations in Bicol Region. At first, the location of each weather station was plotted using their respective latitude and longitude coordinates, and then corresponding annual rainfall values recorded in CY 2000 from each weather station were linked with their plotted spatial location. Then, a Kriging Interpolation, a geostatistical technique often used to interpolated climate fields, was performed using Spatial Analyst to estimate the rainfall values across the study site. Kriging method was used because it is one of the most appropriate spatial interpolation methods to estimate rainfall values (Biau et al., 1999). After that, a Zonal Analysis was executed using administrative boundaries as zones in order to generate the mean annual rainfall for every city/municipality.

Fourthly, the study computed for river density as a proxy indicator for access to water by executing Line Density Analysis using Spatial Analyst from an existing river polyline dataset. Thereafter, a Zonal Analysis was carried to compute for mean river density in every municipality within the pilot site. Moreover, to represent access to road infrastructure, the study calculated road density from the road polyline dataset by performing Kernel Density Analysis in ArcGIS. Kernel density analysis was selected as the appropriate method because it allows greater weight in the computation of road density to major roads than to minor streets in the study area. Subsequently, mean road density of every municipality was computed using Zonal Analysis to allow for comparison with local poverty incidence.

Lastly, to represent proximity to major markets, the study estimated the nearest distance of each town center to major cities within the pilot site as proxy indicator. For this purpose and as mentioned above, Legaspi and Naga Cities were considered as the major cities in this analysis. Rather than selecting the geographical center of each municipality, town center locations were considered more appropriate representation of the geographical center because they serve as the heart of economic, political and cultural life of each locality. To do this, town centers were identified with the aid of Landsat imageries and shapefiles of urban villages known as *barangays*, and thereafter, these were represented as point data in GIS. Based from the existing road network, the distance of each town center to the nearest major city was calculated by taking advantage of the capability of Network Analyst in ArcGIS. Table 16.2 summarizes the different variables/indicators both spatial, i.e. derived from GIS, and socio-economic variables, and their corresponding unit of measurement that were further used in the succeeding statistical analysis.

16.2.5 Regression Analysis

After performing a series of spatial analysis for the different spatial variables to derive auxiliary data that would be comparable with poverty incidence in every locality, the study carried out a Multiple Regression Analysis with the aid of Statistical Packages for Social Sciences (SPSS) software in order to find out which of the different variables if combined altogether significantly affects poverty measured in terms of its incidence. The full regression model for this study can be mathematically stated as follows:

Table 16.2 Unit of measurement for each variable indicator

Variable/indicator	Unit of measurement
Poverty incidence	In percent
Mean elevation	In meters
Slope with agricultural constraint	In percent of total area
Soil with agricultural constraint	In percent of total area
Mean annual rainfall	In millimeters
Mean river density	In sq. m./sq. km.
Mean road density	In sq. m./sq. km.
Distance to major cities	In kilometres
Agrarian reform accomplishment rate	In percent
Internal revenue allotment share	In percent of total amount for 2 provinces
Population growth (1990–2000)	In percent

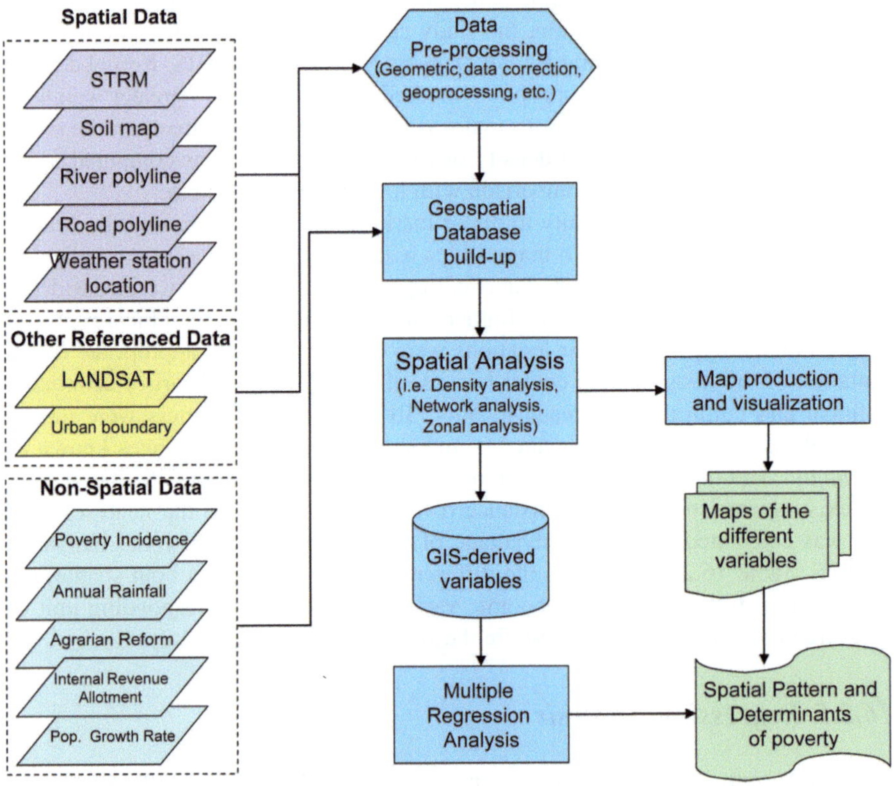

Fig. 16.2 Diagram of research methodology using GIS

$$Y \approx \beta_0 + \beta_1 x_1 + \beta_2 x_2 + \beta_3 x_3 + \beta_4 x_4 + \beta_5 x_5 + \beta_6 x_6 + \beta_7 x_7 + \beta_8 x_8 + \beta_9 x_9 + \beta_{10} x_{10} + \varepsilon$$

where $Y \approx$ poverty, $\beta_0 \approx$ constant, β_1 to $\beta_{10} \approx$ coefficients, $x_1 \approx$ Elevation, $x_2 \approx$ Slope, $x_3 \approx$ Soil texture, $x_4 \approx$ Rainfall distribution, $x_5 \approx$ Access to river resources, $x_6 \approx$ Access to road infrastructure, $x_7 \approx$ Proximity to major markets, $x_8 \approx$ Land distribution, $x_9 \approx$ Fiscal decentralization, $x_{10} \approx$ Population growth, and $\varepsilon \approx$ error.

The diagram (Fig. 16.2) illustrates a summary of all the steps and procedures that were undertaken in this study and its expected outputs.

16.3 Discussion of Results and Analysis

The results of the multiple regression analysis show that eight (8) out of ten (10) variables in this study are statistically significant determinants to the incidence of poverty (see Table 16.3). Furthermore, four (4) out of ten (10) variables (marked as *) are strongly and statistically significant at 0.01. These include access to road infrastructure, fiscal decentralization policy, land distribution program, population growth. That is to say, these four variables appear to be the best predictors for the incidence of poverty within the study site. In the same way, the remaining four (4) variables – proximity to major markets, elevation, slope with agricultural limitations and rainfall – are statistically significant predictor of poverty at 0.05 (marked as **). Of all the variables, mean road density is the strongest influential determinant to poverty as reflected by its high standardized coefficient. This is followed by share in IRA. In contrast, soil with agricultural limitations and access to river resources did not have a statistically significant effect on poverty incidence.

R^2 generated a value of 0.832 which means that this regression model could explain 83.2% of variation in poverty incidence in the study area (see Appendix 1-a). In addition, results from the Analysis of Variance (ANOVA) shows the goodness of fit of the model, and in this case it is statistically significant at 0.001 (Appendix 1-b). In other words, the regression model has the ability to explain the determinants of poverty. Meanwhile, the histogram and the P-P plot of the residual (Appendix 1-c and 1-d) suggest that the residuals are plausibly normally distributed. An existence of collinearity was also checked for every variables using bivariate correlation analysis (see Appendix 2 for the correlation matrix). Correlation results show that none of the variables have a correlation coefficient greater than 0.8. This means that none of the independent variables is correlated or similar to one another, hence they could all be considered in the model. The succeeding portion of this section discusses in more detail each of the variables and how do they affect the incidence of poverty within the study site.

16.3.1 Elevation

As discussed above, elevation is one of statistically significant determinant variables to poverty. Elevation was assumed to have adverse effect on poverty. As such, highly

Table 16.3 Multiple regression results

Variables	Indicator	Unstandardized coefficients		Standardized coefficients	*T*	Sig.
		B	S.E.	Beta		
	Constant	1.469	0.416		3.530	0.001*
Elevation	Mean elevation	−0.0002	0.000	−0.215	−2.221	0.032**
Slope	Percentage of slope with agri. limitations	0.101	0.044	0.248	2.267	0.028**
Soil	Percentage of soil with agri. limitations	−0.0218	0.027	−0.070	−0.797	0.430
Rainfall	Mean annual rainfall	−0.0002	0.000	−0.181	−2.145	0.037**
River access	Mean river density	−0.0532	0.063	−0.082	−0.841	0.405
Road access	Mean road density	−0.111	0.024	−0.490	−4.660	0.001*
Proximity to major markets	Distance to major cities	0.0007	0.000	0.186	2.139	0.038**
Land distribution	Agrarian reform accomplishment rate	0.141	0.049	0.212	2.847	0.007*
Fiscal decentralization	Share in internal revenue allotment	−1.798	0.411	−0.314	−4.370	0.001*
Population growth	Population growth rate	−0.0427	0.015	−0.208	−2.827	0.007*

Note: Poverty incidence as dependent variable
*Significant at 0.01
**Significant at 0.05

elevated areas, especially in developing countries, were perceived to be poorer, thence higher poverty incidence, than low lying lands. However, this assumption appears to contradict the results because of the negative value of its regression coefficient of −0.0002 (Table 16.3). But if its bivariate correlation alone is examined independently, that is without considering other variables, its correlation coefficient generated a positive value of 0.106 (Appendix 2). Such inconsistency could be explained by merely comparing the map of poverty incidence versus elevation (Fig. 16.3a and 16.3b). Areas that have relatively high incidence of poverty have mountainous features – the Caramoan Peninsula, Ragay Coast, and the western coast. However, some low lying municipalities, particularly those located in the Bicol plain also have relatively high poverty incidence while others have low poverty rate. This suggests that the effect of elevation to the level of poverty incidence varies widely by locality. Most likely, elevation has a more significant effect to those municipalities lying in mountain ranges where their topographic features limit their ability to engage in various agricultural and economic activities.

16.3.2 Slope

Secondly, regression result also revealed that the percentage of slope with agricultural constraints, i.e. slope above 8%, is statistically significant determinant to poverty at 95% confidence level. The positive regression coefficient of 0.101 implies that areas covered by land with a slope greater than 8% usually displays high incidence of poverty. Because of their rolling to hilly and steep slope features, they exhibit agricultural development constraints, such as easy access to irrigation. On the other hand, lands with slope 0–8% characterized by flat lands to rolling hills and terrain which are mostly located in the Bicol valley have relatively low level of poverty incidence. These can also be clearly seen by merely comparing slope map and poverty incidence (Fig. 16.3a and 16.3c). In a gist, the results suggest that those municipalities having more percentage of slopes with agricultural constraints generally exhibit higher incidence of poverty.

16.3.3 Soil

Unexpectedly, regression results show that soil texture, especially soil possessing agricultural limitations, is not a statistically significant determinant to the incidence of poverty. This means that soil is not a factor that influences the condition of poverty within the study site. However, by simply comparing the soil and poverty map (Fig. 16.3a and 16.3d), it appears that the poorest section of Caramoan Peninsula possesses a geographical disadvantage because its soil characteristics belong to those classified with agricultural constraint – mountain soil, complex soil and sandy loam; thence, soil is a significant variable in this case. Perhaps, the suitability of crops planted according to the type of soil and its productivity would be more influential to poverty than the soil texture alone.

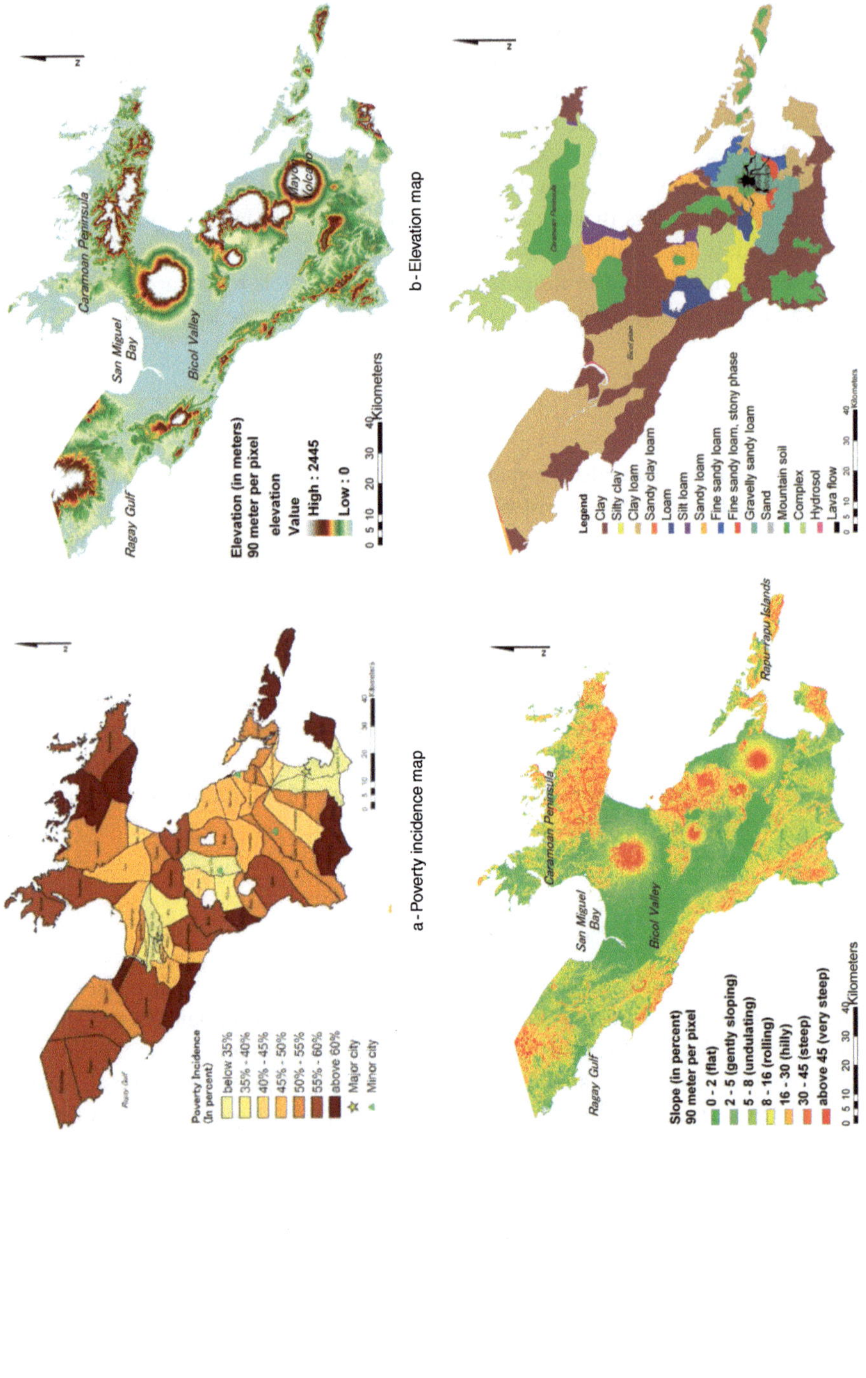

Fig. 16.3 Maps of the different variables (part 1–3)

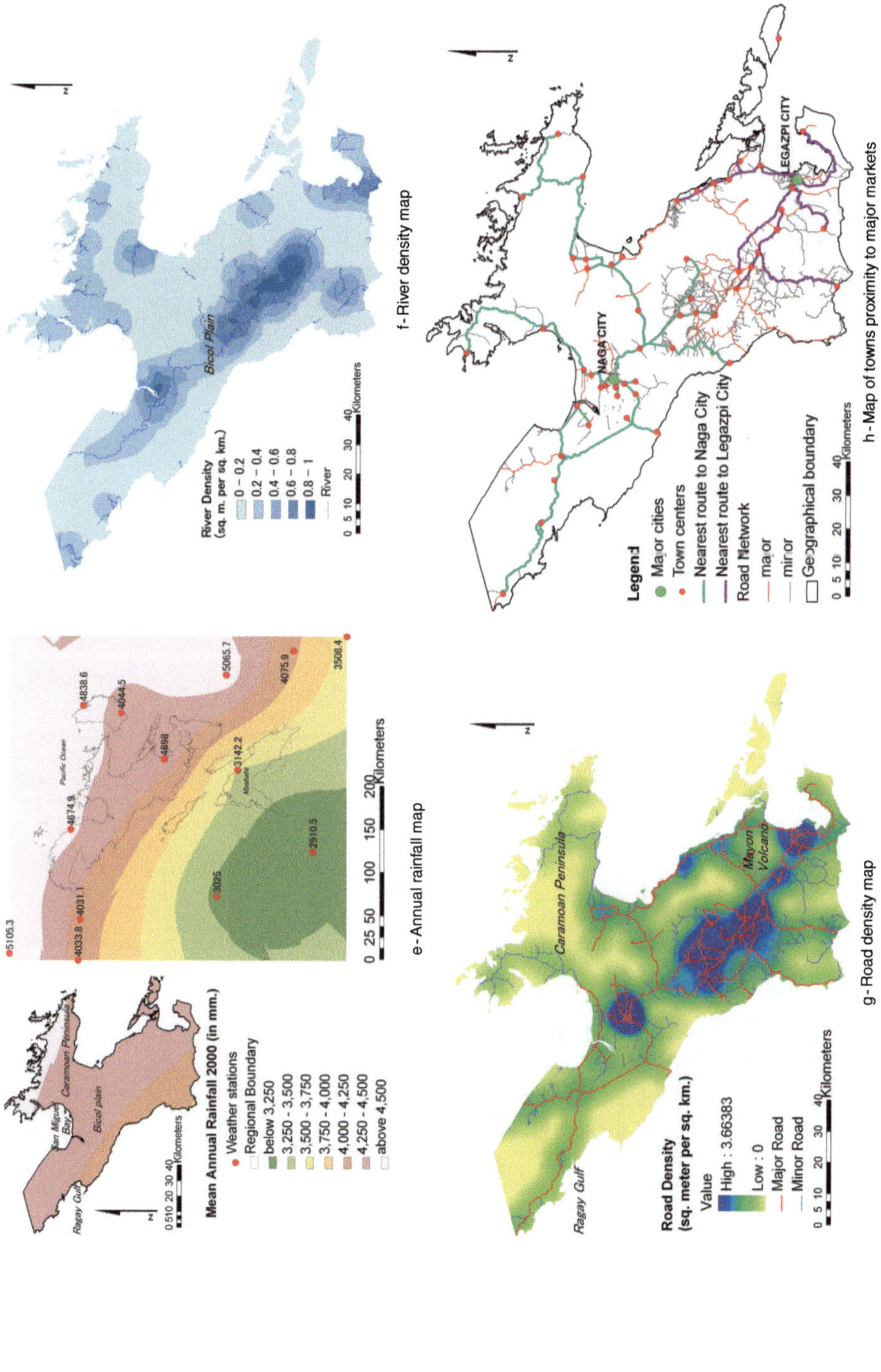

Fig. 16.3 (continued)

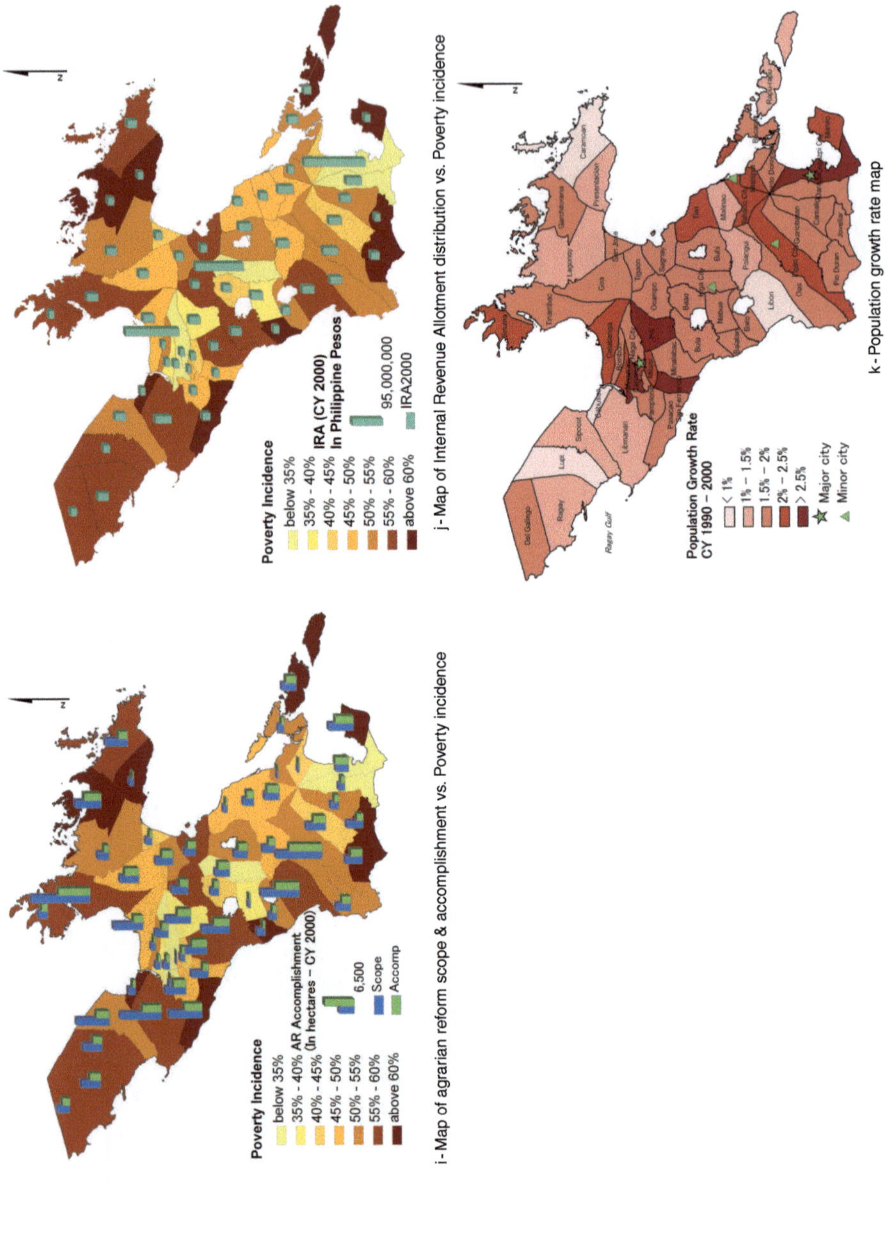

Fig. 16.3 (continued)

16.3.4 Rainfall

Fourthly, the effect of rainfall was found to be statistically significant determinant to the incidence of poverty. Its regression coefficient generated a value of −0.0002 which means that more rainfall leads to lower poverty incidence in a municipality. More rainfall would mean more water for irrigation of crops. However, if other variables are isolated, mean annual rainfall has a positive correlation coefficient with poverty (Appendix 2). This may imply that the effect of rainfall to poverty varies widely. Municipalities in the western coast, particularly in Caramoan Peninsula, which have less amount of annual rainfall exhibit high poverty incidence (see Fig. 16.3a vs. Fig. 16.3e). Rainfall would be beneficial for farmers in this area since they have little access to irrigation facilities as compared to most of the farms in the central plain. On the contrary, municipalities in the northeastern part and certain areas in the central plain with higher rainfall amount also suffer from high poverty incidence. In fact, the study site is frequently battered by typhoons and heavy rains annually. Too much rainfall in these areas may have caused floods, which typically damaged crops and properties.

16.3.5 Access to River

As expected, when river density was computed, it is higher in the central plain with large tracks of agricultural farmlands. Municipalities with relatively higher mean river density generally exhibit low poverty rate and vice versa (see Fig. 16.3a vs Fig. 16.3f). This implies that areas with better access to river resources have a geographical advantage because river is a source of water for irrigation and fishing livelihood. Surprisingly, this assumption is not conclusive as revealed by the results of the multiple regression analysis. River density was not a statistical determinant to poverty. Perhaps this is because, even if river is a potential source of irrigation, about 70% of farmlands in these two provinces have no access to irrigation facilities according to the Census of Agriculture (NSO, 2002).

16.3.6 Access to Road

Mean road density was found to be a strong significant determinant to the incidence of poverty as revealed by its unstandardized regression coefficient of −0.111. Moreover, it is the strongest influential determinant to poverty as reflected in its high standardized coefficient value of −4.660. This means that municipalities with low road density are extremely poor, whereas those with high density are less poor. Figure 16.3a, g clearly shows that high road density is concentrated in major cities of Naga, Legaspi and their peripheries, which are characterized by relatively low poverty rate, while low road density are found mostly in localities with high poverty incidence. These findings suggest that limited access to road infrastructure affects the mobility of people, goods and services. Lack of roads will not spur economic investment that will generate employment and income for the local people. Apparently, more access to road infrastructure notably lowers poverty rate.

16.3.7 Proximity to Major Markets

When travel distance of each town centre was calculated, sixteen (16) municipalities were closest to Legaspi City while the rest were nearest to Naga City (Fig. 16.3h). As such, distance between each town center and their nearest major city was also found to be a statistically significant determinant to poverty. Its positive regression coefficient of 0.0007 suggest that as land travel distance further away from the two major cities, Legaspi and Naga, poverty incidence increases. Apparently, since most businesses, enterprises, offices, industries, including higher educational institutions are located in Legaspi and Naga, the state of development within these cities influences the state of poverty in their surrounding municipalities. This may explain why towns closer to Naga and Legaspi have relatively low poverty incidence (Fig. 16.3a vs. Fig. 16.3 h). This finding implies that most localities in the two provinces are still highly dependent on major cities being the major markets in Bicol region.

16.3.8 Influence of the Agrarian Reform Program

Similar to the above, the accomplishment rate in land distribution which was measured by dividing the total accomplishment as of CY 2000 to the total land distribution target in every municipality, has exhibited a statistically significant determinant to poverty with a positive coefficient value of 0.141. This means that the Comprehensive Agrarian Reform Program (CARP) of the government significantly affects the condition of poverty within the study site. Figure 16.3i shows spatial pattern of lands covered and distributed under CARP. Generally, poorest areas have more targeted agricultural lands for distribution. However, its positive coefficient contradicts the previous assumption because it implies that as the rate of lands distributed to landless poor increases, poverty incidence also increases. Meanwhile, if all other variables are disregarded, its bivariate correlation analysis yielded a negative coefficient value of −0.208 which supports the earlier notion (Appendix 2). Such findings may suggest that, it is not quite clear if agrarian reform has truly contributed to poverty reduction. Conceivably, the lack of support services, particularly rural infrastructure, credit market and agricultural extension, among others, is equally important to make poverty reduction more effective.

16.3.9 Influence of the Fiscal Decentralization Policy

Figure 16.3j shows the distribution of IRA with respect to poverty incidence. Three cities which have low level of incidence gets a higher share of IRA relative to the rest of the municipalities. Regression results also show that the amount of Internal Revenue Allotment (IRA) by municipality significantly affects the incidence of poverty as indicated by its regression coefficient value of −1.798. In fact, it is the second most significant determinant to poverty based from its generated standardized coefficient value of −4.370. This means that municipalities with less funding support from IRA have high poverty incidence. Lower share in IRA suggest meager funds for development, consequently, high poverty rate. It further implies an extreme disparity and geographical bias of the national government to the cities in the allocation of the national wealth.

16.3.10 Influence of Population Growth

Finally, results of the multiple regression analysis revealed a negative regression coefficient value of −0.0427, similar to its bivariate correlation coefficient of −0.460 (Appendix 2). These findings disprove the prior assumption that population growth exacerbates poverty. When mapped with GIS and compared with poverty map (Fig. 16.3a vs. Fig. 16.3k), the map shows that the two major cities and municipalities adjacent to them posted a relatively higher population growth rate as compared to the rest of the study site. This may imply that normally, population grew faster in better-off areas, especially in key cities and nearby municipalities, than in poorest areas; hence its growth rate is higher. Migration could be one of the major reasons. People from the poorest areas would most likely migrate to more affluent areas thereby inducing higher population growth rate in the latter. Furthermore, this finding raised the issue of causality. While population growth is often considered as an original cause of poverty, it may by itself a consequence of poverty rather than its cause.

There may be other demographic as well as socio-economic factors that were not considered in this study, e.g. types of employment, level of economic activities, transportation and communication flow, etc. This could be accounted by the errors accorded in the regression model.

16.4 Summary and Conclusion

This study attempted to explore the determinants of poverty in the Philippines using GIS. It also demonstrated the use of GIS in poverty analysis. It examined the case of Albay and Camarines Sur provinces as pilot areas. It accounted for the various GIS and statistical methodology that were undertaken to solve the problem at hand. The use of GIS for poverty analysis had enabled the incorporation of spatial variables, i.e. agro-climatic factors and accessibility, based from a specified modifiable areal unit, i.e. municipal boundary, to derive auxiliary datasets that could be combined with socio-economic data, i.e. influence of policy and population, to perform multiple regression analysis. The regression model conceptualized in this study was found to have the ability to explain the determinants of poverty.

In conclusion, the results of the study show that the spatial variation in the incidence of poverty is mainly caused by disparities on access to road infrastructure which is further exacerbated by loopholes and geographical bias in fiscal decentralization priorities and deficiency in agrarian reform implementation. Moreover, proximity to major cities where there is a high concentration of development and economic activities and differences in agro-climatic features, particularly, elevation, slope, and rainfall, also proved to be significant determinants to poverty and suggest the presence of geographically disadvantageous areas within the pilot site. All of these findings suggest that geography and facets of public policy have a strong impact on the state of poverty. Finally, the use of GIS proves to be a valuable tool to incorporate spatial factors in poverty analysis.

16.5 Appendix 1

a: Multiple regression model summary

R	R^2	Adjusted R^2	Standard error of the estimate
0.912	0.832	0.794	0.0478069

b: ANOVA table of multiple regression analysis

	Sum of squares	Degrees of freedom	Mean square	F	Sig.
Regression	0.499	10	0.050	21.812	0.001
Residual	0.101	44	0.002		
Total	0.599	54			

c: Histogram of standardized residual

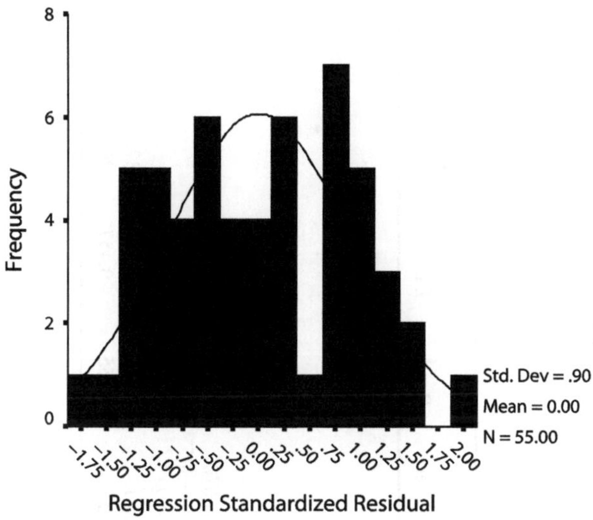

d: Normal P-P Plot of residuals

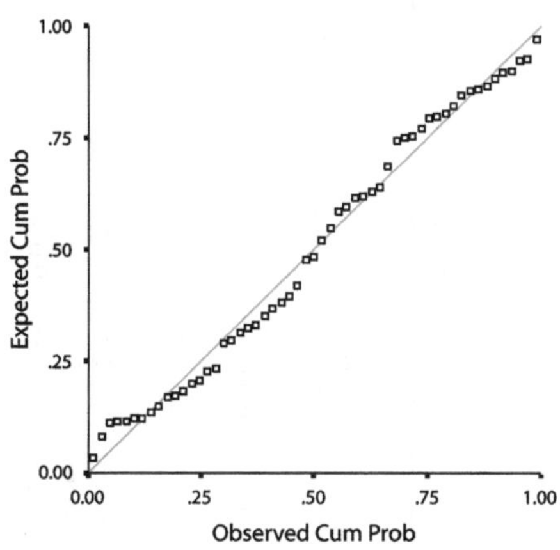

16.6 Appendix 2: Correlation Matrix for all of the Variables Used in Multiple Regression Analysis

Variables	Incidence	Elevation	Slope	Soil	Rainfall	River	Road	Distance	Agrarian reform	IRA	Pop. growth
Poverty incidence	1										
Mean elevation	0.106	1									
Slope above 8%	0.530*	0.658*	1								
Soil with constraint	0.029	0.474*	0.413*	1							
Mean annual rainfall	0.024	0.265	0.135	0.556*	1						
Mean river density	−0.458*	−0.528*	−0.559*	−0.242	−0.287**	1					
Mean road density	−0.746*	−0.322*	−0.56*	−0.195	−0.300**	0.670*	1				
Distance to major cities	0.582*	0.204	0.481*	0.304**	0.357*	−0.305**	−0.512*	1			
Land distribution rate	−0.208	0.130	0	0.236	0.019	0.247	0.319**	−0.127	1		
Percentage of IRA share	−0.530*	0.073	−0.052	0.135	−0.023	0.137	0.379*	−0.199	0.372*	1	
Population growth rate	−0.460*	0.024	−0.244	−0.096	−0.096	0.212	0.255	−0.457*	0.268**	0.128	1

Note: Poverty incidence as independent variable
*Significant at 0.01
**Significant at 0.05

References

Albert, J. R. G., & Collado, P. M. (2004). *Profile and determinants of poverty in the Philippines*. Paper presented at the 9th National Convention on Statistics.

Amarasinghe, U., Samad, M., & Anputhas, M. (2005). Spatial clustering of rural poverty and food insecurity in Sri Lanka. *Food Policy*, *30*, 493–509.

Balisacan, A. M. (2007). Why does poverty persist in the Philippines? Facts, fancies and policies. *SEARCA Agriculture and Development Discussion Paper Series* 1–32.

Balisacan, A. M., Edillon, R. G., & Ducanes, G. M. (2002). *Poverty mapping and targeting for KALAHI-CIDDS Phase I: Final report prepared for the Department of Social Welfare and Development (DSWD), Republic of the Philippines*.

Balisacan, A. M., & Fuwa, N. (2004). Going beyond crosscountry averages: Growth, inequality and poverty reduction in the Philippines. *World Development*, *32*, 1891–1907.

Benson, T., Chamberlin, J., & Rhinehart, I. (2005). An investigation of the spatial determinants of the local prevalence of poverty in rural Malawi. *Food Policy*, *30*, 532–550.

Biau, G., Zorita, E., von Storch, H., & Wackernagel, H. (1999). Estimation of Precipitation by kriging in the EOF space of the sea level pressure field. *Journal of Climate, 12*(4), 1070–1085.

Bigman, D., & Deichmann, U. (2000). Geographic targeting: A review of different approaches. In D. Bigman& H. Fofack (Eds.), *Geographical targeting for poverty alleviation: Methodology and applications* (pp. 43–73). Washington, DC: The World Bank.

Bigman, D., & Fofack, H. (2000). Introduction and overview. In D. Bigman & H. Fofack (Eds.), *Geographical targeting for poverty alleviation: Methodology and applications* (pp. 1–39). Washington, DC: The World Bank.

Coudouel, A., Hentschel, J. S., & Wodon, Q. T. (2002). Poverty measurement and analysis. In J. Klugman (Ed.), *The poverty reduction strategies paper (PRSP) sourcebook* (pp. 29–74). Washington DC: The World Bank.

Davis, B. (2003). *Choosing a method for poverty mapping*. Rome: Food and Agriculture Organization of the United Nations.

Deichmann, U. (1999). *Geographic aspects of poverty*. Retrieved June 21, 2010, from http://siteresources.worldbank.org/INTPGI/Resources/Pro-Poor-Growth/5319_povmap.pdf

Domingo, E. V. (2003). *Poverty mapping in the Philippines*. Paper presented at the UNESCAP Ad Hoc Expert Group Meeting on Poverty Mapping and Monitoring Using Information Technology.

Estudillo, J. P., Sawada, Y., & Hossain, M. (2005). Socioeconomic changes and the decline in poverty: A view from three villages in the Philippines, 1985–1997. *Asian Economic Journal*, *19*(4), 383–406.

Fischer, G., van Velthuizen, H., Nachtergaele, F., & Medow, S. (2000). *Global agro-ecological zones*. Retrieved January 2007, from Food and Agriculture Organization of the United Nations and International Institute for Applied Systems Analysis: http://www.iiasa.ac.at/Research/LUC/GAEZ/index.htm

Fuwa, N. (2007). Pathways out of rural poverty: A case study in socio-economic mobility in the rural Philippines. *Cambridge Journal of Economics*, *31*, 123.

Glasmeier, A. K. (2002). One nation, pulling apart: The basis of persistent poverty in the USA. *Progress in Human Geography*, *26*, 155–173.

Henninger, N., & Snel, M. (2002). *Where are the poor? Experiences with the development and use of poverty maps*. Washington, DC and Arendal: World Resources Institute and UNEP-GRID.

Hyman, G., Larrea, C., & Farrow, A. (2005). Methods, results and policy implications of poverty and food security mapping assessments. *Food Policy*, *30*, 453–460.

Jarvis, A., Reuter, H. I., & Guevarra, E. (2006). *Hole-filled seamless SRTM data V3*. Retrieved October 2006, from International Centre for Tropical Agriculture: http://srtm.csi.cgiar.org

Kam, S. -P., Hossain, M., Bose, M. L., & Villano, L. S. (2005). Spatial patterns of rural poverty and their relationship with welfare-influencing factors in Bangladesh. *Food Policy*, *30*, 551–567.

Kristjanson, P., Radeny, M., Baltenweck, I., Ogutu, J., & Notenbaert, A. (2005). Livelihood mapping and poverty correlates at a meso-level in Kenya. *Food Policy, 30*, 568–583.

Legg, C., Kormawa, P., Maziya-Dixon, B., Okechukwu, R., Ofodile, S., & Alabi, T. (2005). *Report on mapping livelihoods and nutrition in Nigeria using data from the National Rural Livelihoods Survey and the National Food Consumption and Nutrition Survey*. Ibadan, Nigeria: International Institute for Tropical Agriculture.

Minot, N., Bob, B., & Epprecht, M. (2006). Poverty mapping and inequality in Vietnam: Spatial patterns and geographic determinants *International Food Policy Research Institute Research Report, 146*, 1–80.

NASA Landsat Program. (2003). *Landsat ETM+*. Retrieved October 2006, from United States Geological Survey.

NSCB. (2004). *2004 Regional and social economic trends in Bicol Region*. Legaspi City: National Statistical Coordination Board, Regional Division V.

NSCB. (2005). *Estimation of local poverty in the Philippines*. Makati City: National Statistical Coordination Board and The World Bank.

NSCB. (2006). *Overview of the Bicol region*. Retrieved June 21, 2010, from http://www.nscb.gov.ph/ru5/overview/default.html

NSO. (2002). *CY 2002 census of agriculture*. Retrieved June 21, 2010, from http://www.census.gov.ph/data/sectordata/ag0411504.htm

NSO. (2000a). *2000 census of population and housing*. Retrieved June 21, 2010, from http://www.census.gov.ph/data/sectordata/datapop.html

NSO. (2000b). *Final results 2000 family income and expenditures survey*. Retrieved June 21, 2010, from http://www.census.gov.ph/data/sectordata/2000/ie00fr09.htm

Rogers, D., Emwanu, T., & Robinson, T. (2006). Poverty mapping in Uganda: An analysis using remotely sensed and other environmental data. *Pro-poor Livestock Policy Initiative Working Paper, 36*, 1–60.

Index

Y. Murayama, R.B. Thapa (eds.), *Spatial Analysis and Modeling in Geographical Transformation Process,* The GeoJournal Library 100, DOI 10.1007/978-94-007-0671-2,

MIX
Papier aus verantwortungsvollen Quellen
Paper from responsible sources
FSC® C105338

If you have any concerns about our products,
you can contact us on
ProductSafety@springernature.com

In case Publisher is established outside the EU,
the EU authorized representative is:
Springer Nature Customer Service Center GmbH
Europaplatz 3, 69115 Heidelberg, Germany

Printed by Libri Plureos GmbH
in Hamburg, Germany